Studies in Logic
Volume 70

Proceedings of the International Conference
Philosophy, Mathematics, Linguistics: Aspects of Interaction, 2012

Euler International Mathematical Institute

St. Petersburg, May 22—25, 2012

Volume 59
The Psychology of Argument. Cognitive Approaches to Argumentation and Persuasion
Fabio Paglieri, Laura Bonelli and Silvia Felletti, eds.

Volume 60
Absract Algebraic Logic. An Introductory Textbook
Josep Maria Font

Volume 61
Philosophical Applications of Modal Logic
Lloyd Humberstone

Volume 62
Argumentation and Reasoned Action. Proceedings of the 1st European Conference on Argumentation, Lisbon 2015. Volume I
Dima Mohammed and Marcin Lewiński, eds.

Volume 63
Argumentation and Reasoned Action. Proceedings of the 1st European Conference on Argumentation, Lisbon 2015. Volume II
Dima Mohammed and Marcin Lewiński, eds.

Volume 64
Logic of Questions in the Wild. Inferential Erotetic Logic in Information Seeking Dialogue Modelling
Paweł Łupkowski

Volume 65
Elementary Logic with Applications. A Procedural Perspective for Computer Scientists
D. M. Gabbay and O. T. Rodrigues

Volume 66
Logical Consequences. Theory and Applications: An Introduction.
Luis M. Augusto

Volume 67
Many-Valued Logics: A Mathematical and Computational Introduction
Luis M. Augusto

Volume 68
Argument Technologies: Theory, Analysis, and Applications
Floris Bex, Floriana Grasso, Nancy Green, Fabio Paglieri and Chris Reed, eds.

Volume 69
Logic and Conditional Probability: A Synthesis
Philip Calabrese

Volume 70
Proceedings of the International Conference
Philosophy, Mathematics, Linguistics: Aspects of Interaction, 2012 (PhML-2012)
Oleg Prosorov, ed.

Studies in Logic Series Editor
Dov Gabbay   dov.gabbay@kcl.ac.uk

Proceedings of the International Conference

Philosophy, Mathematics, Linguistics: Aspects of Interaction, 2012

Euler International Mathematical Institute

St. Petersburg, May 22—25, 2012

Edited by

Oleg Prosorov

© Individual author and College Publications 2017
All rights reserved.

ISBN 978-1-84890-256-5

College Publications
Scientific Director: Dov Gabbay
Managing Director: Jane Spurr

http://www.collegepublications.co.uk

Printed by Lightning Source, Milton Keynes, UK

All rights reserved. No part of this publication may be reproduced, stored in a retrieval system or transmitted in any form, or by any means, electronic, mechanical, photocopying, recording or otherwise without prior permission, in writing, from the publisher.

Contents

Preface vii

Kant's Logic Revisited 1
Theodora Achourioti and Michiel van Lambalgen

Implicit Dynamic Function Introduction and Ackermann-Like Function Theory 23
Marcos Cramer

Locology and Localistic Logic: Mathematical and Epistemological Aspects 35
Michel De Glas

Secure Multiparty Computation without One-way Functions 63
Dima Grigoriev and Vladimir Shpilrain

IF Logic and Linguistic Theory 83
Jaakko Hintikka

Commentary on Jaakko Hintikka's "IF Logic and Linguistic Theory" 95
Gabriel Sandu

Foundations as Superstructure (*Reflections of a practicing mathematician*) 101
Yuri I. Manin

Classical and Intuitionistic Geometric Logic 117
Grigori Mints

Commentary on Grigori Mints' "Classical and Intuitionistic Geometric Logic" 127
Roy Dyckhoff and Sara Negri

The Historical Role of Kant's Views on Logic 133
Andrei Patkul

The Logic for Metaphysical Conceptions of Vagueness 157
Francis Jeffry Pelletier

A Note on the Axiom of Countability 177
Graham Priest

**Topologies and Sheaves Appeared as Syntax and Semantics
of Natural Language** 183
Oleg Prosorov

Venus Homotopically 245
Andrei Rodin

**On a Combination of Truth and Probability:
Probabilistic IF Logic** 267
Gabriel Sandu

Towards Analysis of Information Structure of Computations 277
Anatol Slissenko

Horizons of Scientific Pluralism: Logics, Ontology, Mathematics . . 301
Vladimir L. Vasyukov

Preface

The international conference *Philosophy, Mathematics, Linguistics: Aspects of Interaction 2012* (PhML-2012) was held on May 22–25, 2012 at the Euler International Mathematical Institute (EIMI), which is a part of the St. Petersburg Department of Steklov Mathematical Institute (PDMI). This conference was the second sequel in the PhML series of conferences intended to provide a forum for the presentation of current researches, and to stimulate an interdisciplinary dialogue between philosophers, mathematicians, logicians, computer scientists and linguists.

The first PhML conference was held in 2009 and was initially conceived as a part of the *World Philosophy Day in Russia* proclaimed for 2009 by UNESCO. The first conference clearly revealed the need for the scientific community in a broad dialogue between representatives of natural sciences and humanities because of the increasing mathematisation of scientific knowledge. The project was successfully continued with PhML conferences in 2012 and 2014.

As it often happens in science, similar tendencies appear almost simultaneously in different research centers. It is noteworthy that in recent years, numerous international conferences with interdisciplinary topics are organized in many countries. Apparently, the reason for this lies in the fact that the contemporary social development is characterized by intensive contacts between different cultures, in which the understanding of human phenomena and the knowledge of the world are achieved in the interaction of various forms of collective consciousness and professional scientific activity – in art and literature, in humanities and natural sciences.

It is generally recognized that the globalization is a characteristic feature of the contemporary stage of civilization development. In the field of material production, the globalization manifests itself in the international division of labour, in which the most of high-tech products include components made in several countries. In the field of culture, the globalization manifests itself in the free availability of works of literature, music, fine arts, cinema, no matter where they were created. But in sciences, the globalization is manifested not only in the free availability of scientific research results offered in open access publications on the Internet, but also in the wide development of interdisciplinary researches, in which previously isolated sciences fruitfully interact, and where fundamental research and applied research are complementary.

The most important sign of this phenomenon becomes the increasing mathematisation of all sciences, which began in the Modern Ages. In this regard, in the preface to his *Metaphysical Foundations of Natural Science*, I. Kant asserted in 1786 that "in any special doctrine of nature there can be only as much proper science as there is mathematics therein."[1]

The process of mathematisation is obviously manifested in the constant expansion of application areas of mathematics invading today not only in the natural and technical sciences, but also in the humanities. Now, the scientists, who are working in the humanities, and mathematicians, who are interested in the expansion of the field of applications of mathematics, have a serious need for interdisciplinary dialogue and personal contacts to discuss various aspects of interplay between mathematics and humanities. That is why in 2011, the Scientific Council of the EIMI decided to hold on May 22–25, 2012 the international interdisciplinary conference PhML-2012, the subject of which would constitute different aspects of interaction of philosophy, mathematics, linguistics, and the main goal would be to broaden the dialogue between mathematicians, logicians, computer scientists, philosophers and linguists.

The PhML series of conferences was organized with the intention to reveal that the "effectiveness of mathematics in the natural sciences",[2] pointed out in 1960 by Eugene Wigner, is also manifested in such humanities as philosophy and linguistics. Following Wigner: "The miracle of the appropriateness of the language of mathematics for the formulation of the laws of physics is a wonderful gift which we neither understand nor deserve."[3] A possible way to explain the applicability of mathematics was suggested by Alain Connes in reflections on the nature of mathematics he expressed during the talk *Espace-temps, nombres premiers, deux défis pour la géométrie*[4] given on November 12, 2010 at the Institut Henri Poincaré, Paris, where he said in particular:

> C'est qu'il faut essayer de comprendre quand les gens vous posent la question : « À quoi les mathématiques sont-elles utiles ? » En fait, les mathématiques, c'est sans doute l'usine la plus performante pour fabriquer des concepts, et des concepts, après, qui servent partout.[5] (03:42 – 03:59)

Concerning the universality of concepts, Émile Durkheim wrote in 1912: "conversation and intellectual dealings among men consist in an exchange of

---

[1] Metaphysical Foundations of Natural Science. In H. Allison and P. Heath, eds., *Immanuel Kant. Theoretical Philosophy after 1781*, p. 185. Cambridge University Press, UK, 2002.

[2] E. P. Wigner. The Unreasonable Effectiveness of Mathematics in the Natural Sciences. *Communications in Pure and Applied Mathematics*, 13(1):1–14, 1960.

[3] Ibid., p. 14.

[4] Watch this talk on Vimeo at https://vimeo.com/24504403, on the page maintained by the Société Mathématique de France.

[5] Our translation of this quotation is: "It is that you should try to understand when people ask you the question: 'For what mathematics is useful?' In fact, the mathematics is probably the most efficient factory to produce concepts, and such concepts that, afterwards, serve everywhere."

concepts. The concept is, in essence, an impersonal representation. By means of it, human intelligences communicate."[6]

For such an exchange of concepts, the PhML conferences provide a forum for philosophers, linguists, and especially for working mathematicians and logicians who are interested in philosophy and linguistics, and who have something to say in these domains of knowledge on the basis of their own professional practice where mathematics intervenes in philosophy/linguistics as a factory of concepts and/or as a factory of structures. And this is done for the purpose to analyze the philosophical/linguistic problems in mathematical terms, to draw philosophical/linguistic conclusions from philosophical/linguistic hypotheses by a mathematical proof, and to construct mathematical models in which one can study philosophical/linguistic problems. This means that the conferences of PhML series aim to promote mathematical methods in philosophy and linguistics. But the conferences of PhML series are not limited in this issue; they are interested in the philosophy of mathematics and logic, as well as they are interested in other related issues mentioned on their CfP webpages.

Moreover, the organizers considered as important to invite for participation in the conference PhML-2012 the scientists of the last wave of Russian emigration. As a result of extensive preparatory work, the conference PhML-2012 had brought together 37 registered participants from 11 countries. Apart from plenary sessions intended to present invited papers, there were also parallel sessions of thematic sections to present contributed papers. In addition, the conference featured a panel discussion entitled *The Heritage of Kant and Contemporary Formal Logic*.

It is worth noting that the conference PhML-2012 aroused great interest among researches of the PDMI, graduate students, faculty and staff members of the St. Petersburg State University, members of the St. Petersburg Mathematical Society, members of the St. Petersburg Philosophical Society, and among the wide scientific community of St. Petersburg. Some talks had gathered such an audience that the EIMI conference-hall was overcrowded. Following to the general opinion of organizers and to numerous comments of participants, this second conference in the series PhML was highly successful.

The conference PhML-2012 is presented on the website of the Euler International Mathematical Institute at http://www.pdmi.ras.ru/EIMI/2012/PhML/index.htm. The working languages of the conference were Russian and English, but for a better understanding between the participants, when giving a talk or speaking in discussions, all used English. With this in mind, in printed

---

[6]É. Durkheim. *The Elementary Forms of Religious Life*, p. 435. The Free Press, NY, 1995.

announcements about the schedule and the program, and in announcements on the Internet about the conference, organizers used the spelling of personal names and titles of talks in English.

At the conference PhML-2012, there were presented 23 invited talks in plenary sessions and 6 contributed talks in parallel sessions of two thematic sections *Mathematics and Language* and *Logic and Semantics*. The contents of most of the papers are outlined in brief abstracts presented on the EIMI website at http://www.pdmi.ras.ru/EIMI/2012/PhML/index_files/abstracts_PhML_2012.htm. This table of abstracts lists all the talks given at the conference. The actual volume of the Proceedings of PhML-2012 provides an opportunity for readers to acquaint with a selection of expanded papers those were presented during the PhML-2012 conference.

Since the *Studies in Logic* is a reviewed series, all the papers have had to be passed through the peer reviewing process of the type "Single Blind Review", that is, a paper is sent to reviewers in such a form as it is prepared for publication by the author(s), but the names of reviewers are hidden from the author(s). In result, all papers of the present Proceedings were peer reviewed, in most cases by more than one referee, and sometimes by four referees. The published versions of original talks given at the conference have benefited greatly due to the cooperative work of authors and reviewers during such an editorial process. I am very grateful to all reviewers, without whose work the present Proceedings of PhML-2012 would not have been possible to realize in its actual form.

The only papers being kept intact in the editorial process are the papers of Jaakko Hintikka and Grigori Mints, who were passed away while these Proceedings were been prepared. Their original papers are accompanied by detailed—however indispensable—commentaries written for this volume by Gabriel Sandu, Roy Dyckhoff and Sara Negri, to whom I am very grateful.

The project to publish the Proceedings of PhML-2012 dates back to the August 2013 when Grigori Mints accepted my proposal to head the prospective editorial board. We planned to start editorial work after the next conference PhML-2014 were completed in April 2014. To my deepest sorrow and regret, he suddenly died on May 29, 2014, and the work could not start as planned. One year later, I had returned to the project. Now, after two years of intensive editorial work, I propose the present volume to the attention of researchers who are interested in those aspects of the interaction of philosophy, mathematics, and linguistics that were discussed during the conference PhML-2012.

Oleg Prosorov
St. Petersburg Department of
Steklov Mathematical Institute

Kant's Logic Revisited

Theodora Achourioti

ILLC, Amsterdam University College
University of Amsterdam
Science Park 113, 1098XG Amsterdam, The Netherlands
`T.Achourioti@uva.nl`

Michiel van Lambalgen

ILLC, Department of Philosophy
University of Amsterdam
Oude Turfmarkt 141, 1012GC Amsterdam, The Netherlands
`M.vanLambalgen@uva.nl`

Abstract: Kant considers his *Critique of Pure Reason* to be founded on the act of judging and the different forms of judgement, hence, take pride of place in his argumentation. The consensus view is that this aspect of the *Critique of Pure Reason* is a failure because Kant's logic is far too weak to bear such a weight. Here we show that the consensus view is mistaken and that Kant's logic should be identified with geometric logic, a fragment of intuitionistic logic of great foundational significance.

1. Preview

Below the reader will find a condensed revisionist account of Kant's so-called 'general logic', usually thought to be substandard, even when compared with the traditional logic of his day [4].[1] Ultimately our interest is in the formalisation of Kant's 'transcendental logic' (for which see [1]), but since transcendental logic takes its starting point in the judgement forms listed in the Table of Judgement (most of which have their origin in general logic) we must take a close look at the actual logical forms of these judgements. The result of this investigation is

---

We are grateful to the referees for insightful comments.

[1] Not to mention the scathing verdicts from the standpoint of modern logic which we take to have started with Frege and Strawson.

that Kant's general logic is not monadic, not finitary, not classical, and perhaps linear rather than intuitionistic. We will here not elaborate on the last point[2] but we will restrict ourselves to stating a completeness theorem identifying Kant's general logic with a fragment of intuitionistic logic.

2. Validity in general logic

The key to any insightful formalisation of Kant's logic is the observation that judgements in Kant's sense participate in two kinds of logics: *general* logic and *transcendental* logic. Here is how Kant introduces 'general logic' in the first *Critique* [7]:

> [G]eneral logic abstracts from all the contents of the cognition of the understanding and of the difference of its objects, and has to do with nothing (A55-6/B80) but the mere form of thinking. (A54/B78)

And later, with a slightly different emphasis:

> General logic abstracts [...] from all content of cognition, i.e. from any relation of it to the object, and considers only the logical form in the relation of cognitions to one another, i.e. the form of thinking in general. (A55/B79)

So what is the 'mere form of thinking'?

The first two paragraphs of the *Jäsche Logik* [5] marvel at the fact that all of nature, including ourselves, is bound by rules. It continues:

> Like all our powers, the understanding is bound in its actions to rules [...] Indeed, the understanding is to be regarded in general as the source and the faculty for thinking rules in general [...] [T]he understanding is the faculty for thinking, i.e. for bringing the representations of the senses under rules.

From this it derives a characterisation of logic:

> Since the understanding is the source of rules, the question is thus, according to what rules does it itself operate? [...] If we now put aside all cognition that we have to borrow from objects and merely reflect on the use just of the understanding, we discover those of its rules which are necessary without qualification, for any purpose and without regard to any particular objects, because without them we would not think at all. [...] [T]his science of the necessary laws of the understanding and of reason in general, or what

---

[2]Grigori Mints was planning on studying the connection between Kant's disjunctive judgement and multiplicative linear logic.

is one and the same, of the mere form of thought as such, we call *logic*. [5, pp. 527-8] (cf. also A52/B76)

To appreciate the real import of this passage, one must resist the temptation to consider logic as consisting of a motley set of inference rules, such as modus ponens and syllogistic inferences, even though the *Jäsche Logik* will later list these too. Two definitions are pertinent here:

§58 A rule is an assertion under a universal condition. [5, p. 615]

Here it is important to bear in mind Kant's notion of *universal representation* as 'a representation of what is common in several objects' [5, §1, p. 589]. A rule is, therefore, applicable to a domain of indefinite extension.

The second definition is that of an *inference of reason*:

§56 An inference of reason is the cognition of the necessity of a proposition through the subsumption of its condition under a given universal rule. [5, p. 614]

At this point we will not yet provide an elaborate explanation of the notion of 'condition', but the reader is invited to take modus ponens as a concrete example. We then have the following sequence of ideas: (i) the understanding operates according to rules, (ii) the understanding's operations are necessary insofar as they pertain to the formal features of rules, and (iii) the most general formal principle is rule-application (or rule composition – as we shall see the distinction was not always made in those days). Thus Kant's logic has a general and constructive definition of validity, a consequence of the meaning of 'rule'. The *Jäsche Logik* will give concrete instances of this most general principle, such as modus ponens, but the full force of the principle will only become apparent when we come to discuss the true logical form of Kant's 'judgements'. We must note here that the general inference principle limits logic to judgements that can be seen as rules. We view Kant's emphasis on rules and their structural properties as marking the 'formal' character of his general logic. The definition of validity just given should be contrasted with the Bolzano-Tarski definition of validity: 'an argument is valid if its conclusion is true whenever its premises are' – for in this part of Kant's logic (what he calls 'general logic') there is no truth yet, there are only rules. A different kind of logic, 'transcendental logic' will introduce truth.

3. Three definitions of judgement and a Table ...

Any modern logic textbook makes a strict separation between syntax, semantics and consequence relation, and makes no reference at all to psychological processes that may be involved in a concrete case of asserting a syntactically well-formed sentence. These processes are studied in psycholinguistics, and start from the assumption that there are specific syntactic and semantic binding processes at work in the brain. For logical theorising such psycholinguistic approaches are deemed to be irrelevant. For Kant they are in fact of the essence, and his definitions of judgement also contain a cognitive component.

But the reader trying to piece together Kant's views on logic may be forgiven a sense of bewilderment when she finds not one but three seemingly very different definitions of 'judgement', none of which specifies a syntactic form, together with a 'Table of Judgement' which specifies some syntactic forms (for example, categorical, hypothetical, disjunctive, with various other subdivisions), without an indication of how these forms relate to the three definitions. Lastly, there are the examples of judgements that Kant uses in various works, whose logical forms do not fit easily in the Table of Judgement. This looks unpromising material, but we shall show that Kant's logic is nevertheless coherent and surprisingly relevant to modern concerns.

Let us begin with the three definitions of judgement:

> A judgement is the representation of the unity of the consciousness of various representations, or the representation of their relation insofar as they constitute a concept. [5, p. 597]

> A judgement is nothing but the manner in which given cognitions are brought to the objective unity of apperception. That is the aim of the copula **is** in them: to distinguish the objective unity of given representations from the subjective [. . .] Only in this way does there arise from this relation a judgement, i.e. a relation that is objectively valid [. . .][3] (B141-2)

> Judgements, when considered merely as the condition of the unification of representations in a consciousness, are rules. (*Prol.* §23; see [8])

Even for those unfamiliar with Kant's technical vocabulary it will be obvious that 'unity' plays a central role in all three definitions. These are different

---

[3]Where 'objectively valid' means 'having relation to an object', which is not the same as 'true of the object'.

Kant's Logic Revisited

ways of saying that the expressions occurring in a judgement must be bound together so that they can be simultaneously present to consciousness. The first definition posits unity simply as a requirement. The second says that unity in a judgement is achieved if the judgement has 'relation to an object'. The third definition links unity to the meaning of a judgement. Just as an example: if for a hypothetical judgement $\varphi \to \psi$ there exists a rule transforming a proof of φ into a proof of ψ, then that judgement is unified. If the hypothetical is a truth functional material implication, then antecedent and consequent are independent, hence this is not a unified representation. The presence of a notion of unity of representation raises three questions: (i) what has this got to do with formal logic?, (ii) is there a relation between the unity and the reference to objects occurring in the second definition? and (iii) what is the relation between unity and the concrete forms of judgement given in the Table of Judgement?

3.1. Objects, concepts and general logic

Categorical judgements are composed of concepts, and objects 'fall under' concepts,[4] in a sense hinted at in the following note:

> *Refl.* **3042** Judgement is a cognition of the unity of given concepts: namely, that B belongs with various other things x, y, z under the same concept A, or also: that the manifold which is under B also belongs under A, likewise that the concepts A and B can be represented through a concept B. [9, p. 58]

It appears that both concepts and objects may fall under a given concept C. The given concept is therefore *transitive* in the sense that if (concept) M belongs to C (by being a subconcept) and (object) a belongs under M, then a belongs under C. Kant uses this semantics for concepts in his 'principle for categorical inferences of reason':

> *What belongs to the mark of a thing also belongs to the thing itself.* [5, p. 617]

The next note supplies more information about these objects 'in the logical sense' (so called because they make a cameo appearance in the section 'The logical employment of the understanding' (A68-9/B93)).

> *Refl.* **4634** We know any object only through predicates that we can say or think of it. Prior to that, whatever representations are found in us are to be counted only as materials for cognition but not as cognition. Hence

---

[4] Kant also uses the phrases 'object a belongs under concept C' and 'C belongs to a'.

> an object is only a something in general that we think through certain
> predicates that constitute its concept. In every judgment, accordingly, there
> are two predicates that we compare with one another, of which one, which
> comprises the given cognition of the object, is the logical subject, and the
> other, which is to be compared with the first, is called the logical predicate.
> If I say: a body is divisible, this means the same as: Something x, which I
> cognize under the predicates that together comprise the concept of a body,
> I also think through the predicate of divisibility. [9, p. 149]

What this *Reflexion* tells us is that an object is generic (or most general) for the 'predicates that constitute its concept', and that the quantifier 'something x' ranges over such generic objects only.

The same idea is prominent in the section of CPR entitled 'On the logical use of the understanding in general':

> [T]he understanding can make no other use of concepts than that of judging
> by means of them. Since no representation pertains the object immediately
> except intuition alone, a concept is thus never immediately related to an
> object, but is always related to some other representation of it (whether
> that be an intuition or itself already a concept). Judgement is therefore the
> mediate cognition of an object, hence the representation of a representation
> of it. (A68/B93)

An object is therefore rather like what logicians call a *type*: i.e. a set[5] $p(x)$ of formulas containing at least the free variable x;[6] free variables not identical to x can be replaced by formal parameters representing objects, hence specified by a type. As an example, consider the predicate 'body' and the type 'x is a massive body which orbits star y' – which can be used to defined the predicate 'planet', by existential quantification over y or by replacing y by a formal parameter (representing the Sun, say). Let T be the theory of the relevant concepts. If M is a concept, we say that $M(x)$ belongs to $p(x)$ if $T, p(x) \vdash M(x)$. For example, if T contains

$$\forall x (A(x) \wedge \exists y B(x, y) \rightarrow M(x)),$$

then $p(x) = \{A(x), \exists y B(x, y)\}$ belongs to $M(x)$. It is technically convenient to introduce suitable constants witnessing a type: if $p(x)$ is a (consistent) type,

---

[5] In our context a finite set.

[6] Relations enter Kant's logic especially in connection with the hypothetical judgement (see section 3.3.2); furthermore, as Hodges observed in [4], traditional logic allowed relations in syllogisms.

let a_p be a new constant satisfying $p(a_p)$.[7] These constants correspond to the 'objects in general' that we encountered in *Reflexion* **4634**. One may then view $p(x)$ and a_p as determining the same object; and in this formal sense we have that M belongs to a_p.

The next question to consider is whether Kant's theory of concepts puts a bound on the complexity of concepts, i.e. the complexity of the types belonging under the concept. The $p(x)$ given in the previous paragraph can be viewed as a single *positive primitive* formula:

Definition 1. A formula is *positive primitive* if it is constructed from atomic formulas using only $\vee$, (infinite) $\bigvee, \wedge, \exists, \bot$.

Suppose M, P are concepts all of whose subconcepts can be defined using positive primitive types (equivalently, formulas). The judgement 'all M are P' – or in the language of *Reflexion* **4634**: 'To everything x, to which M belongs, also P belongs – may then be expressed as

$$\bigwedge_{p \in M} \bigvee_{q \in P} \forall x (p(x) \to q(x)),$$

which is equivalent to

$$\forall x (\bigvee_{p \in M} p(x) \to \bigvee_{q \in P} q(x)),$$

and this formula satisfies the definition of a *geometric implication*:

Definition 2. A formula is *geometric* or a *geometric implication* if it is of the form $\forall \bar{x}(\theta(\bar{x}) \to \psi(\bar{x}))$, where θ and ψ positive primitive.

As it turns out, Kant's theories of concepts and of judgements contain the resources to restrict the complexity of $p(x)$ to positive primitive. The reason for this is that the complexity of the relation '$M(x)$ belongs to $p(x)$' is at most that of geometric implications. For the proof we must refer the reader to [1]; but a sketch will be given in section 4.

Geometric logic – the inferential relationships between geometric formulas – is therefore naturally suggested by Kant's theory of concepts. We will see that the logical form of Kant's own examples of judgements (in so far as they are 'objectively valid' (see section 3.2)) is that of geometric implications. As a consequence, we can show by means of 'dynamical proofs' of geometric implications that judgements can be viewed as rules:

---

[7]The constant a_p implicitly depends on the parameters and free variables (x excluded) occurring in $p(x)$.

Judgements, when considered merely as the condition of the unification of representations in a consciousness, are rules. (*Prol.* §23; see [8])

3.2. Unity, objects and transcendental logic

The second characterisation of judgement maintains that if a judgement has a certain kind of unity (the 'objective unity of apperception') then it relates to an object – has 'objective validity' – and can express a truth or falsehood of that object; it is 'truth-apt', in modern terminology. This is the domain of *transcendental logic*, which Kant defines as follows:

> [...] a science of pure understanding and of the pure cognition of reason, by means of which we think objects completely a priori. Such a science, which would determine the origin, the domain, and the objective validity of such cognitions, would have to be called transcendental logic since it has to do merely with the laws of the understanding and reason, but solely insofar as they are related to objects a priori and not, as in the case of general logic, to empirical as well as pure cognitions of reason without distinction. (A57/B81-2)

For Kant, perceiving objects about which judgements can be made is an instance of what would now be called the binding problem: objects are always given as a 'manifold' of parts and features, which have to be bound together through a process of *synthesis*. What is very distinctive about Kant's treatment here is that the binding that binds expressions in judgement together at the same time binds parts and features together with a view toward constructing an object out of sensory material that relates to the judgement. Therefore the binding process, necessary to bring separately perceived parts and features together, is in the end a complex logical operation, described by transcendental logic:

> Transcendental logic is the expansion of the elements of the pure cognition of the understanding and the principles without which no object can be thought at all (which is at the same time a logic of truth). For no cognition can contradict it without at the same time losing all content, i.e. all relation to any object, hence all truth. (A62-3/B87)

In the *Critique*, transcendental logic is not recognisably presented as a logic, and it is commonly thought that it cannot be so presented. The article [1] shows otherwise, mainly by focussing on the semantics of transcendental logic. There is a vast difference between the notion of object as it occurs in first order models, and in Kant's logic. In the former, objects are mathematical entities supplied by the metatheory, usually some version of set theory. These objects have no internal structure, at least not for the purposes of the model theory. Kant's notions of object, as they occur in the semantics furnished by transcendental logic, are very different. For instance, there are 'objects of experience', somehow constructed out of sensory material; transcendental logic deals with *a priori* and completely general principles which govern the construction of such objects, and relate judgements to objects so that we may come to speak of *true* judgements.

3.3. The Table of Judgement (A70/B95)

The three definitions describe judgement either in terms of certain cognitive operations ('unity of representations') or in terms of a function that a judgement has to perform (establishing 'relation to an object'). There is no hint of a specific form of judgement here. We find such hints in the Table of Judgement, but there we do not find a comparison with definitions of judgement; e.g. the *Critique*'s definition occurs only at (B141-2), way after the Table of Judgement is introduced. This raises the problem of how we know that the forms proposed in the Table satisfy the three definitions, and conversely, how for instance the functional characterisation given at (B141-2) leads to specific forms of judgement.

We now turn to the forms of judgement listed in the Table of Judgement, and we discuss (some of) the inferences in which these judgements participate, in part to emphasise the many differences between Kant's logic and modern logic[8] We will also comment on the relation between the Table of Judgement and the Table of Categories (A80/B106), although a full treatment is beyond the scope of this paper.

We will begin our discussion with the title 'Relation' (A70/B95), where we find

Relation
Categorical
Hypothetical
Disjunctive

3.3.1. Categorical judgements

These are judgements in subject-predicate form, combined with quantifiers and optional negation, which can occur on the copula and on the concepts occur-

---
[8]See note 1.

ring in the judgement. The Table of Judgement further specifies categorical judgements with regard to Quantity and Quality:

Quantity
Universal
Particular
Singular

In the Table of Categories we find a corresponding list of 'pure concepts of the understanding':

Of Quantity
Unity
Plurality
Totality

The precise correspondence between judgement forms and Categories is a matter of controversy. Here we argue on logical grounds that Kant intended a correspondence between the universal judgement and Unity, between the particular judgement and Plurality, and between the singular judgement and Totality.[9]

As explained in section 3.1, the universal judgement 'all M are P', or as Kant would have it 'To everything x to which M belongs, also P belongs', should not be interpreted as the classical $\forall x(M(x) \to P(x))$, but as

$$\forall x(\bigvee_{p \in M} p(x) \to \bigvee_{q \in P} q(x));$$

and because the subject is maintained 'assertorically', not 'problematically', we require that the types in M do not contain $\bot$. These types are therefore satisfiable – meaning that the (nonempty) collection of M's is given as that which the judgement is about, and the quantifier 'To everything x' is restricted to M, not to some universe of discourse.

The association 'universality – unity' is motivated by the fact that in the universal judgement 'all M are P' the predicate P makes no distinctions among the things falling under the subject M. Relative to P, M can hence be taken as a unit.

The things falling under M form a plurality that is not a unity (with respect to the predicate P) if there are true particular judgements 'some M are P' and 'some M are not P'.

---
[9] See Frede and Krüger [3] for a different correspondence linking the singular judgement and Unity.

In an unpublished note about the relation between universal and singular judgement, Kant writes:

> *Refl.* **3068** In the universal concept the sphere [=extension] of a concept is entirely enclosed in the sphere of another concept; [...] in the singular judgement, a concept that has no sphere at all is consequently merely enclosed as a part under the sphere of another concept. Thus singular judgements are to be valued equally with the universal ones, and conversely, a universal judgement is to be considered a singular judgement with regard to the sphere, much as if it were only one by itself. [9, p. 62]

Now consider (B111), where we read 'Thus **allness** (totality) is nothing other than plurality considered as a unity [...]'

Taking a plurality M to be a totality involves considering M as a unity, which means that a pair of judgements 'some M are P' and 'some M are not P' is replaced by one of 'all M are P' and 'all M are not P'. M is thus totally determined with respect to the available predicates. Since M cannot be divided using a predicate, this means that the concept M is used singularly, and hence a universal judgement 'all M are P' can equivalently be regarded as the singular judgement 'M is P', whence the correspondence between the singular judgement and totality.

Quality
Affirmative
Negative
Infinite

There is no need for our present purposes to dwell extensively on this Category, except to say that Kant makes a distinction between sentence negation as in the negative particular judgement 'some A are not B' and predicate negation, represented by the infinite judgement 'some A are non-B', which is affirmative but requires infinitary logic for its formalisation: $\bigvee_{B \cap C = \emptyset}$ (some A are C). Hence Kant's logic is not finitary. The difference with classical first order logic will only increase as we go on.

3.3.2. Hypothetical judgements

It would be a mistake to identify Kant's hypothetical judgements with a propositional conditional $p \to q$, let alone material implication as defined by its truth table: a material implication need not have any rule-like connection between antecedent and consequent. Here is the definition in the *Jäsche Logik*:

> The matter of hypothetical judgements consists of two judgements that are connected to each other as ground and consequence. One of these judgements, which contains the ground, is the *antecedent*, the other, which is related to it as consequence, is the *consequent*, and the representation of this kind of connection of two judgements to one another for the unity of consciousness is called the *consequentia* which constitutes the form of hypothetical judgements. [5, p. 601, par. 59][10]

This definition seems to say that the hypothetical is a propositional connective, and some of Kant's examples fall into this category:

> If there is perfect justice, then obstinate evil will be punished. (A73/B98)

However, other examples exhibit a more complex structure, involving relations, variables and binding. In the context of a discussion of the possible temporal relations between cause and effect Kant writes in *CPR*:

> If I consider a ball that lies on a stuffed pillow and makes a dent in it as a cause, it is simultaneous with its effect. (A203/B246)

The hypothetical that can be distilled from this passage is:

> If a ball lies on a stuffed pillow, it makes a dent in that pillow.

From this we see that (i) the antecedent and consequent need not be closed judgements but may contain variables, and (ii) antecedent and consequent may contain relations and existential quantifiers.

We now give an extended quote from the *Prolegomena* §29 [8] which provides another example of a hypothetical judgement whose logical structure likewise exhibits the features listed in (i) and (ii) above:

> It is, however, possible that in perception a rule of relation will be found, which says this: that a certain appearance is constantly followed by another (though not the reverse); and this is a case for me to use a hypothetical judgement and, e.g., to say: If a body is illuminated by the sun for long enough, it becomes warm. Here there is of course not yet the necessity of connection, hence not yet the concept of cause. But I continue on, and say: if the above proposition, which is merely a subjective connection of perceptions, is to be a proposition of experience, then it must be regarded as necessarily and universally valid. But a proposition of this sort would be: The sun through its light is the cause of the warmth. The foregoing empirical rule is now regarded as a law, and indeed as valid not merely of appearances, but of them on behalf of a possible experience, which requires

---

[10]Here it is of interest to observe that in the same paragraph *consequentia* is also used to refer to an inference.

universally and therefore necessarily valid rules [...] the concept of a cause indicates a condition that in no way attaches to things, but only to experience, namely that experience can be an objectively valid cognition of appearances and their sequence in time only insofar as the antecedent appearance can be connected with the subsequent one according to the rule of hypothetical judgements. [8, p. 105]

The logical form of the first hypothetical (a 'judgement of perception') is something like:

If x is illuminated by y between time t and time s and $s - t > d$ and the temperature of x at t is v, then there exists a $w > 0$ such that the temperature of x at s is $v + w$ and $v + w > c$,

where d is the criterion value for 'long enough' and c a criterion value for 'warm'. We find all the ingredients of polyadic logic here: relations and quantifier alterations. The causal connection which transforms the judgement into a 'judgement of experience' arises when the existential quantifiers are replaced by explicitly definable functions.

We now move on to the logical properties of the hypothetical judgement. Here it is of some importance to note that the term *consequentia*, characterising the logical form of the hypothetical, is also used to describe the inferences from the hypothetical:

The *consequentia* from the ground to the grounded, and from the negation of the grounded to the negation of the ground, is valid. [5, p. 623]

Furthermore, the negation of a hypothetical is not defined.[11] This strongly suggests that the hypothetical judgement is really a license for inferences. Indeed, in the *Jäsche Logik* Kant characterises inferences such as modus ponens and modus tollens as immediate inferences and as such needing only one premise, not two premises [5, p. 623]. Modern proof systems conceive of modus ponens as a two-premise inference, p implies q and p, therefore q. But Kant does not think of it in this way. He thinks of it as an inference with premise p, conclusion q, which is governed by a license for inference. This strongly suggests that Kant does not have a single entailment relation, as in modern logic,[12] but only local entailment relations defined by specific inferences. We end this discussion of the hypothetical judgement with a further

---

[11] Note that the negation of a categorical judgement is defined, although its properties do seem to be weaker than classical negation: 'some A are not B' is the negation of 'All A are B', but it is a moot point whether the negative particular judgement has existential import. Its infinitive counterpart does have existential import.

[12] See Hodges [4] for relevant discussion.

twist: its logical properties change when it is considered in a causal context, i.e. in transcendental logic:

> When the cause has been posited, the effect is posited ‹*posita causa ponitur effectus*› already flows from the above. But when the cause has been cancelled, the effect is cancelled ‹*sublata causa tollitur effectus*› is just as certain; when the effect has been cancelled, the cause is cancelled ‹*sublato effectu tollitur causa*› is not certain, but rather the causality of the cause is cancelled ‹*tollitur causalitas causae*›. [6, p.336-7]

3.3.3. Disjunctive judgements

These are again not what one would think, judgements of the form $p \lor q$. The *Jäsche Logik* provides the following definition:

> A judgement is disjunctive if the parts of the sphere of a given concept determine one another in the whole or toward a whole as complements [...] [A]ll disjunctive judgements represent various judgements as in the community of a sphere [...] [O]ne member determines every other here only insofar as they stand together in community as parts of a whole sphere of cognition, outside of which, in a certain relation, nothing may be thought.(*Jäsche Logik*, §27, 28) [5, pp. 602-3]

As examples Kant provides:

> Every triangle is either right-angled or not right-angled.
> A learned man is learned either historically, or in matters of reason.

Thus the logical form is something like $\forall x(C(x) \to A(x) \lor B(x))$, where C represents the whole, A, B its parts; here it is not immediately clear whether the parts can be taken to exist outside the context of the whole. But actually the situation is much more complicated. The *Jäsche Logik* equivocates between concepts and judgements making up the whole, and this is intentional, as we read in the *Vienna Logic*:

> The disjunctive judgment contains the relation of different judgment insofar as they are equal, as *membra dividentia*, to the *sphaera* of a *cognitio divisa*. E.g., All triangles, as to their angles, are either right-angled or acute or obtuse. I represent the different members as they are opposed to one another and as, taken together, they constitute the whole *sphaera* of the *cognitio divisa*. This is in fact nothing other than a logical division, only in the division there does not need to be a *conceptus divisus*; instead, it can be a *cognitio divisa*. E.g., If this is not the best world, then God was not able or did not want to create a better one. This is the division of the *sphaera* of the cognition that is given to me. [5, p. 374-5]

So it is not just concepts that can be divided in the familiar way, also cognitions (*Erkenntnisse*), including judgements, can be so divided. What this means for the complexity of Kant's logic can be seen if we look at the expanded example in the *Dohna-Wundlacken Logic*:

> If this world is not the best, then God either was unfamiliar with a better [one] or did not wish to create it or could not create [it], etc. Together these constitute the whole *sphaera*. [5, p. 498]

It will be instructive to formalise this example. Let w_0 be the actual world, G a constant denoting God, let $B(w_0, w)$ represent 'w is a better world than w_0', and let $\mathit{Uf}(G,w)$, $\mathit{Uw}(G,w)$, $\mathit{Uc}(G,w)$ represent: 'God was unfamiliar with w', 'God was unwilling to create w' and 'God was unable to create w', respectively. We then get the combined hypothetical-disjunctive judgement:

$$\exists w B(w_0, w) \to \forall w (B(w_0, w) \to (\mathit{Uf}(G, w) \vee \mathit{Uw}(G, w) \vee \mathit{Uc}(G, w))).$$

It is to be noted that this hypothetical-disjunctive judgement consists entirely of relations, and that the division is formulated in terms of singular judgements containing a parameter ('God') and a variable. As in the case of the hypothetical judgement, the negation for a disjunctive judgement is not defined, which suggests that it is actually a license for inferences, using quantified forms of the disjunctive syllogism, for example:

1. Starting from the premise 'God is familiar with a better world' (which is taken to imply $\exists w(B(w_0, w) \wedge \neg \mathit{Uf}(G, w)))$ now introduces the positive primitive formula $\exists w(B(w_0, w) \vee (\mathit{Uw}(G, w) \vee \mathit{Ua}(G, w)))$.

2. Similarly the premise 'God is familiar with all better worlds' yields the formula $\forall w(B(w_0, w) \to (\mathit{Uw}(G, w) \vee \mathit{Ua}(G, w)))$.

Kant evidently believes these inferences are perfectly proper cases of the disjunctive syllogism, but the present-day reader may well ask whether his general logic has the resources to break down these inferences in smaller steps. But if the hypothetical and the disjunctive judgement are licenses for inferences, this means that they can be taken as given as far as general logic is concerned (much like a Prolog program is taken as given and is used only to derive atomic facts). This somewhat eases the burden on general logic, in the sense that it need not have the resources to prove hypothetical and disjunctive judgements.

As we did for the hypothetical judgement, we will also look at the intended transcendental use of the disjunctive judgement:

> The same procedure of the understanding when it represents to itself the sphere of a divided concept, it also observes in thinking of a thing as divisible; and just as in the first case the members of the division exclude each other, and yet are connected in one sphere, so in the latter case the understanding represents to itself the parts of the latter as being such that existence pertains to each of them (as substances) exclusively of the others, even while they are combined together in one whole. (B113)

The disjunctive judgement is said to involve the cognitive act of dividing a thing, while keeping the resulting parts simultaneously active in one representation. Here we are concerned with the logical principles that Kant's disjunction satisfies. Kant gives as inferences valid for a disjunctive judgement $C \to A \vee B$, the two halves of the so-called disjunctive syllogism:

C and $\neg A$ implies B
C and A implies $\neg B$.

These inference rules are considerably weaker than those that are valid for the classical or intuitionistic disjunction, and remind one of the multiplicative disjunction of linear logic. Can one impose stronger inference rules on the disjunction? That is doubtful. For example, the standard right disjunction rule in sequent calculus:

$$\frac{\Gamma \Rightarrow A, \Delta}{\Gamma \Rightarrow A \vee B, \Delta}$$

is invalid for Kant, because it allows the addition of an arbitrary B to A, without the guarantee that A, B constitute a whole.

An additional consideration is the connection with divisibility; here the parts must be present simultaneously, which is what the rule just given expresses. This formulation lends some credibility to Kant's association of the disjunctive judgement with the category of simultaneity in the third Analogy of Experience. However, the new formulation raises the issue of what one should say if A and B are identical. Kant makes an important distinction between two kinds of identity in 'On the amphiboly of concepts of reflection':

> If an object is presented to us several times, but always with the same inner determinations, then it is always exactly the same if it counts as an object of pure understanding, not many but only one thing; but if it is appearance, then [...] however identical everything may be in regard to [concepts], the difference of the places of these appearances at the same time is still an adequate ground for the numerical difference of the object (of the senses) itself. Thus, in the case of two drops of water one can completely abstract from all inner difference (of quality and quantity), and it is enough that they

be intuited in different places at the same time for them to be held to be numerically different. (A263-4/B319-20)

Suppose one has a 'whole' that is divided into spatially distinct parts that have 'the same inner determinations'. This hypothetical situation suggests that a logic for Kant's disjunction does not include a rule for (right) contraction:

$$\frac{\Gamma \Rightarrow A, A, \Delta}{\Gamma \Rightarrow A, \Delta}$$

But in that case also the standard rule for left disjunction introduction:

$$\frac{\Gamma, A \Rightarrow \Delta \quad \Gamma, B \Rightarrow \Delta}{\Gamma, A \vee B \Rightarrow \Delta}$$

must be dropped because otherwise right contraction becomes derivable. Instead, one would have a rule like:

$$\frac{\Gamma, A \Rightarrow \Delta \quad \Gamma, B \Rightarrow \Delta'}{\Gamma, A \vee B \Rightarrow \Delta, \Delta'}$$

3.4. Logical form of judgements

Looking back at our examples we see that, with one exception (the negative particular judgement, which, as discussed in [1] was meant by Kant to be purely negative), they are all geometric judgements. Geometric logic, i.e. the logic of geometric formulas, plays an important role in several branches of mathematics, Euclidean geometry being one but not the only example. More germane to our purposes is a result in [1], which shows that all objectively valid judgements in the sense of (B141-2) must be finite conjunctions of geometric implications.

3.5. 'Functions of unity in judgements': dynamical proofs

In a dynamical proof one takes a geometric theory[13] as defining a consequence relation holding between two sets of facts. An example, taken from Coquand [2], illustrates the idea. The theory is:[14]

1. $P(x) \wedge U(x) \rightarrow Q(x) \vee \exists y R(x, y)$

2. $P(x) \wedge Q(x) \rightarrow \bot$

---
[13] We assume the geometric implications in the theory have antecedents consisting of conjunctions of atomic formulas only.
[14] We omit the universal quantifiers.

3. $P(x) \wedge R(x,y) \to S(x)$

4. $P(x) \wedge T(x) \to U(x)$

5. $U(x) \wedge S(x) \to V(x) \vee Q(x)$

And here is an example of a derivation of $V(a_0)$ from $P(a_0), T(a_0)$:

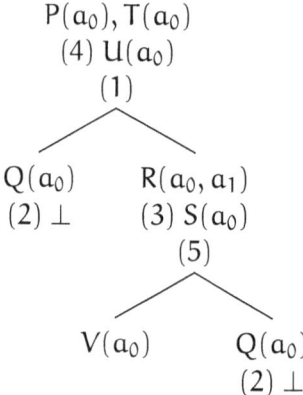

We give some comments on the derivation. The dynamical proof just given can also be taken to prove $\forall x(P(x) \wedge T(x) \to V(x))$, where the proof is the link between antecedent and consequent, hence a 'function of unity'. Furthermore, the geometric theory defines the consequence relation, hence the geometric implications occurring in it can be seen as inference rules. Disjunctions lead to branching of the tree, as we see in (1) and (5). The existential quantifier in formula (1) introduces a new term in the proof, here a_1, which appears in the right branch of (1). This constant is the 'object in general' of *Reflexion* **4634**. Lastly, a fact is derivable if it appears on every branch not marked by $\bot$, which leaves $V(a_0)$. If X is a collection of facts whose terms are collected in I, F a fact with terms in I, and T a geometric theory, then there exists a dynamical proof of F from X if and only if $T, X \vdash F$ in intuitionistic logic.

It is clear how a dynamical proof of a geometric implication from a geometric theory proceeds: if T is the geometric theory and $\forall \bar{x}(\tau(\bar{x}) \to \theta(\bar{x}))$ the geometric implication (τ is a conjunction of atomic formulas, and for simplicity take θ an existentially quantified conjunction θ' of atomic formulas; we interpret θ' as a set), choose new terms not occurring in either T or $\forall \bar{x}(\tau(\bar{x}) \to \theta(\bar{x}))$, plug these terms into τ and construct a dynamical proof tree with the sets θ' at the leaves. There may occur terms in θ' not in τ; these have to be quantified existentially. Introduce any other existential quantifiers on θ' as required by θ. The result is an intuitionistic derivation of $\forall \bar{x}(\tau(\bar{x}) \rightarrow \theta(\bar{x}))$ from T. Conversely, if there is an intuitionistic derivation of $\forall \bar{x}(\tau(\bar{x}) \rightarrow \theta(\bar{x}))$ from T, then there exists a dynamical proof in the sense just sketched.

Dynamical proofs as a semantics for geometric implications can explain Kant's characterisation of judgements as rules, as well as 'a unity of the consciousness of various representations'; after all, the diagram represents 'unity' as a single spatial representation. What remains to be done is to situate a judgement's 'objective validity' relative to its other properties.

4. Completeness of the Table of Judgement

In [1] it is argued that (i) Kant's implied semantics for logic is radically different from that of classical first order logic, (ii) the implied semantics, centered around Kant's three different notions of object, can be given a precise mathematical expression, thus leading to a formalised transcendental logic, and (iii) on the proposed semantics, Kant's formal logic turns out to be geometric logic.

It is not appropriate to repeat the technical exposition here, so we will follow a different strategy starting from Kant's most fundamental characterisation of judgement:

> A judgement is nothing but the manner in which given cognitions are brought to the objective unity of apperception. (B141)

A judgement is the act of binding together mental representations; this is what the term 'unity' refers to. The aim of judgement is indicated by means of the word 'objective', which is Kant's terminology for 'having relation to an object'. But for Kant, objects are not found in experience, but they are constructed ('synthesised') from sensory matter under the guidance of the Categories, which are defined as 'concepts of an object in general, by means of which the intuition of an object is regarded as determined in respect of one of the logical functions of judgement' (B128). It is here that judgement plays an all-important role, since Kant's idea is that objects are synthesised through the act of making judgements about them.

Technically, these acts of synthesis are modelled as a kind of possible worlds structure (an 'inverse system'), where the possible worlds are finite first order models whose elements are partially synthesised objects, except for the unique top-world (the 'inverse limit') which represents (the idea of) fully synthesised objects. Bringing a (formal) judgement φ to the 'objective unity of apperception' is now characterised by the property: for any such possible worlds structure, if φ is true on all worlds, then φ is also true on the top-world. That is to say, if φ is true for all stages of synthesis of an object, then φ is true of some fully synthesised object. Kant calls judgements φ satisfying this conditional property 'objectively valid.' It turns out that the objectively valid formulas are exactly the geometric formulas. It follows that no judgement whose logical form is more complex than that allowed by the Table of Judgement can be objectively valid, i.e. this Table is complete.

It is of some interest that the key idea in the proof sheds light on Kant's logical reinterpretation of the Categories of Quantity as constraints on concepts (B113-6):

> In every cognition of an object there is, namely, **unity** of the concept, which one can call **qualitative unity** insofar as by that only the unity of the comprehension of the manifold of cognition is thought, as, say, the unity of the theme in a play, a speech, or a fable. Second, truth in respect of the consequences. The more true consequences from a given concept, the more indication of its objective reality. One could call this the **qualitative plurality** of the marks that belong to a concept as a common ground ... Third, finally, **perfection**, which consists in plurality conversely being traced back to the unity of the concept, and agreeing completely with this one and no other one, which one can call **qualitative completeness** (totality).

The phrase 'unity of the theme in a play' is probably a reference to Aristotle's 'unity of action' in tragedy, where

> the structural union of the parts [must be] such that, if any one of them is displaced or removed, the whole will be disjointed and disturbed. For a thing whose presence or absence makes no visible difference, is not an organic part of the whole (*Poetics*, VIII).

Hence we read 'qualitative unity' as the requirement that the concept under consideration is integrated with other concepts by means of a theory, and is invariant under structure-preserving mappings (homomorphisms). The latter requirement forces all subconcepts of the given concept to have the same logical complexity. We are now in a position to spell out the logical meaning of B113-6 in formal terms.

Let C be a concept which satisfies 'qualitative unity' and let T be the first order theory witnessing 'qualitative unity'. Define a 'qualitative plurality' Σ

by
$$\Sigma(x) = \{\theta(x) \mid T \models \forall x(C(x) \to \theta(x)), \theta \text{ pos. prim.}\}.$$

Because we may have, for each θ, 'some θ aren't C', for all we know Σ could be a proper plurality. But 'qualitative completeness' now becomes provable:

$$\Sigma(x), T \models C(x),$$

hence by compactness there is positive primitive $\tau(x)$ such that

$$T \models \forall x(\tau(x) \leftrightarrow C(x)).$$

It follows that, as announced in section 3.1, universal judgements 'all M are P' can be expressed as geometric implications, provided the concepts M, P satisfy 'qualitative unity'.

In summary, we have shown that after formalisation, Kant's general logic turns out to be at least as rich as geometric logic, while it coincides with it when taking into account the semantics of judgements dictated by 'transcendental logic'.[15] This latter result is but one example of interesting metalogical theorems that may be proved about Kant's logic; B113-6, formally reinterpreted as a theorem about definability of concepts, is another.

References

[1] T. Achourioti and M. van Lambalgen. A formalisation of Kant's transcendental logic. *Review of Symbolic Logic*, 4(2):254–289, 2011. DOI: http://dx.doi.org/10.1017/S1755020310000341.

[2] T. Coquand. A completeness proof for geometric logic. Technical report, Computer Science and Engineering Department, University of Gothenburg, 2002. Retrieved September 29, 2010, from http://www.cse.chalmers.se/~coquand/formal.html.

[3] M. Frede and L. Krüger. Ueber die Zuordnung der Quantitaeten des Urteils und der Kategorien der Groesse bei Kant. *Kant-Studien* 61, 28–49, 1970.

[4] W. Hodges. Traditional Logic, Modern Logic and Natural Language. *Journal of Philosophical logic*, 38:589–606, 2009. DOI: http://dx.doi.org/10.1007/s10992-009-9113-y.

[5] I. Kant. *Lectures on Logic; translated from the German by J. Michael Young*. The Cambridge edition of the works of Immanuel Kant. Cambridge University Press, Cambridge, 1992.

---

[15] We simplify here and do not consider multiplicative disjunction.

[6] I. Kant. *Lectures on Metaphysics; edited by Karl Ameriks and Steve Naragon.* Cambridge University Press, Cambridge, 1997.

[7] I. Kant. *Critique of pure reason; translated from the German by Paul Guyer and Allen W. Wood.* The Cambridge edition of the works of Immanuel Kant. Cambridge University Press, Cambridge, 1998.

[8] I. Kant. *Theoretical philosophy after 1781; edited by Henry Allison and Peter Heath.* The Cambridge edition of the works of Immanuel Kant. Cambridge University Press, Cambridge, 2002.

[9] I. Kant. *Notes and fragments; edited by Paul Guyer.* The Cambridge edition of the works of Immanuel Kant. Cambridge University Press, Cambridge, 2005.

[10] M. Wolff. Vollkommene Syllogismen und reine Vernunftschluesse: Aristoteles und Kant. Eine Stellungnahme zu Theodor Eberts Gegeneinwaenden. Teil 2. *Journal for General Philosophy of Science*, 2010. DOI: http://dx.doi.org/10.1007/s10838-010-9136-7.

Received: 31 July 2016

Implicit Dynamic Function Introduction and Ackermann-like Function Theory

Marcos Cramer

University of Luxembourg
6, rue Richard Coudenhove-Kalergi, 1359 Luxembourg
`marcos.cramer@uni.lu`

Abstract: We discuss a feature of the natural language of mathematics – the implicit dynamic introduction of functions – that has, to our knowledge, not been captured in any formal system so far. If this feature is used without limitations, it yields a paradox analogous to Russell's paradox. Hence any formalism capturing it has to impose some limitations on it. We sketch two formalisms, both extensions of Dynamic Predicate Logic, that innovatively do capture this feature, and that differ only in the limitations they impose onto it. One of these systems is based on *Ackermann-like Function Theory*, a novel foundational theory of functions that is inspired by Ackermann Set Theory and that interprets ZFC.

Keywords: dynamic predicate logic, function introduction, Ackermann set theory, function theory.

1. Dynamic predicate logic

Dynamic predicate logic (*DPL*) [7] is a formalism whose syntax is identical to that of standard first-order predicate logic (PL), but whose semantics is defined in such a way that the dynamic nature of natural language quantification is captured in the formalism:

1. If a farmer owns a donkey, he beats it.
2. PL: $\forall x \, \forall y \, (farmer(x) \land donkey(y) \land owns(x,y) \rightarrow beats(x,y))$.
3. DPL: $\exists x \, (farmer(x) \land \exists y \, (donkey(y) \land owns(x,y))) \rightarrow beats(x,y)$.

In PL, 3 is not a sentence, since the rightmost occurrences of x and y are free. In DPL, a variable may be bound by a quantifier even if it is outside its scope.

The semantics is defined in such a way that 3 is equivalent to 2. So in DPL, 3 captures the meaning of 1 while being more faithful to its syntax than 2.

1.1. DPL semantics

We present DPL semantics in a way slightly different but logically equivalent to its definition by Groenendijk and Stokhof in [7]. Structures and assignments are defined as for PL: A structure S specifies a domain $|S|$ and an interpretation a^S for every constant, function or relation symbol a in the language. An S-assignment is a function from variables to $|S|$. Let G_S denote the set of S-assignments. Given two assignments g, h, we define $g[x]h$ to mean that g differs from h at most in what it assigns to the variable x. Given a DPL term t, we recursively define

$$[t]_S^g = \begin{cases} g(t) & \text{if } t \text{ is a variable,} \\ t^S & \text{if } t \text{ is a constant symbol,} \\ f^S([t_1]_S^g, \ldots, [t_n]_S^g) & \text{if } t \text{ is of the form } f(t_1, \ldots, t_n). \end{cases}$$

Groenendijk and Stokhof [7] define an interpretation function $[\![\cdot]\!]_S$ from DPL formulae to subsets of $G_S \times G_S$. We instead recursively define for every $g \in G_S$ an interpretation function $[\![\cdot]\!]_S^g$ from DPL formulae to subsets of G_S:[1]

1. $[\![\top]\!]_S^g := \{g\}$.
2. $[\![t_1 = t_2]\!]_S^g := \{h | h = g \text{ and } [t_1]_S^g = [t_2]_S^g\}$.[2]
3. $[\![R(t_1, \ldots, t_2)]\!]_S^g := \{h | h = g \text{ and } ([t_1]_S^g, \ldots, [t_2]_S^g) \in R^S\}$.
4. $[\![\neg \varphi]\!]_S^g := \{h | h = g \text{ and there is no } k \in [\![\varphi]\!]_S^h\}$.
5. $[\![\varphi \wedge \psi]\!]_S^g := \{h | \text{there is a } k \text{ s.t. } k \in [\![\varphi]\!]_S^g \text{ and } h \in [\![\psi]\!]_S^k\}$.
6. $[\![\varphi \to \psi]\!]_S^g := \{h | h = g \text{ and for all } k \text{ s.t. } k \in [\![\varphi]\!]_S^h, \text{ there is a } j \text{ s.t. } j \in [\![\psi]\!]_S^k\}$.
7. $[\![\exists x \, \varphi]\!]_S^g := \{h | \text{there is a } k \text{ s.t. } k[x]g \text{ and } h \in [\![\varphi]\!]_S^k\}$.

$\varphi \vee \psi$ and $\forall x \, \varphi$ are defined to be a shorthand for $\neg(\neg \varphi \wedge \neg \psi)$ and $\exists x \, \top \to \varphi$ respectively.

---

[1] This can be viewed as a different currying of the uncurried version of Groenendijk and Stokhof's interpretation function.

[2] The condition $h = g$ in cases 2, 3, 4 and 6 implies that the defined set is either $\emptyset$ or $\{g\}$.

2. Implicit dynamic introduction of function symbols

Functions are often dynamically introduced in an implicit way in mathematical texts. For example, [10] introduces the additive inverse function on the reals as follows:

(a) For each a there is a real number $-a$ such that $a + (-a) = 0$. [10, p. 1]

Here the natural language quantification "there is a real number $-a$" *locally* (i.e. inside the scope of "For each a") introduces a new real number to the discourse. But since the choice of this real number depends on a and we are universally quantifying over a, it *globally* (i.e. outside the scope of "For each a") introduces a function "$-$" to the discourse.

The most common form of implicitly introduced functions are functions whose argument is written as a subscript, as in the following example:

(b) Since f is continuous at t, there is an open interval I_t containing t such that $|f(x) - f(t)| < 1$ if $x \in I_t \cap [a, b]$. [10, p. 62]

If one wants to later explicitly call the implicitly introduced function a function, the standard notation with a bracketed argument is preferred:

(c) Suppose that, for each vertex v of K, there is a vertex $g(v)$ of L such that $f(\operatorname{st}_K(v)) \subset \operatorname{st}_L(g(v))$. Then g is a simplicial map $V(K) \to V(L)$, and $|g| \simeq f$. [8, p. 19]

When no uniqueness claims are made about the object locally introduced to the discourse, implicit function introduction presupposes the existence of a choice function, i.e. presupposes the Axiom of Choice. We hypothesise that the naturalness of such implicit function introduction in mathematical texts contributes to the wide-spread feeling that the Axiom of Choice must be true.

Implicitly introduced functions generally have a restricted domain and are not defined on the whole universe of the discourse. In the example (c), g is only defined on vertices of K and not on vertices of L. Implicit function introduction can also be used to introduce multi-argument functions, but for the sake of simplicity and brevity, we restrict ourselves to unary functions in this paper.

If the implicit introduction of functions is allowed without limitations, one can derive a contradiction:

(d) For every function f, there is a natural number $g(f)$ such that

$$g(f) = \begin{cases} 0 & \text{if } f \in \text{dom}(f) \text{ and } f(f) \neq 0, \\ 1 & \text{if } f \notin \text{dom}(f) \text{ or } f(f) = 0. \end{cases}$$

Then g is defined on every function, i.e. $g(g)$ is defined. But from the definition of g, $g(g) = 0$ iff $g(g) \neq 0$.

This contradiction is due to the *unrestricted function comprehension* that is implicitly assumed when allowing implicit introductions of functions without limitations. Unrestricted function comprehension could be formalised as an axiom schema as follows:

Axiom Schema 1 (Unrestricted function comprehension). For every formula $\varphi(x,y)$, the following is an axiom: $\forall x \, \exists y \, \varphi(x,y) \to \exists f \, \forall x \, \varphi(x, f(x))$.

The inconsistency of unrestricted function comprehension is analogous to the inconsistency of unrestricted set comprehension, i.e. Russell's paradox.

Russell's paradox led to the abandonment of unrestricted comprehension in set theory. Two radically different approaches have been undertaken for restricting set comprehension: Russell himself restricted it through his Ramified Theory of Types, which was later simplified to Simple Type Theory (STT), mainly known via Church's formalisation in his simply typed lambda calculus [2]. On the other hand, the risk of paradoxes like Russell's paradox also contributed to the development of ZFC (Zermelo-Fraenkel set theory with the Axiom of Choice), which allows for a much richer set theoretic universe than the universe of simply typed sets. Since all the axioms of ZFC apart from the Axiom of Extensionality, the Axiom of Foundation and the Axiom of Choice are special cases of comprehension, one can view ZFC as an alternative way to restrict set comprehension.

Similarly, the above paradox must lead to the abandonment of unrestricted function comprehension. The type-theoretic approach is easily adapted to functions, so we will first sketch the system that formalises this approach, *Typed Higher-Order Dynamic Predicate Logic*. For an untyped approach, there is no clear way to transfer the limitations that ZFC puts onto set comprehension to the case of function comprehension. However, there is an axiomatization of set theory (with classes) called *Ackermann set theory* that is a conservative extension of ZFC. It turns out that the limitations that Ackermann set theory poses on set comprehension can be transferred to the case of function comprehension, and hence to the case of implicit dynamic function introduction.

The need to deal with implicit function introduction arose for us in the context of the *Naproche project*, a project aiming at automatic formalisation of natural language mathematics [3, 5, 6]. It has been implemented in the Naproche system using type restrictions as in Typed Higher-Order Dynamic Predicate Logic, and we plan to implement it using the less strict restrictions of the untyped Higher-Order Dynamic Predicate Logic in a future version of the system.

3. Typed higher-order dynamic predicate logic

In this section, we extend DPL to a system called *Typed Higher-Order Dynamic Predicate Logic* (THODPL), which formalises implicit dynamic function introduction, and also allows for explicit quantification over functions. THODPL has variables typed by the types of STT. In the below examples we use x and y as variables of the basic type i, and f as a variable of the function type $i \to i$. A complex term is built by well-typed application of a function-type variable to an already built term, e.g. $f(x)$ or $f(f(x))$.

The distinctive feature of THODPL syntax is that it allows not only variables but any well-formed terms to come after quantifiers. So (1) is a well-formed formula:

$$\forall x\ \exists f(x)\ R(x, f(x)), \tag{1}$$

$$\forall x\ \exists y\ R(x, y), \tag{2}$$

$$\exists f\ (\forall x\ R(x, f(x))). \tag{3}$$

The semantics of THODPL is to be defined in such a way that (1) has the same truth conditions as (2). But unlike (2), (1) dynamically introduces the function symbol f to the context, and hence turns out to be equivalent to (3).

We now sketch how these desired properties of the semantics can be achieved. In THODPL semantics, an assignment assigns elements of $|S|$ to variables of type i, functions from $|S|$ to $|S|$ to variables of type $i \to i$ etc. Additionally, an assignment can also assign an object (or function) to a complex term. For example, any assignment in the interpretation of $\exists f(x)\ R(x, f(x))$ has to assign some object to $f(x)$. The definition of $g[x]h$ can now naturally be extended to a definition of $g[t]h$ for terms t. The definition of $[\![t]\!]_S^g$ has to be adapted in the natural way to account for function variables.

Just as in the case of DPL semantics, we recursively define an interpretation $[\![\cdot]\!]_S^g$ from DPL formulae to subsets of G_S (the cases 1-5 of the recursive definition are as in Section 1.1):

6. $[\![\varphi \to \psi]\!]_S^g := \{h | h$ differs from g in at most some function variables $f_1, \ldots, f_n$ (where this choice of function variables is maximal), and there is a variable x such that for all $k \in [\![\varphi]\!]_S^g$, there is an assignment $j \in [\![\psi]\!]_S^k$ such that $j(f_i(x)) = h(f_i)(k(x))$ for $1 \leq i \leq n$, and if $n > 0$ then $k[x]g\}$.

7. $[\![\exists t\ \varphi]\!]_S^g := \{h |$ there is a k s.t. $k[t]g$ and $h \in [\![\varphi]\!]_S^k\}$.

In order to make case 6 of the definition more comprehensible, let us consider its role in determining the semantics of (1), i.e. of $\exists x\ \top \to \exists f(x)\ R(x, f(x))$: First note that $[\![\exists f(x)\ R(x, f(x))]\!]_S^k$ is the set of assignments j satisfying $R(x, f(x))$ (i.e. for which $[\![R(x, f(x))]\!]_S^j$ is non-empty) such that $j[f(x)]k$. Furthermore note that $[\![\exists x\ \top]\!]_S^g$ is the set of assignments k such that $k[x]g$. So by case 6 with $n = 1$,

$[\![\exists x\ \top \to \exists f(x)\ R(x, f(x))]\!]_S^g = \{h | h[f]g$ and there is a variable x such that for all k such that $k[x]g$, there is an assignment j satisfying $R(x, f(x))$ such that $j[f(x)]k$ and $j(f(x)) = h(f)(k(x))$, and $k[x]g\}$

$= \{h | h[f]g$ and for all k such that $k[x]g$, there is an assignment j satisfying $R(x, f(x))$ such that $j[f(x)]k$ and $j(f(x)) = h(f)(k(x))\}$

$= \{h | h[f]g$ and for all k such that $k[x]h$, k satisfies $R(x, f(x))\}$

$= [\![\exists f\ (\forall x\ R(x, f(x)))]\!]_S^g$.

The type restrictions THODPL imposes may be too strict for some applications: Mathematicians sometimes do make use of functions that do not fit into the corset of strict typing, e.g. a function defined on both real numbers and real functions. To overcome this restriction, we will introduce an untyped variant HODPL in Section 6. But for this, we require some foundational preliminaries.

4. Ackermann set theory

Ackermann set theory [1] postulates not only sets, but also proper classes which are not sets.[3] The sets are distinguished from the proper classes by a unary predicate M (from the German word "Menge" for "set").

Ackermann presented a pure version of his theory without urelements, and a separate version with urelements, which we will present here. The language of Ackermann set theory contains three predicates: A binary predicate $\in$, a unary predicate M and a unary predicate U for urelements. We introduce $L(x)$ ("x is limited") as an abbreviation for $M(x) \vee U(x)$. The idea is that sets and urelements are objects of limited size, and are distinguished from the more problematic classes of unlimited size.

The axioms of Ackermann set theory with urelements are as follows:

- *Extensionality axiom:* $\forall x \, \forall y \, (\forall z \, (z \in x \leftrightarrow z \in y) \rightarrow x = y)$.

- *Class comprehension axiom schema:* Given a formula $F(y)$ (possibly with parameters[4]) that does not have x among its free variables, the following is an axiom:
$\forall y \, (F(y) \rightarrow L(y)) \rightarrow \exists x \, \forall y \, (y \in x \leftrightarrow F(y))$.

- *Set comprehension axiom schema:* Given a formula $F(y)$ (possibly with parameters that are limited[5]) that does not have x among its free variables and does not contain the symbol M, the following is an axiom:
$\forall y \, (F(y) \rightarrow L(y)) \rightarrow \exists x \, (M(x) \wedge \forall y \, (y \in x \leftrightarrow F(y)))$.

- *Elements and subsets of sets are limited:*
$\forall x \, \forall y \, (M(y) \wedge (x \in y \vee \forall z \, (z \in x \rightarrow z \in y)) \rightarrow L(y))$.

---

[3]Note, however, that unlike the more well-known class theory NBG, Ackermann set theory also allows for proper classes that contain proper classes.

[4]This means that F may actually be of the form $F(\bar{z}, y)$, and that these parameters are universally quantified in the axiom:
$\forall \bar{z} \, (\forall y \, (F(\bar{z}, y) \rightarrow M(y)) \rightarrow \exists x \, \forall y \, (y \in x \leftrightarrow F(\bar{z}, y)))$.

[5]Formally, with the parameters made explicit, the set comprehension axiom schema reads as follows:
$\forall z_1, \ldots, z_n \, (L(z_1) \wedge \cdots \wedge L(z_n) \rightarrow (\forall y \, (F(z_1, \ldots, z_n, y) \rightarrow L(y)) \rightarrow \exists x \, (M(x) \wedge \forall y \, (y \in x \leftrightarrow F(z_1, \ldots, z_n, y)))))$.

So unlimited set comprehension is replaced by two separate comprehension schemata, one for class comprehension and one for set comprehension. In both cases, the comprehension is restricted by the constraint that only limited objects satisfy the property that we are applying comprehension to. But for set comprehension, we have the additional constraint that the property may not be defined using the setness predicate or using a proper class as parameter. Ackermann justified this approach by appeal to a definition of "set" from Cantor's work [1].

If an Axiom of Foundation for sets is added, Ackermann set theory turns out to be – in what it says about sets – precisely equivalent to ZF [9]. But this equivalence is not a triviality: It is especially hard to establish Replacement for the sets of Ackermann set theory.

5. Ackermann-like function theory

Now we transfer the ideas of a comprehension limited in this way from set comprehension to function comprehension. For this a dichotomy similar to that between sets and classes has to be imposed on functions. We propose the terms *function* and *map* respectively for this dichotomy, and call the theory resulting from these limitations on function comprehension *Ackermann-like Function Theory* (AFT). AFT can be shown to be equiconsistent with Ackermann set theory and hence with ZFC (see Theorem 4 below).

The language of Ackermann-like function theory (L_{AFT}) contains

- a unary predicate F for functions,

- a unary predicate U for urelements,

- a constant symbol u for undefinedness, and

- a binary function symbol a for function application.

Instead of $a(f, t)$ we usually simply write $f(t)$. We write $L(x)$ instead of $U(x) \vee F(x)$. The undefinedness constant u is needed for formalising the idea that a function is only defined for certain values and undefined for others. In this language, the unrestricted function comprehension schema would be as follows:

Axiom Schema 2 (Unrestricted function comprehension in L_{AFT}). Given a variable z and formulae $P(z)$ and $R(z,x)$ (possibly with parameters), the following is an axiom: $\forall z\, (P(z) \to \exists x\, R(z,x)) \to \exists f\, (\neg U(f) \wedge \forall z\, ((P(z) \to R(z, f(z))) \wedge (\neg P(z) \to f(z) = u)))$.

Analogously to the case of Ackermann set theory, AFT has separate comprehension schemata for maps and functions. The restriction that is imposed on both schemata now is $\forall z\, \forall x\, (R(z,x) \to L(z) \wedge L(x))$. In the function comprehension schema, in which F(f) appears among the conclusions we may draw about f, the additional restriction is that the formula $R(z,x)$ may not contain the symbol F and may not have unlimited objects as parameters.

Additionally to these comprehension schemata, AFT has

- a function extensionality axiom,

- an axiom stating that any value a function takes and any value a function is defined at is limited, and

- an axiom stating that submaps of functions are functions.

In AFT one can interpret Ackermann set theory with Foundation, and hence ZFC (see Theorems 1 and 3 below). Since the map and function comprehension schemata presuppose the existence of choice maps and choice functions, the Axiom of Choice naturally comes out true in these interpretations.

We now state the main theorems about AFT. Their proofs can be found in the author's PhD thesis [5, pp. 58–62].

Theorem 1 (Theorem 4.2.7 in [5, p. 58]). *AFT interprets Ackerman set theory with urelements and the Axiom of Choice.*

Theorem 2 (Theorem 4.2.20 in [5, p. 61]). *Ackermann set theory with the Axiom of Foundation and the Axiom of Global Choice interprets AFT.*

Theorem 3 (Theorem 4.2.8 in [5, p. 59]). *AFT interprets ZFC.*

Theorem 4 (Corollary in [5, p. 62]). *AFT is equiconsistent with ZFC.*

6. Higher-order dynamic predicate logic

Now we are ready to sketch the untyped *Higher-Order Dynamic Predicate Logic* (HODPL). The restriction we impose on implicit function introduction are those imposed by AFT. AFT gives us untyped maps, which always have a restricted domain. So instead of using types to syntactically restrict the possible arguments for a given function term, we implement a semantic restriction on function application by integrating a formal account of presuppositions into the HODPL.[6] HODPL syntax thus allows for any term to be applied to any number of arguments to form a new term.

Besides the binary "=", HODPL has two unary logical relation symbols, U for urelements and F for functions. HODPL syntax does not depend on a signature, as we do not allow for constant, function and relation symbols other than "=", U and F. These can be mimicked by variables that respectively denote a non-function, denote a normal function or denote a function that only takes two predesignated urelements ("booleans") as values.

The domain of a structure always has to be a model of AFT. The possibility of presupposition failure is implemented in HODPL semantics by making the interpretation function partial rather than total. For conveniently talking about partial functions, we use the notation $\text{def}(f(x))$ to abbreviate that f is defined on x.

We define the partial interpretation function $[\![\cdot]\!]_S^g \subseteq G_S \times G_S$ by specifying its domain and its values trough a simultaneous recursion (the cases 3-8 of the second part are as in THODPL):

- Domain of $[\![\cdot]\!]_S^g$:

  1. $\text{def}([\![U(t)]\!]_S^g)$ iff $[t]_S^g \neq u^S$.
  2. $\text{def}([\![F(t)]\!]_S^g)$ iff $[t]_S^g \neq u^S$.
  3. $\text{def}([\![\top]\!]_S^g)$.
  4. $\text{def}([\![t_1 = t_2]\!]_S^g)$ iff $[t_1]_S^g \neq u^S$ and $[t_2]_S^g \neq u^S$.
  5. $\text{def}([\![\neg\varphi]\!]_S^g)$ iff $\text{def}([\![\varphi]\!]_S^g)$.
  6. $\text{def}([\![\varphi \wedge \psi]\!]_S^g)$ iff $\text{def}([\![\varphi]\!]_S^g)$ and for all $h \in [\![\varphi]\!]_S^g$, $\text{def}([\![\psi]\!]_S^h)$.
  7. $\text{def}([\![\varphi \to \psi]\!]_S^g)$ iff $\text{def}([\![\varphi]\!]_S^g)$ and for all $h \in [\![\varphi]\!]_S^g$, $\text{def}([\![\psi]\!]_S^h)$.
  8. $\text{def}([\![\exists t\ \varphi]\!]_S^g)$ iff for all h s.t. $h[t]g$, $\text{def}([\![\varphi]\!]_S^h)$.

- Values of $[\![\cdot]\!]_S^g$:

  1. $[\![U(t)]\!]_S^g := \{h | g = h \text{ and } [t]_S^g \in U^S\}$.
  2. $[\![F(t)]\!]_S^g := \{h | g = h \text{ and } [t]_S^g \in F^S\}$.

---
[6]See [4] for an introduction to presuppositions in mathematical texts.

One can define a sound proof system for HODPL that can prove everything provable in AFT: In the author's PhD thesis, a proof system for an extension of HODPL is defined [5, pp. 108–113] and proven to be sound [5, pp. 147, 148] and complete [5, pp. 156–176]. The details of this proof system are beyond the scope of this paper.

7. Conclusion

We have studied a feature of the natural language of mathematics that has previously not been studied by other logicians or linguists, the *implicit dynamic function introduction*, exemplified by constructs of the form "for every x there is an $f(x)$ such that ...". If this feature is used without limitations, it yields a paradox analogous to Russell's paradox. Hence any formalism capturing it has to impose some limitations on it. We have sketched two higher-order extensions of Dynamic Predicate Logic, *Typed Higher-Order Dynamic Predicate Logic* (THODPL) and *Higher-Order Dynamic Predicate Logic* (HODPL), which capture this feature, and which differ only in the limitations they impose onto it. HODPL is based on *Ackermann-like Function Theory*, a novel foundational theory of functions that is inspired by Ackermann Set Theory and that interprets ZFC.

References

[1] W. Ackermann. Zur Axiomatik der Mengenlehre. *Mathematische Annalen*, 131:336–345, 1956.

[2] A. Church. A formulation of the simple theory of types. *Journal of Symbolic Logic*, 5:56–68, 1940.

[3] M. Cramer, B. Fisseni, P. Koepke, D. Kühlwein, B. Schröder, and J. Veldman. The Naproche Project – Controlled Natural Language Proof Checking of Mathematical Texts. In *CNL 2009 Workshop, LNAI 5972*, pages 170–186, 2010.

[4] M. Cramer, D. Kühlwein, and B. Schröder. Presupposition Projection and Accommodation in Mathematical Texts. In *Semantic Approaches in Natural Language Processing: Proceedings of the Conference on Natural Language Processing 2010 (KONVENS)*, pages 29–36. Universaar, 2010.

[5] Marcos Cramer. *Proof-checking mathematical texts in controlled natural language*. PhD thesis, Universität Bonn, 2013.

[6] Marcos Cramer. The Naproche system: Proof-checking mathematical texts in controlled natural language. *Sprache und Datenverarbeitung – International Journal for Language Data Processing*, 38(1–2):9–33, 2014.

[7] J. Groenendijk and M. Stokhof. Dynamic Predicate Logic. *Linguistics and Philosophy*, 14(1):39–100, 1991.

[8] M. Lackenby. Topology and Groups. Retrieved December 1, 2016, from `http://people.maths.ox.ac.uk/lackenby/tg050908.pdf`, 2008. Lecture Notes.

[9] W. Reinhardt. Ackermann's set theory equals ZF. *Annals of Mathematical Logic*, 2:189–249, 1970.

[10] W. Trench. *Introduction to Real Analysis*. Prentice Hall, 2003.

Received: 6 December 2016

Locology and Localistic Logic: Mathematical and Epistemological Aspects

Michel De Glas

SPHERE, CNRS-Université Paris Diderot
5, rue Thomas Mann 75205 Paris Cedex 13 France
`michel.deglas@univ-paris-diderot.fr`

Abstract: The object of this paper is to present and thoroughly study a new logic, called localistic logic, the essential features of which are as follows. First, it relies upon a rejection of the positive paradox axiom and a weakening of the deduction theorem. Second, the localistic logic provides locology with a logical framework. Third, the concepts of prelocus and locus provide logic (and locology) with a categorical substratum.

Keywords: locology, localistic logic, prelocus, locus, constructivism.

Introduction

One may point out, in modern mathematics, many mathematical, logical and philosophical oppositions to Cantor's transfinite "paradise". As is well known, Kronecker, Poincaré, Brouwer, Weyl, Feferman, and some others are particularly reluctant to accept Cantor's conception of the continuum ("The actual infinite is not required for the mathematics of the physical world", Feferman says).

Surprisingly enough, topology has never really been touched by the criticisms on set theory and actual infinity, although it incorporates many problematic notions of set theory. For instance, unless one chooses to consider non-T_1 topological spaces (i.e. spaces of little mathematical significance and of practically no use in applied domains), boundaries are lines of Lebesgue-measure zero. Next, contrary to what intuition suggests, the operators of interior and closure are idempotent. Moreover, the concept of neighbourhood, which is supposed to model the notion of proximity or nearness, is somehow transitive. It is not difficult to prove that all these counterintuitive and mathematically hard to accept situations are immediate consequences of the actual infinity, particularly of the atomic nature of the continuum.

The various attempts to generalize point-set topology take place in the course of "point elimination". It is on this road that one can meet abstract spaces (first studied by Hausdorff who took the notion of open set as a primitive in the study of continuity in such spaces), Heyting algebras (which arose from the epistemological deliberations of Brouwer), pointless topology (where open-set lattices are taken as primitive notions, irrespective of whether they are composed of points), point-wise, or formal, topology (an intuitionistic approach to topology, based upon Martin-Löf's type theory, which proves to be slightly more restrictive than pointless topology). A further step in the process of (pointless) abstraction may be taken by considering the category of locales (whose objects are complete lattices equipped with the infinite distributive law, and whose morphisms are maps preserving finite meets and arbitrary joins), which, according to many category-theorists is the structure within which pointless topology must be developed. Whatever one may think of the latter assertion, an essential feature of the results available is that they all invoke non constructive principles: the localistic framework allows to give classical theorems of topology constructive proofs. What one can gain by doing constructive topology is that there are contexts in which one may like to do "topology" but one does not wish to assume the law of excluded middle or the axiom of choice. Such contexts are called topoï. Apart from this alleged constructive aspect, there are nevertheless several results which say that, from one point of view (i.e., when one works with spatial locales), working with locales is doing nothing more than a disguised version of classical point-set topology. One may, of course, consider non spatial locales but very little work has been done on specific applications of such tools. Furthermore, large parts of the theory of locales can be internalized in any topos and a topos is nothing but a category which is sufficiently "like" the category of sets for one to carry out set-theoretical constructions inside it. Point-wise (or formal) topology is related to pointless topology by the adjunction (in the category-theoretic sense of the word) between the category of locales and the category of topological spaces. In the case of spatial locales, the adjunction reduces to an adjoint equivalence between the category of spatial locales and the category of sober topological spaces. Thus the point-wise and the pointless approaches are essentially equivalent as soon as one wishes to deal with spatiality.

The overall gain possibly provided by pointless or point-wise topology is thus quite limited. The basic reason is that, despite the generalization provided by the "elimination" of points and whatever the level of abstraction is, the algebraic structures implied by these approaches are essentially the same as those defined in a point-set topological framework.

Locology [1, 2, 5–7] has been elaborated by the author as an alternative to topology in order to provide new mathematically and philosophically acceptable and fruitful solutions to the above mentioned problems. It allows in particular, from the giving of a reflexive (and possibly symmetric) relation, which may be seen as a relation of resemblance or as the measure of a granularity over some carrier set, to redefine most concepts of topology in a more satisfactory way: the concepts of *core* and *shadow*, which are substituted for that of interior and closure, are not idempotent; to any subset in a locological space may be associated its *frontier* and its *boundary* (the former being divided into its inner and outer parts), the distinction between the two entities being of prime importance both from a mathematical and an epistemological viewpoint (mathematically speaking, a frontier has a certain "thickness"; epistemologically speaking, it allows to distinguish between punctuality and indivisibility); the relevant algebraic structure is that of a complete and complemented, but not distributive, lattice with a semi-implication. The distinction, in locology, between boundaries (which have no analogue in the world of "real" entities) and frontiers allows, in particular, to revisit some fundamental problems left open by topology (and mereotopology): that of contiguity and contact [2]. These problems originate from the set-theoretical and topological definition of the continuum and the consecutive failure in the treatment of boundaries. It also leads to formalize, in an essentially new way, the key concepts of categorization [7]. Locological spaces encompass Poincaré-Zeeman tolerances spaces [9, 10, 13], Choquet's pretopological spaces [3, 12], and mathematical morphology. The study of locological concepts and the structure thus implied allow to understand why these three (independent) streams of research have not been followed up.

The anti-realism at the root of the rejection of actual infinity and the Cantorian conception of the continuum is, as is well known, intimately related with the anti-realism (or anti-platonism) in logic which leads to substructural logics, in particular intuitionistic logic. However, the topology/locology alternative, sketched above, suggests, first, that the criticisms addressed to topology translate to intuitionistic logic, and, second that a new logic of which locology would be the "geometric" counterpart is needed.

Localistic logic, the definition and the study of which are the object of Section 2, meets this requirement. We first prove that, contrary to the intuitionists' claim, the law of excluded middle is in no way a principle of omniscience and is perfectly compatible with a constructive view of logic and mathematics, and that the excluded middle and the reductio ad absurdum are not in general mutually dependent. Next, localistic logic allows to revisit the question of the admissibility, from a constructivist viewpoint, of the thinning on the left (also called the positive paradox axiom), i.e. $A \to (B \to A)$ and the thinning on the right, i.e. $A \to (\sim A \to B)$, which are admitted by both classical and intuitionistic logics. It is worth noticing that this question was raised by the first intuitionists: Kolmogorov rejected $A \to (\sim A \to B)$ but accepted $A \to (B \to A)$; Glivenko, whose axiomatization was the one adopted by Gentzen, raised the same question but eventually followed Heyting in keeping both thinning on the right and thinning on the left. This question has also been tackled by relevant logics, the first axiomatization of which was actually proposed by another Russian intuitionist, Orlov. The rejection of both thinning on the right and thinning on the left by relevant logics is the main departure from intuitionistic logic. However, the various versions of relevant logics fall short of an interesting semantics (i.e. a semantics where the truth-values may be expressed in terms of classes of objects).

Localistic logic leads to rejection of $A \to (B \to A)$ on the basis that, if A may be derived from a set Γ of hypotheses ($\Gamma \vdash A$) then there is no reason that, for any B, $B \to A$ may be derived *gratis prodeo* ($\Gamma \vdash B \to A$). However, this may hold for some Γ's, in particular for $\Gamma = \emptyset$: if $\vdash A$ then $\vdash B \to A$ (a theorem may be derived from anything). What relevantists did not actually realize is that a theorem is more than a formula deduced from an empty set of hypotheses. As far as thinning on the right is concerned, the localistic argument is as follows: $A \to (\sim A \to B)$ being related to the reductio ad absurdum, its (in)admissibility depends on some further assumptions. A study of propositional and predicate logics is performed in Section 2. An essential feature is, as alluded to above, *the weakening of the deduction theorem*.

Section 3 is devoted to the study of the categorical substratum of localistic logic. It is shown that the *theory of (pre)loci* provides localistic logic with a category-theoretic basis. However, the role played by localistic logic for loci theory is not quite analogous to that played by topoï theory for intuitionistic logic. Indeed localistic logic cannot be seen as an internal logic of a locus. On the contrary, it may be seen as *emerging* from a (pre)locus. We may prove however the equivalence between locus-validity and localistic provability.

1. Locology

Let X be a set and let λ be a reflexive relation on X: $x\lambda x$, for any x in X. The relation λ is to be thought of as a resemblance or an indistinguishability relation on X. The set $\lambda[x] = \{y : x\lambda y\}$ of λ-relatives of x, the elements of which may be seen as being close to (or resembling, or being indistinguishable from) x, is called the halo of x. As λ may be defined as a map $X \longrightarrow \wp(X)$, $x \longmapsto \lambda[x]$, we denote by $\lambda(A)$ the set

$$\lambda(A) = \bigcup_{x \in A} \lambda[x]$$

so that $\lambda[x] = \lambda(\{x\})$. Next, one defines the following two operators h and s which associate to any A in $\wp(X)$ its *core* and its *shadow* respectively. More precisely, let $h : \wp(X) \longrightarrow \wp(X)$ be the operator which associates to any A its core

$$h(A) = \{x \in X : \lambda[x] \subseteq A\}.$$

Immediate properties follow:

(1) $h(A) \subseteq A, h(X) = X$,
(2) If $A \subseteq B$ then $h(A) \subseteq h(B)$,
(3) $h \circ h(A) \subseteq h(A)$,
(4) $h(A \cup B) \supseteq h(A) \cup h(B)$,
(5) $h(\bigcap_i A_i) = \bigcap_i h(A_i)$.

It is worth emphasizing that, contrary to the properties of an interior operator in topology, h is not idempotent (unless λ is assumed to be transitive). On the other hand, property (5) holds for infinite intersections.

The *shadow* operator is defined in a dual way. It associates to any $A \in \wp(X)$ its shadow $s(A)$ defined by

$$s(A) = \{x \in X : \lambda[x] \cap A \neq \varnothing\}.$$

The operators h and s are interdefinable since clearly

$$s(A) = \overline{h(\overline{A})},$$

where $\overline{A}$ denotes the complement of A. Immediate properties of s then come out:

(1) $s(A) \supseteq A$, $s(\emptyset) = \emptyset$,
(2) If $A \subseteq B$, then $s(A) \subseteq s(B)$,
(3) $s \circ s(A) \supseteq s(A)$,
(4) $s(A \cap B) \subseteq s(A) \cap s(B)$,
(5) $s(\bigcup_i A_i) = \bigcup_i s(A_i)$.

Like h, the operator s is not idempotent and, unlike topology, equality (5) holds for infinite unions.

The idea of using a reflexive (and symmetric) relation to recapture the intuitive notion of indistinguishability is not new. That of having recourse to non idempotent "interior" or "closure" operators is not without predecessors either. The former idea can be traced back to Poincaré's works on the physical continuum. As claimed by Poincaré [9], "the raw result of experience may be expressed by the relation $A = B, B = C, A < C$, which may be taken as a formula of the physical continuum". Here, $A = B$ is to be understood as "A and B are indistinguishable", and $A = B$ is then a reflexive and symmetric relation over the collection of entities under study. This approach was exploited by Zeeman [13] in his works on tolerance spaces.

The idea of a non idempotent closure operator can be traced back to Choquet's paper on pretopology [3]. Such an operator is nowadays referred to as a Čech closure operator [12]. Depending upon the properties it is equipped with (isotony, accretivity, sub-linearity, . . .) the resulting spaces are called extended topologies [8], neighbourhood spaces [4], Smyth spaces [11], or pretopologies.

However these two streams of research have not been followed up. The basic reasons seems to be the following. The Poincaré-Zeeman approach is essentially geometric and is then deprived of an algebraic (and a logical) content: there is nothing, in tolerance spaces, which can play the role of the lattice of open sets in a topological space. In a symmetric way, the approaches pertaining to the stream initiated by Choquet have a poor geometric content. Furthermore, they lead to very poor algebraic structures too: as a generalization of the Kuratowski closure algebra, the algebra $\{cl(A) : A \subset X\}$, where cl denotes the generalized closure operator, fails, for instance, to be sup-complete, so that the disjunction of objects of the algebra cannot be defined. If A and B are "closed" sets, nothing can be said of the entity "A and B" (apart from the fact that, in general, $A \cap B$ is not "closed"). These limitations are insurmountable.

Owing to the property (5) of h and s, the corresponding algebras, as will be seen below, have much stronger properties. Many results may then be derived, most of which are not derivable in a generalized closure space or in a tolerance space.

We consider the two families:

$$\mathcal{L} = \{h(A) : A \subseteq X\},$$

$$\mathcal{K} = \{s(A) : A \subseteq X\}.$$

It is clear that $\mathcal{L} = \{\bar{A} : A \in \mathcal{K}\}$ and $\mathcal{K} = \{\bar{A} : A \in \mathcal{L}\}$. In view of the properties of h, $(\mathcal{L}, \cap)$ is a complete and bounded inf-semi-lattice. However, for A and B in $\mathcal{L}$, $A \cup B$ may not be an element of $\mathcal{L}$. Indeed, given A and B there may not exist $C \in \wp(X)$ such that $h(C) = A \cup B$. Hence $(\mathcal{L}, \cup)$ is not a sup-semi-lattice. We may however define, for A and B in $\mathcal{L}$:

$$A \sqcup B = \bigcap\{C \in \mathcal{L} : C \supseteq A, C \supseteq B\},$$

the existence of which is guaranteed by the inf-completeness and the boundedness of $\mathcal{L}$. $A \sqcup B$ is thus the least upper bound of the set of objects of $\mathcal{L}$ which contain $A \cup B$. Thus $(\mathcal{L}, \cap, \sqcup)$ is a complete lattice. But it fails to be distributive. Indeed we may have $A \cap B = A \cap C$ and $A \sqcup B = A \sqcup C$ and $B \neq C$, the equality $B = C$ being a necessary and sufficient condition for a lattice to be distributive. Furthermore

$$h \circ \lambda(A) = \bigcap\{B \in \mathcal{L} : B \subseteq A\},$$

for any $A \subseteq X$. Hence

$$\lambda \circ h(A) \subseteq A \subseteq h \circ \lambda(A),$$

where the equality holds on the right-hand side iff $A \in \mathcal{L}$, and then

$$A \sqcup B = h \circ \lambda(A \cup B),$$

for any $A, B \in \mathcal{L}$.

The lattice $(\mathcal{L}, \cap, \sqcup)$ is called a *locology* and X a *locological space*. Despite the differences between locology and topology (non-distributivity of $\mathcal{L}$, completeness of $\mathcal{L}$, non-idempotency of h and s), the consequences of which are of prime importance, it is quite clear that the objects of $\mathcal{L}$ bear some resemblance with open sets in a topological space. However, $\mathcal{L}$ may be defined as

$$\mathcal{L} = \{A \subseteq X : A = h \circ \lambda(A)\}.$$

The operator $h \circ \lambda$ is accretive, order-preserving, and idempotent. Hence $h \circ \lambda$ is an algebraic closure operator, i.e. a closure operator in $\wp(X)$ viewed as an ordered set. Of course, it is not a topological closure operator since e.g. $h \circ \lambda(A \cup B) \neq h \circ \lambda(A) \cup h \circ \lambda(B)$. This means that the objects of $\mathcal{L}$ have something in common with *closed sets* in topology.

A dual analysis may be carried out for the algebra

$$\mathcal{K} = \{s(A) : A \subseteq X\}.$$

Indeed, if one defines, for $A, B \in \mathcal{K}$,

$$A \sqcap B = \bigcup \{C \in \mathcal{K} : C \subseteq A, C \subseteq B\},$$

then $(\mathcal{K}, \sqcap, \cup)$ is a complete, but not distributive lattice, where, furthermore:

$$\overline{h \circ \lambda(\bar{A})} = \bigcup \{C \in \mathcal{K} : B \subseteq A\},$$

$$A \sqcap B = \overline{h \circ \lambda(\bar{A} \sqcap \bar{B})},$$

and, if λ is symmetric,

$$\overline{h \circ \lambda(A)} = \lambda \circ h(A)$$

$$A \sqcap B = \lambda \circ h(A \sqcap B)$$

The fact that $\mathcal{K}$ may be rewritten as

$$\mathcal{K} = \{A \subseteq X : A = \overline{h \circ \lambda(\bar{A})}\} = \{A \subseteq X : \bar{A} = h \circ \lambda(\bar{A})\}$$

and, if λ is symmetric, as

$$\mathcal{K} = \{A \subseteq X : A = \lambda \circ h(A)\}$$

shows that the objects of $\mathcal{K}$ have a superficial similarity with closed sets in a topological space and a deeper resemblance with *open* sets in a topological space.

This may seem paradoxical at first glance. It is, however, a key point of locology. To more accurately specify this, one first has to revisit the concept of a boundary. The critical analysis of the concept of a boundary in topology leads us to actually define two different concepts: to any region A, one associates its *frontier* and its *boundary*, the former being divided into its inner and outer parts.

To any $A \subseteq X$, one associates its *inner frontier* $\partial_{in}(A) = \lambda(\bar{A}) \cap A$ and its *outer frontier* $\partial_{out}(A) = \lambda(A) \cap \bar{A}$. The *frontier* of A is then $\partial(A) = \partial_{in}(A) \cup \partial_{out}(A) = \lambda(A) \cap \lambda(\bar{A})$. Among many properties, which follow from their definition, a remarkable property is the idempotency of ∂_{in} and ∂_{out} which follows from

$$h[\partial_{in}(A)] = h[\partial_{out}(A)] = \emptyset.$$

The epistemological significance of these equalities is that locology allows us to distinguish between punctuality and indivisibility (these two notions being unduly identified to each other in topology). Indeed, $\partial_{in}(A)$ and $\partial_{out}(A)$ may be considered, on the one hand, as indivisible since they have empty cores and, on the other hand, as having a certain "thickness" (unless λ coincides with the diagonal of the carrier set X).

One may now define the concept of a boundary. The *boundary* of $A \subseteq X$ is defined to be the core of its frontier

$$\beta(A) = h(\partial(A)).$$

Since $\partial(A) = \lambda(A) \cap \lambda(\bar{A})$, one has

$$\beta(A) = h \circ \lambda(A) \cap h \circ \lambda(\bar{A}).$$

Hence

(1) $\beta(A) = \beta(\bar{A}) \in \mathcal{L}$,
(2) $A \smallsetminus \beta(A) = \lambda \circ h(A)$,
(3) $A \cup \beta(A) = h \circ \lambda(A)$,
(4) $A \in \mathcal{L}$ iff $\beta(A) \subseteq A$,

(5) $A \in \mathcal{K}$ iff $A \cap \beta(A) = \emptyset$.

Thus, as already alluded to above, objects of $\mathcal{L}$ and $\mathcal{K}$ have properties in common with closed sets and open sets in topology, respectively: any A in $\mathcal{L}$ contains its boundary; any A in $\mathcal{K}$ is disjoint from its boundary.

This shows that locology allows us, not only, to define purely locological concepts (the core h, the shadow s, the frontiers ∂_{in} and ∂_{out}) which have no counterpart in topology, but also concepts that may be seen as "quasi topological" (the operators $h \circ \lambda$ and $\lambda \circ h$, the boundary β) with common features and essential differences with their topological pendants. It may also be shown that topology is the limit case of locology corresponding to an infinitely small granularity.

Given a locological space X, one may define in $\mathcal{L}$ the unary operator $\neg$ by setting $\neg A = h(\bar{A})$, the core of the complement of A. The operator $\neg$ clearly satisfies

(1) $A \cap \neg A = \emptyset$,
(2) $A \cup \neg A \neq X$,
(3) $A \sqcup \neg A = X$,
(4) $\neg\neg A = A$ iff λ is symmetric.

Hence, $\neg$ is a complementation in $\mathcal{L}$ and an orthocomplementation iff λ is symmetric (these properties being not equivalent in a non distributive lattice). Anticipating on the next section, this translates into logical terms as follows. First, (1)–(3) show that, contrary to the intuitionists' claim, *the law of excluded middle is in no way a principle of omniscience*: for any $A \in \mathcal{L}$, A and $\neg A$ are disjoint and their disjunction $A \sqcup \neg A$ covers the universe, but there may exist objects of X that do not belong to either of A and $\neg A$. Second, (3) and (4) show that the law of excluded middle and the reductio ad absurdum may not be interdependent.

Next, for any A and B in $\mathcal{L}$, let $\Rightarrow$ be the binary operator defined by

$$A \Rightarrow B = h(\bar{A} \cup B),$$

the essential properties of which are

(1) $A \Rightarrow B = X$ iff $A \subseteq B$,

Locology and Localistic Logic 45

(2) $A \Rightarrow \emptyset = \neg A$,
(3) $A \cap (A \Rightarrow B) \subseteq B$,
(4) $(A \Rightarrow B) \cap (A \Rightarrow C) \subseteq A \Rightarrow (B \cap C)$,
(5) $(A \Rightarrow B) \cap (B \Rightarrow C) \subseteq A \Rightarrow C$,
(6) $\dfrac{C \subseteq A \Rightarrow B}{A \cap C \subseteq B}$.

However, the reciprocal to (6), i.e.

$$\frac{A \cap C \subseteq B}{C \subseteq A \Rightarrow B}$$

holds only if $\mathcal{L}$ is distributive (in which case $\mathcal{L}$ is a Boolean algebra) i.e. only if λ is transitive. Therefore, $\Rightarrow$ is not, strictly speaking, an implication. It implies, in particular, that

$$\frac{A = X}{B \Rightarrow A = X}$$

holds but

$$A \Rightarrow (B \Rightarrow A) \ne X.$$

Anticipating, once again, on the following section, this inequality translates into the non-validity of the *positive paradox axiom* (thinning on the left).

Similarly, although

$$\frac{A \Rightarrow C = X \quad B \Rightarrow C = X}{(A \sqcup B) \Rightarrow C = X}$$

holds, one generally has

$$(A \Rightarrow C) \cap (B \Rightarrow C) \nsubseteq (A \sqcup B) \Rightarrow C.$$

The properties of $\Rightarrow$ (called, from now on, a *semi-implication*) and of the disjunction $\sqcup$, as compared to those enjoyed by the corresponding operators in a Boolean or a Heyting algebra (hence in classical and intuitionistic logics), are essential features of the locological framework from a logico-algebraic viewpoint.

One may then define, as a natural abstraction of the above algebraic structure, the concept of Λ-algebra. A Λ-*algebra* is a 5-tuple $(\mathcal{L}, \wedge, \vee, \neg, \Rightarrow)$ such that

(1) $(L, \wedge, \vee)$ is a lattice with a least element 0 and a greatest element 1.

(2) $(L, \Rightarrow)$ satisfies
$a \Rightarrow b = 1$ iff $a \leq b$,
$a \wedge (a \Rightarrow b) \leq b$,
$(a \Rightarrow b) \wedge (b \Rightarrow c) \leq (a \Rightarrow c)$,
$(a \Rightarrow b) \wedge (a \Rightarrow c) \leq a \Rightarrow (b \wedge c)$.

(3) $(L, \neg)$ satisfies
$\neg a = a \Rightarrow 0$,
$\neg \neg a = a$.

where $a \leq b$ iff $a \wedge b = a$.

Theorem 1.1. *In a Λ-algebra L, the following properties hold:*

(a) $\neg$ *is an order-reversing involution.*

(b) $a \wedge \neg a = 0$.

(c) $\neg(a \vee b) = \neg a \wedge \neg b$; $\neg(a \wedge b) = \neg a \vee \neg b$.

(d) $a \Rightarrow b$ *is increasing wrt a and decreasing wrt b.*

(e) $a \vee \neg a = 1$.

(f) $(a \Rightarrow b) \wedge (a \Rightarrow c) = a \Rightarrow (b \wedge c)$.

(g) $(a \vee b) \Rightarrow c \leq (a \Rightarrow c) \wedge (b \Rightarrow c)$.

(h) *if* $c \leq a \Rightarrow b$ *then* $a \wedge c \leq b$.

(i) *if* $a \wedge c \leq b$ *then* $1 \Rightarrow c \leq a \Rightarrow b$. □

From now on, we will consider locologies whose underlying relation λ is symmetric, i.e., locologies that are orthocomplemented, as lattices.

Theorem 1.2. (a) *Any locology is a Λ-algebra.* (b) *Any Λ-algebra $(L, \wedge, \vee, \neg, \Rightarrow)$ may be embedded into a locology on some set.*

Proof. (a) is obvious. To prove (b), consider the MacNeille completion L^* of L, i.e.

$$L^* = \{A^* : A \in L\} \subset \wp(L)$$

where A^* is defined, for any $A \in L$, by

$$M_A = \{m : a \leq m, \text{ for any } a \in A\},$$

$$A^* = \{x : x \leq m, \text{ for any } m \in A\}.$$

If $A = \emptyset$ then $M_A = L$ and $A^* = 0$ where 0 is the least element of A. Hence $(L^*, \subseteq, \cap)$ is a complete inf-semi-lattice with a least element 0. It may then be equipped with the structure of a complete lattice $(L^*, \subseteq, \cap, \vee^*)$ by setting

$$A^* \vee^* B^* = \bigcap \{C \in L^* : C \supseteq A^*, C \supseteq B^*\}.$$

The image under the transformation $A \longmapsto A^*$ of a singleton a of L is the subset $\{b \in L : b \leq a\}$ which will be denoted $(a \downarrow)$. Let $\mathcal{T}$ be the set $\{(a \downarrow) : a \in L\}$. The mapping $a \longmapsto (a \downarrow) \in \mathcal{T}$ is obviously one-to-one and onto. Since $a \leq b$ in L iff $(a \downarrow) \subseteq (b \downarrow)$ in $\mathcal{T}$, identifying $\{a\}$ with a leads to considering a mapping $f : L \longrightarrow \mathcal{T} \subseteq L^*$, $a \longmapsto (a \downarrow)$. One may then easily prove that f is a monomorphism. $\square$

2. Localistic logic

The language of propositional localistic logic (LL for short) has an alphabet consisting of proposition symbols: $p_0, p_1, \ldots$, connectors: $\wedge, \vee, \rightarrow, \leftrightarrow, \bot$, and auxiliary symbols: (,). The set Φ of formulas is the smallest set X such that
(1) $p_i \in X$, $i \in \mathbb{N}$, $\bot \in X$,
(2) If $\phi, \psi \in X$, then $\phi \wedge \psi, \phi \vee \psi, \phi \rightarrow \psi \in X$.

The axioms and the inference rules for propositional LL are instances of one of the following forms, where $\sim \phi$ stands for $\phi \rightarrow \bot$ and $\phi \leftrightarrow \psi$ stands for $(\phi \rightarrow \psi) \wedge (\psi \rightarrow \phi)$:

Axioms

A1 $\bot \rightarrow \phi$

A2a $\phi \wedge \psi \rightarrow \phi$

A2b $\phi \wedge \psi \rightarrow \psi$

A3 $((\phi \rightarrow \psi) \wedge (\phi \rightarrow \chi)) \rightarrow (\phi \rightarrow (\psi \wedge \chi))$

A4 $(\phi \to \psi) \wedge (\psi \to \chi) \to (\phi \to \chi)$

A5a $\phi \to \phi \vee \psi$

A5b $\psi \to \phi \vee \psi$

A6 $(\phi \wedge (\phi \to \psi)) \to \psi$

A7 $\sim\sim \phi \leftrightarrow \phi$

Inference rules

R1 $\dfrac{\phi \quad \phi \to \psi}{\psi}$

R2 $\dfrac{\phi}{\psi \to \phi}$

R3 $\dfrac{\phi \to \chi \quad \psi \to \chi}{(\phi \vee \psi) \to \chi}$

A formula is said to be *provable*, denoted $\vdash_{LL} \phi$ or simply $\vdash \phi$, iff there exists a sequence $\phi_1, \phi_2, \ldots, \phi_n$ of formulas such that $\phi_n = \phi$ and, for any $i \leq n$, ϕ_i is either an axiom or follows form earlier formulas in the sequence by a rule of inference from {R1, R2, R3}. A *valuation* v is a mapping $v : \phi \longrightarrow (L, \wedge, \vee, \neg, \Rightarrow)$, where L is a Λ-algebra, such that

$$v(\bot) = 0,$$

$$v(\phi \wedge \psi) = v(\phi) \wedge v(\psi),$$

$$v(\phi \vee \psi) = v(\phi) \vee v(\psi),$$

$$v(\phi \to \psi) = (\phi) \Rightarrow v(\psi),$$

$$v(\sim \phi) = \neg v(\phi).$$

Let L be a Λ-algebra. A formula $\phi \in \Phi$ is L-valid iff, for any algebra v, $v(\phi) = 1$, the greatest element of L.

Theorem 2.1. $\vdash \phi$ *iff* ϕ *is L-valid for any Λ-algebra L.*

The above completeness theorem deals only with the equivalence between *provability* and validity in a Λ-algebra. One now has to consider deducibility from a set Γ of formulas (which, as usual, may be thought of as hypotheses). Unlike the classical and the intuitionistic cases, the extension from provability ($\vdash \phi$) to deducibility from Γ ($\Gamma \vdash \phi$) is far from obvious. It leads, in particular, to the non-validity of the deduction theorem.

We say that ϕ is *deducible* from a set Γ of formulas, denoted $\Gamma \vdash \phi$, iff either ϕ is provable ($\vdash \phi$) or there exists a sequence $\phi_1, \ldots, \phi_n = \phi$ of formulas such that each ϕ_i is either an axiom or a formula of Γ or follows from earlier formulas in the sequence by the inference rule R1.

We say that ϕ is Γ-valid, denoted $\Gamma \models \phi$ iff, for any L and any valuation v, there exists $\Gamma_0 \subseteq \Gamma, \Gamma_0$ finite, such that

$$\bigwedge_{\gamma \in \Gamma_0} v(\gamma) \leq v(\phi).$$

Theorem 2.2. *If $\phi \vdash \psi$, then $\vdash \phi \rightarrow \psi$.*

Proof. Two cases have to be considered. If $\vdash \psi$ then $\vdash \phi \rightarrow \psi$ by R2. If $\not\vdash \psi$ then $\phi \vdash \psi$ means that ψ may be deduced from ϕ, A1-A7 plus R1. As R1 is the only inference rule, unless ψ and ϕ are the same formula (in which case the statement of the theorem is trivially true), $\phi \vdash \phi \rightarrow \psi$. But this holds iff $\vdash \phi \rightarrow \psi$. □

We seem to be well on the way to a proof of the deduction theorem. Indeed, from theorem 2.2 which asserts

$$\frac{\phi \vdash \psi}{\vdash \phi \rightarrow \psi}$$

and consequently

$$\frac{\phi_1, \phi_2, \ldots, \phi_n \vdash \psi}{\vdash (\phi_1 \wedge \phi_2 \wedge \cdots \wedge \phi_n) \rightarrow \psi}$$

we might expect that, for any Γ,

$$\frac{\Gamma \vdash \phi}{\Gamma \vdash \psi \rightarrow \phi}.$$

However, we have the

Theorem 2.3. *$\Gamma \vdash \phi$ does not entail $\Gamma \vdash \psi \rightarrow \phi$.*

Proof. Suppose that $\Gamma \vdash \phi$ implies $\Gamma \vdash \psi \to \phi$ for any ψ. Then, applying theorem 2.2 above and theorem 2.7 below, $\Gamma \vDash \phi$ implies $\Gamma \vDash \psi \to \phi$ for any ψ. A necessary and sufficient condition is that $v(\psi) \Rightarrow v(\phi) \geq v(\phi)$ for any valuation v. Such an inequality generally does not hold. $\square$

We must emphasize the contrast between the validity of

$$\frac{\psi \vdash \phi}{\vdash \psi \to \phi}$$

and the non-validity of

$$\frac{\Gamma, \psi \vdash \phi}{\Gamma \vdash \psi \to \phi}$$

i.e. of the deduction theorem. The validity of the former means that either $\vdash \phi$ and then $\vdash \psi \to \phi$ (a theorem may be deduced from anything) or $\nvdash \phi$ in which case $\psi \vdash \phi$ iff $\vdash \psi \to \phi$, i.e. iff $\psi \to \phi$ has already been proved. The validity of the latter would mean that if ϕ may be deduced from $\Gamma \cup \{\psi\}$ then, irrespective of whether ψ (or $\psi \to \phi$) appears or not in the deduction of ϕ, $\psi \to \phi$ may be deduced, *gratis prodeo*, from Γ. Such a scheme is not allowed in a localistic framework. Although highly non constructive, this derivation is possible in intuitionistic logic.

It must be clear that the weakening of the deduction theorem (or, equivalently, the non-validity of the positive paradox axiom $\phi \to (\psi \to \phi)$) is the logical counterpart of the non-distributivity of a Λ-algebra, which itself is the translation in algebraic terms of the non-idempotency of the shadow and the core operators in a locological space.

A set Γ of formulas is said to be consistent if $\Gamma \nvdash \bot$. Otherwise, it is said to be inconsistent. Γ is said to be complete iff, for any formula ϕ, $\Gamma \vdash \phi$ or $\Gamma \vdash \sim \phi$.

Theorem 2.4. (a) *If $\Gamma \cup \{\phi\}$ is consistent then $\Gamma \nvdash \sim \phi$.* (b) *If Γ is complete, then the reciprocal to (a) holds.*

Proof. (a) Let $\Gamma \cup \{\phi\}$ be consistent and suppose that $\Gamma \vdash \sim \phi$. Then $\Gamma, \phi \vdash \sim \phi$. But, since $\Gamma, \phi \vdash \phi$, then $\Gamma, \phi \vdash \bot$, a contradiction.

(b) If $\Gamma \nvdash \sim \phi$, since Γ is complete, $\Gamma \vdash \phi$. Suppose that $\Gamma, \phi \vdash \bot$. Then $\Gamma \vdash \bot$, and $\Gamma \vdash \sim \phi$, a contradiction. $\square$

Theorem 2.5. *Γ is consistent iff Γ is satisfiable.*

Proof. (a) If Γ is satisfiable, then there exists a valuation v such that $v(\gamma) = 1$ for any $\gamma \in \Gamma$. Suppose Γ is not consistent. Then $\Gamma \vdash \bot$ and therefore $\Gamma_0 \vdash \bot$ for some $\Gamma_0 \subseteq \Gamma$, Γ_0 finite. Setting $\Gamma_0 = \{\gamma_1, \gamma_2, \ldots, \gamma_n\}$ leads to $\gamma_1, \gamma_2, \ldots, \gamma_n \vdash \bot$, i.e. $\vdash \sim(\gamma_1 \wedge \gamma_2 \wedge \cdots \wedge \gamma_n)$ and $\vDash \sim (\gamma_1 \wedge \gamma_2 \wedge \cdots \wedge \gamma_n)$. Thus, $v(\gamma_1 \wedge \gamma_2 \wedge \cdots \wedge \gamma_n) = 0$, a contradiction.

(b) Let Γ be consistent. From part (a) of theorem 2.4, if γ is consistent then $\nvdash \sim \gamma$ and, consequently, $\nvDash \sim \gamma$. Suppose Γ is not satisfiable, in which case there exists $\gamma \in \Gamma$ such that, for any v, $v(\gamma) < 1$. By induction on the length of γ, $v(\gamma) = 0$ for any v, i.e. $v(\sim \gamma) = 1$, for any v. Thus $\vDash \sim \gamma$, a contradiction. □

Theorem 2.6. *If $\Gamma \vdash \phi$ then $\Gamma \vDash \phi$.*

Proof. If $\vdash \phi$ then $\vDash \phi$, whence $\Gamma \vDash \phi$. If $\Gamma \vdash \phi$ (and $\nvdash \phi$), then there exists $\Gamma_0 = \{\gamma_1, \gamma_2, \ldots, \gamma_n\} \subseteq \Gamma$ such that $\Gamma_0 \vdash \phi$. Then $\vdash (\gamma_1 \wedge \gamma_2 \wedge \cdots \wedge \gamma_n) \to \phi$ and $\vDash (\gamma_1 \wedge \gamma_2 \wedge \cdots \wedge \gamma_n) \to \phi$. Thus $\Gamma_0 \vDash \phi$ and $\Gamma \vDash \phi$. □

Theorem 2.7. *If $\Gamma \vDash \phi$, then $\Gamma \vdash \phi$.*

Proof. (a) If Γ is finite, the reciprocal to theorem 2.6 clearly holds. Indeed, if $\Gamma = \{\gamma_1, \gamma_2, \ldots, \gamma_n\}$ then $\gamma_1, \gamma_2, \ldots, \gamma_n \vDash \phi$ entails $\vDash (\gamma_1 \wedge \gamma_2 \wedge \cdots \wedge \gamma_n) \to \phi$ hence $\vdash (\gamma_1 \wedge \gamma_2 \wedge \cdots \wedge \gamma_n) \to \phi$ i.e. $\gamma_1, \gamma_2, \ldots, \gamma_n \vdash \phi$.

(b) If Γ is infinite, there exists $\Gamma_0 \subseteq \Gamma$, Γ_0 finite, such that

$$\bigwedge_{\gamma \in \Gamma_0} v(\gamma) \leq v(\phi),$$

i.e. $\Gamma_0 \vDash \phi$. Applying (a) leads to $\Gamma_0 \vdash \phi$. Thus $\Gamma \vdash \phi$. □

3. Categorical substratum

The aim of this section is to provide localistic logic (and locology) with a categorical substratum which would play to some extent the role played by topoï theory and set theory for intuitionistic and classical logic respectively. As a Λ-algebra is a non distributive lattice the disjunction of which is weaker than the intuitionistic and the classical ones, it is a priori quite clear that the required categorical framework must exhibit a weakened form of exponentiation.

3.1. Preloci

A category $\mathcal{C}$ is said to have semi-exponentiation if

(1) any pair $\langle A, B \rangle$ of objects of $\mathcal{C}$ has product $A \times B$,

(2) for any pair $\langle A, B \rangle$ of objects, there is an object B^A and an arrow $e: B^A \times A \longrightarrow B$ such that, for any $g : C \times A \longrightarrow B$, there exists at most one arrow $\hat{g} : C \longrightarrow B^A$ such that the diagram

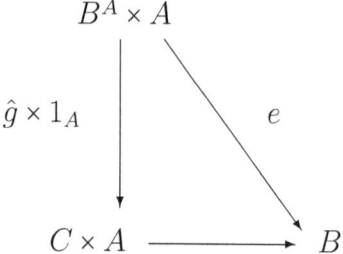

commutes. If, for a given $g, \hat{g}$ exists, we will write

$$\frac{g : C \times A \longrightarrow B}{\hat{g} : C \longrightarrow B^A}$$

where g and $\hat{g}$ may be omitted.

(3) The following rules hold

(a) $\dfrac{A \longrightarrow B}{1 \longrightarrow B^A}$,

(b) $\dfrac{C^B \times B^A \times A \longrightarrow C}{C^B \times B^A \longrightarrow C^A}$,

(c) $\dfrac{B^A \times C^A \times A \longrightarrow B \times C}{B^A \times C^A \longrightarrow (B \times C)^A}$.

The arrow e is called the *evaluation arrow*. The arrow $\hat{g}$, if it exists, is called the *exponential adjoint* of g.

Theorem 3.1. *The correspondence $g \longmapsto \hat{g}$ is bijective.* □

Although the correspondence $g \mapsto \hat{g}$ is bijective, $\mathrm{Hom}(C \times A, B)$ and $\mathrm{Hom}(C, B^A)$ are not isomorphic. As $\hat{g}$ may not exist, the correspondence is not necessarily a total function.

A category $\mathcal{A}$ which has

(1) a terminal object 1,

(2) a pullback and a pushout for each pair of arrows,

(3) semi-exponentiation

will be called a *prelocus* (plural: *preloci*). Clearly, a prelocus has initial object 0, for any pair $\langle A, B \rangle$ of objects, a product defined by an object $A \times B$ together with projections $\pi_{A,B} : A \times B \longrightarrow A$ and $\pi'_{A,B} : A \times B \longrightarrow B$ and a coproduct given by an object $A + B$ and injection arrows $j_{A,B} : A \longrightarrow A + B$ and $j'_{A+B} : B \longrightarrow A + B$.

Theorem 3.2. *In a prelocus $\mathcal{A}$, the following properties hold*
(1) $0 \cong 0 \times A$, *for any object A.*
(2) *If there exists an arrow $A \longrightarrow 0$, then $A \cong 0$ and the arrow is a mono.* □

3.2. The algebra of subobjects in a prelocus

Let $\mathcal{A}$ be a prelocus and let X be an object of $\mathcal{A}$. First recall that a subobject of X is defined as follows. Given two monos $f : A \twoheadrightarrow X$ and $g : B \twoheadrightarrow X$, one sets $f \subseteq g$ iff there exists $h : A \twoheadrightarrow B$ such that $f = g \circ h$. Then, the relation $\simeq$ defined by $f \simeq g$ iff $f \subseteq g$ and $g \subseteq f$ is an equivalence on the set of monos with codomain X. Furthermore, if $f \simeq g$, there exists an iso $k : B \longrightarrow X$ with inverse $h : A \longrightarrow X$ such that $f = g \circ h$ and $g = f \circ k$. The equivalence class of f modulo $\simeq$ is denoted $[f]$ and is said to be a subobject of X. The set $\mathrm{Sub}(X)$ of subobjects of X is thus

$$\mathrm{Sub}(X) = \{[f] : f : A \twoheadrightarrow X, \text{some } A\}.$$

We will usually write "the subobject f" when we mean "the subobject $[f]$".

3.2.1. Conjunction

Let $f : A \twoheadrightarrow X$ and $g : B \twoheadrightarrow X$ be two subobjects of X. The *conjunction* of f and g is defined to be the pullback of f and g, i.e. the subobject $f \cap g$: $A \times_X B \twoheadrightarrow X$ such that

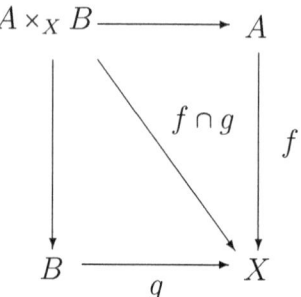

is a pullback square. The subobject $f \cap g$ is thus defined up to isomorphism.

3.2.2. Disjunction

The *disjunction* of two subobjects $f : A \twoheadrightarrow X$ and $g : B \twoheadrightarrow X$ of X is the subobject $f \cup g$ of X such that the diagram

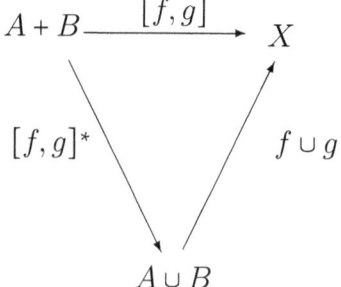

is an epi-mono factorization. In other words, $f \cup g$ is the image of the coproduct arrow $[f,g]$ (the least subobject of X through which $[f,g]$ factors) and $A \cup B \cong [f,g]^*(A+B)$, $[f,g]^*$ being an epi.

Theorem 3.3. $(\mathrm{Sub}(X), \subseteq, \cap, \cup)$ *is a lattice with a least element* 0_X *and a greatest element* 1_X. □

However, $\mathrm{Sub}(X)$ is not, in general, distributive. Indeed, let $f : A \twoheadrightarrow X$ and $g : B \twoheadrightarrow X$ be subobjects of X such that

$$f \cap g \simeq f \cap h \simeq 0_X,$$

Locology and Localistic Logic

$$f \cup g \simeq f \cup h.$$

We then have the following commutative diagrams

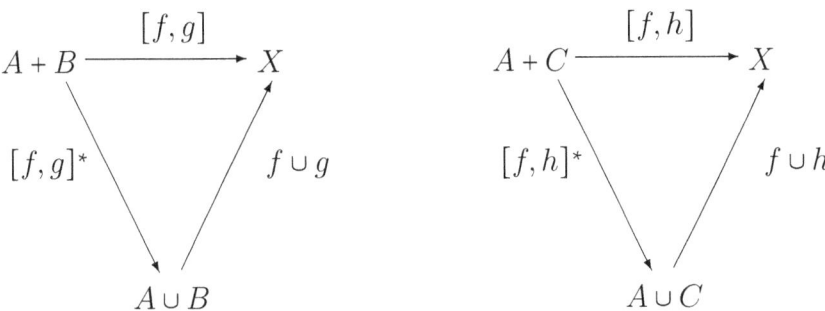

i.e. $f \cup g \simeq [f, g]$ and $f \cup h \simeq [f, h]$. But, $g \not\simeq h$ since $B \not\simeq C$. Thus, in general, Sub(X) is not distributive.

3.2.3. Semi-implication

Let $f : A \twoheadrightarrow X$ and $g : B \twoheadrightarrow X$ be two subobjects of X. We define $f \Rightarrow g : B^A \twoheadrightarrow X$ as a subobject of X such that the diagram

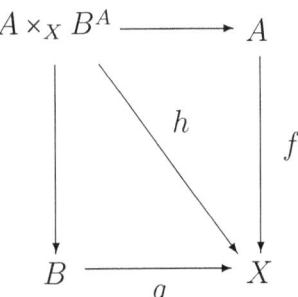

where $h = f \cap (f \Rightarrow g)$, commutes. From the definition of $f \cap g$, we have the following commutative diagram

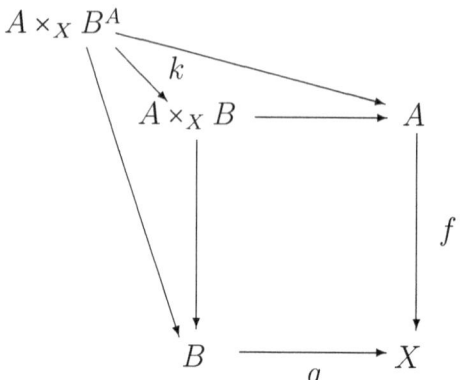

Hence $(f \cap g) \circ k \approx f \cap (f \Rightarrow g)$. Thus

$$f \cap (f \Rightarrow g) \subseteq f \cap g.$$

The existence of such a subobject is guaranteed by the fact that, in any lattice L, for any x, y such that $y \leq x$, there exists $z \in L$ such that $x \wedge z \leq y$. Clearly there are several possible choices.

Theorem 3.4. *The following holds*

(a) *If* $h \subseteq f \Rightarrow g$ *then* $h \cap f \subseteq g$,
(b) *If* $f \Rightarrow g \simeq 1_X$ *then* $f \subseteq g$. □

The converse to (a) does not hold. Thus $f \Rightarrow g$ is not an implication. That is why it is called a *semi-implication*.

Theorem 3.5. *For any subobjects* $f : A \twoheadrightarrow X, g : B \twoheadrightarrow X$ *and* $h : C \twoheadrightarrow X$ *of* X, *the following holds:*

(a) $(f \Rightarrow g) \cap (g \Rightarrow h) \subseteq f \Rightarrow h$,
(b) $(f \Rightarrow g) \cap (f \Rightarrow h) \subseteq f \Rightarrow (g \cap h)$,
(c) *If* $f \subseteq g$ *then* $f \Rightarrow g \simeq 1_X$. □

Let $f : A \twoheadrightarrow X$ be a subobject of X in a prelocus $\mathcal{A}$ and let $x : 1 \longrightarrow X$. If there exists $k : 1 \longrightarrow A$ such that the diagram

Locology and Localistic Logic

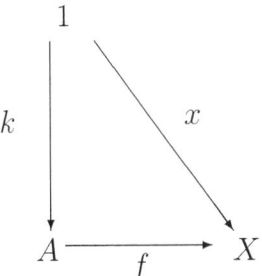

commutes, we say that x is an element of f, denoted $x \in f$.

Theorem 3.6. *In any prelocus, for any object X, we have in $\mathrm{Sub}(X)$, $x \in f \cap g$ iff $x \in f$ and $x \in g$.*

Proof. (a) If $x \in f \cap g$, then there exists k such that $x = (f \cap g) \circ k$. Since $f \cap g \subseteq f$, there exists j such that $f \cap g = f \circ j$. Thus $x = f \circ j \circ k$, i.e. $x \in f$. Similarly, $x \in g$.

(b) Suppose that $x \in f$ and $x \in g$ and consider the diagram

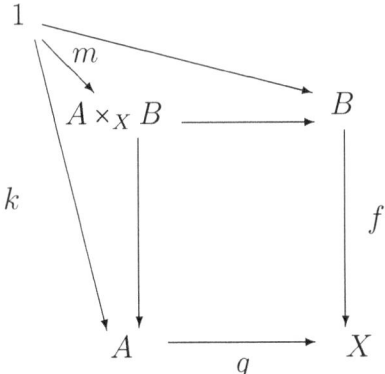

By definition of $f \cap g$, the inner square is a pullback, so the arrow m does exist making the whole diagram commute. Hence $(f \cap g) \circ m = f \circ k = x$. Thus $x \in f \cap g$. □

3.2.4. Complementation

In a prelocus, one can associate to any $f : A \twoheadrightarrow X$ in $\mathrm{Sub}(X)$ the subobject $\neg f : A \twoheadrightarrow X$, defined by $\neg f \simeq f \Rightarrow 0_X$ and called the *complement* of f.

Theorem 3.7. *For any X and any $f : A \twoheadrightarrow X$ and $g : B \twoheadrightarrow X$ in* $\mathrm{Sub}(X)$, *we have*
  (a) $f \cap \neg f \simeq 0_X$,
  (b) *If* $f \subseteq g$ *then* $\neg g \subseteq \neg f$.

Proof. (a) follows from the definition of the semi-implication. (b) Given $f \subseteq g$, then, for any $h : C \twoheadrightarrow X$, $g \Rightarrow h \subseteq f \Rightarrow h$. In particular $g \Rightarrow 0_X \subseteq f \Rightarrow 0_X$ i.e. $\neg g \subseteq \neg f$. □

3.3. Loci

A prelocus is called a *locus* if the commutativity of one of the two diagrams

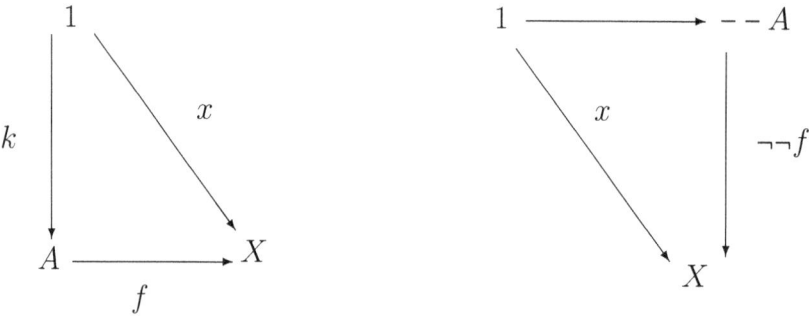

implies that of the other and then of the square

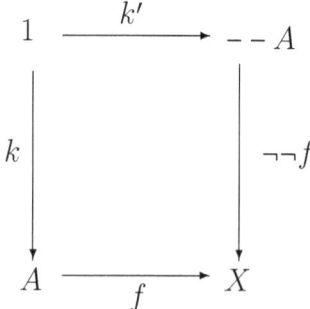

for any object X. In other words, a prelocus $\mathcal{A}$ is a locus iff, for any object X of $\mathcal{A}$, and any $x : 1 \twoheadrightarrow X$, the following equivalence holds true: $(x \in f)$ iff $(x \in \neg\neg f)$.

Theorem 3.8. *For any object X in a locus $\mathcal{A}$ and any $f : A \twoheadrightarrow X$ and $g : B \twoheadrightarrow X$ in $\mathrm{Sub}(X)$, we have*
 (a) $f \simeq \neg\neg f$.
 (b) *If $\neg f \subseteq \neg g$ then $g \subseteq f$.*
 (c) $f \cup \neg f \simeq 1_X$.
 (d) $\neg(f \cup g) \simeq \neg f \cap \neg g$.
 (e) $\neg(f \cap g) \simeq \neg f \cup \neg g$. □

Theorem 3.9. *For any object X in a locus $\mathcal{A}$, $(\mathrm{Sub}(X), \subseteq, \cap, \cup, \neg, \Rightarrow)$ is a Λ-algebra.* □

Remark. In a locus, the following do *not* hold true:
 (a) If $f \cap g \simeq 0_X$ then $g \subseteq \neg f$,
 (b) If $x \in \neg f$ then not $x \in f$,
 (c) If $x \in f \cup g$ then $x \in f$ or $x \in g$,
although the converse implications hold true.

The relationships between loci and Λ-algebras may be made more precise. In a lattice L, when considered as a poset category, there exists an arrow $a \longrightarrow b$ between two elements of L iff $a \leq b$. Since, furthermore, in a Λ-algebra, $x \leq a \Rightarrow b$ entails $x \wedge a \leq b$ (the converse being generally false), the existence of an arrow $x \longrightarrow (a \Rightarrow b)$ implies that of an arrow $x \wedge a \longrightarrow b$. This is reminiscent of the situation in a locus where there is a bijection between a subset of $\mathrm{Hom}(x, b^a)$ and $\mathrm{Hom}(x \times a, b)$. Now, in a Λ-algebra $a \wedge x = x \wedge a$ is the product $x \times a$ and $a \Rightarrow b$ provides us with the exponential b^a. The evaluation arrow $b^a \times a \longrightarrow b$ is the unique arrow $(a \Rightarrow b) \wedge a \longrightarrow b$ which appears in the definition of the semi-implication. Conversely, semi-exponentiation provides semi-implication. Thus, categorically, a Λ-algebra is nothing but a category with a terminal object, with pullbacks and pushouts for any pair of arrows, with products and coproducts for any pair of objects and with a semi-exponentiation. Thus any Λ-algebra is a locus.

3.4. Locus-validity

The above remark on the links between loci and Λ-algebras leads us to consider the concept of locus-validity and its relation with localistic provability. Given a set Φ of formulas, defined via the formation rules given in section 2.1, and a locus $\mathcal{A}$, a formula $\phi \in \Phi$ is said to be $\mathcal{A}$-valid, denoted $\mathcal{A} \vDash \phi$, iff, for any object X of $\mathcal{A}$ and for any valuation $v \colon \phi \longrightarrow \mathrm{Sub}(X), v(\phi) = 1_X$.

Theorem 3.10. $\vdash_{LL} \phi$ iff ϕ is $\mathcal{A}$-valid for any locus $\mathcal{A}$.

Proof. (a) If $\vdash_{LL} \phi$ then ϕ is L-valid for any Λ-algebra L. Then, in particular, ϕ is Sub(X)-valid, for any X in $\mathcal{A}$ and any $\mathcal{A}$, i.e. $\mathcal{A} \vDash \phi$ for any $\mathcal{A}$.
(b) If $\mathcal{A} \vDash \phi$ for any locus $\mathcal{A}$, then ϕ is Sub(X)-valid for any X in $\mathcal{A}$ and for any $\mathcal{A}$. Suppose that $\nvdash_{LL} \phi$. Then $\nvDash_{LL} \phi$ i.e. there exists a Λ-algebra L such that ϕ is not L-valid. But any Λ-algebra being a locus, this leads to a contradiction. □

Clearly, in a locus with exponentiation we have $f \Rightarrow (g \Rightarrow f) \simeq 1_X$ and $f \subseteq g \Rightarrow h$ iff $f \cap g \subseteq h$, for any object X and any $f : A \twoheadrightarrow X$, $g : B \twoheadrightarrow X$ and $h : C \twoheadrightarrow X$ in Sub(X). This, in turn, leads us to compare the locus-theoretic and the topos-theoretic frameworks, in particular from a logico-algebraic viewpoint. The question of whether the algebraic operators $\cap, \cup, \neg$ and $\Rightarrow$ in the algebra Sub(X) of subobjects of some X in a locus (or, equivalently, in view of theorem 3.10, the logical connectors in localistic logic) may be - or should be - internalized is of prime interest. Indeed, a remarkable feature of the above analysis and results is that loci theory makes no use of the concept of subobject classifier and that the logical connectors or the corresponding algebraic operators have no internal counterpart.

Far from being a drawback, the impossibility to internalize the logical connectors and then to consider localistic logic as an internal logic of some (hence any) locus, is a highly desirable result. First, it means that LL must be seen as emerging from a locus-theoretic structure and it asserts the pre-eminence of the (categorical) structure over the logic which emerges from it. The links between topoï theory and intuitionistic logic (IL) convey the opposite - and highly controversial - view. Second, the definition of a subobject classifier in **Set**, which allows to recapture the definition of the Boolean topos **Set** as a special case of the general definition of a topos, is rather artificial. Finally, the equivalence between IL-provability and topos validity hides a situation which seems somehow anomalous. On the one hand, topos validity and IL-provability only depend on the algebraic - hence external to the topos - structure of the algebra Sub(1) of subobjects of the terminal object 1, Sub(1) being not an actual object in a topos. On the other hand, from an internal viewpoint - which should prevail since IL is defined, via truth-arrows, as an internal logic -, what actually matters is Ω^X, for any X, i.e. the internal version of the notion of power set, of which Sub(X) is the external version. But Ω^X plays no role in the definition of the validity/provability. The divorce between the internal and the external versions culminates in a rather counter-intuitive result: there are non-Boolean topoï, i.e. topoï where Sub(Ω) is not a Boolean algebra, which do validate classical logic. The usual claim that topoï theory is to IL what set theory is to classical logic (CL) and, therefore, that topoï theory is the *right* generalization of set theory in some sense, is, to say the least, questionable. Such a situation is just impossible in a locus-theoretic framework. Indeed, if we define a Boolean locus as a locus such that, for any object X, Sub(X) is a Boolean algebra (i.e. such that, for any $f : A \twoheadrightarrow X, g : B \twoheadrightarrow X$ and $h : C \twoheadrightarrow X, f \Rightarrow (g \Rightarrow f) \simeq 1_X$ and $f \cap g \subseteq h$ iff $f \subseteq g \Rightarrow h$), a formula ϕ is CL-valid iff ϕ is valid in any Boolean locus.

References

[1] J.-P. Barthélemy, M. De Glas, J. Desclés, and J. Petitot. Logique et dynamique de la cognition. *Intellectica*, 23:219–301, 1996.

[2] O. Breysse and M. De Glas. A New Approach to the Concept of Boundary and Contact: Toward an Alternative to Mereotopology. *Fondamenta Informaticae*, 78:217–238, 2007.

[3] G. Choquet. Convergences. *Annales de l'Université de Grenoble*, 23:55–112, 1947.

[4] M. M. Day. Convergence, Closure, and Neighborhoods. *Duke Math. J.*, 11:181–199, 1947.

[5] M. De Glas. Locological Spaces. *Cognitiva*, 90:437–445, 1990.

[6] M. De Glas. Subintuitionnistic Logics and Locology. *Logic Colloquium*, 1997.

[7] M. De Glas and J.-L. Plane. *Une approche formelle de la typicité*, volume 20. École Polytechnique, Paris, 2005. ISSN 0984-5100.

[8] P. C. Hammer. Extended Topology: Continuity. *Portugaliae Mathematica*, 25:77–93, 1964.

[9] H. Poincaré. *La science et l'hypothèse*. Flammarion, 1902.

[10] H. Poincaré. *La valeur de la science*. Flammarion, 1905.

[11] M. B. Smyth. Semi-Metric, Closure Spaces, and Digital Topology. *Th. Comp. Sci.*, pages 257–276, 1995.

[12] E. Čech. *Topological Spaces*. Wiley, 1966.

[13] E. C. Zeeman and O. P. Buneman. Tolerance Spaces and the Brain. In C. H. Waddington, editor, *Towards a Theoretical Biology*, pages 140–151. Oxford University Press, 1968.

Received: 14 April 2016

Secure Multiparty Computation without One-way Functions

Dima Grigoriev[*]

CNRS, Mathématiques, Université de Lille
59655, Villeneuve d'Ascq, France
dmitry.grigoryev@math.univ-lille1.fr

Vladimir Shpilrain[†]

Department of Mathematics, The City College of New York
New York, NY 10031, USA
shpil@groups.sci.ccny.cuny.edu

Abstract: We describe protocols for secure computation of the sum, product, and some other functions of two or more elements of an arbitrary constructible ring, without using any one-way functions. One of the new inputs that we offer here is that, in contrast with other proposals, we conceal intermediate results of a computation. For example, when we compute the sum of k numbers, only the final result is known to the parties; partial sums are not known to anybody. Other applications of our method include voting/rating over insecure channels and a rather elegant and efficient solution of the "two millionaires problem".

We also give a protocol, without using a one-way function, for the so-called "mental poker", i.e., a fair card dealing (and playing) over distance.

Finally, we describe a secret sharing scheme where an advantage over Shamir's and other known secret sharing schemes is that nobody, including the dealer, ends up knowing the shares (of the secret) owned by any particular player.

It should be mentioned that computational cost of our protocols is negligible to the point that all of them can be executed without a computer.

---

[*]Research was supported by the RSF grant 16-11-10075.
[†]Research was partially supported by the NSF grant CNS-1117675 and by the ONR (Office of Naval Research) grant N000141512164.

1. Introduction

As a society, we have become dependent on information technology for many aspects of our daily life, and as a consequence, dependent upon cryptography. The need for developing various cryptographic tools to address new challenges in storing and processing information is therefore clear. One of these challenges, namely how to securely and efficiently process information owned by several different parties, is addressed in this paper.

The problem of secure multi-party computation was originally suggested by Yao [19] in 1982. The concept usually refers to computational systems in which several parties wish to jointly compute some value based on individually held secret bits of information, but do not wish to reveal their secrets to anybody in the process. For example, two individuals, each possessing some secret numbers, x and y, respectively, may wish to jointly compute some function $f(x, y)$ without revealing any information about x or y other than what can be reasonably deduced by knowing the actual value of $f(x, y)$.

Secure computation was formally introduced by Yao as secure two-party computation. His "two millionaires problem" (cf. our Section 3) and its solution gave way to a generalization to multi-party protocols, see e.g. [4], [7]. Secure multi-party computation provides solutions to various real-life problems such as distributed voting, private bidding and auctions, sharing of signature or decryption functions, private information retrieval, etc.

In this paper, we showcase several protocols, originally offered in [13], for secure computation of various functions (including the sum and product) of three or more elements of an arbitrary constructible ring, without using encryption or any one-way functions whatsoever. We require in our scheme that there are k secure channels for communication between the $k \geq 3$ parties, arranged in a cycle. We also show that less than k secure channels is not enough.

Unconditionally secure multiparty computation was previously considered in [4] and elsewhere. A new input that we offer here is that, in contrast with [4] and other proposals, we conceal "intermediate results" of a computation. For example, when we compute a sum of k numbers n_i, only the final result $\sum_{i=1}^{k} n_i$ is known to the parties; partial sums are not known to anybody. This is not the case in [4] where each partial sum $\sum_{i=1}^{s} n_i$ is known to at least some of the parties. This difference is important because, by the "pigeonhole principle", at least one of the parties may accumulate sufficiently many expressions in n_i to be able to recover at least some of the n_i other than his own.

Here we show how our method works for computing the sum (Section 2) and the product (Section 2.2) of private numbers. We ask what other functions can be securely computed without revealing intermediate results.

Other applications of our method include voting/rating over insecure channels (Section 2.4) and a rather elegant solution of the "two millionaires problem" (Section 3).

In Section 5, we consider a cryptographic primitive known as "mental poker", i.e., fair card dealing (and playing) over distance. Several protocols for doing this, most of them using encryption, have been suggested, the first by Shamir, Rivest, and Adleman [18], and subsequent proposals include [5] and [9]. As with bit commitment, fair card dealing between just two players over distance is impossible without a one-way function since commitment is part of any meaningful card dealing scenario. However, this turns out to be possible if the number of players is $k \geq 3$. What we require though is that there are k secure channels for communication between players, arranged in a cycle. We also show that our protocol can, in fact, be adapted to deal cards to just 2 players. Namely, if we have 2 players, they can use a "dummy" player (e.g. a computer), deal cards to 3 players, and then just ignore the "dummy"'s cards, i.e., "put his cards back in the deck". An assumption on the "dummy" player is that he cannot generate any randomness, so randomness has to be supplied to him by the two "real" players. Another assumption is that there are secure channels for communication between either "real" player and the "dummy". We believe that this model is adequate for 2 players who want to play online but do not trust the server. "Not trusting" the server exactly means not trusting with generating randomness. Other, deterministic, operations can be verified at the end of the game; we give more details in Section 5.2.

We note that the only known (to us) proposal for dealing cards to $k \geq 3$ players over distance without using one-way functions was published in [1], but their protocol lacks the simplicity, efficiency, and some of the functionalities of our proposal; this is discussed in more detail in our Section 6. Here we just mention that computational cost of our protocols is negligible to the point that they can be easily executed without a computer.

Finally, in Section 7, we propose a secret sharing scheme where an advantage over Shamir's [17] and other known secret sharing schemes is that nobody, including the dealer, ends up knowing the shares (of the secret) owned by any particular players. The disadvantage though is that our scheme is a (k,k)-threshold scheme only.

2. Secure computation of a sum

In this section, our scenario is as follows. There are k parties $P_1, \ldots, P_k$; each P_i has a private element n_i of a fixed constructible ring R. The goal is to compute the sum of all n_i without revealing any of the n_i to any party $P_j, j \neq i$.

One obvious way to achieve this is well studied in the literature (see e.g. [8, 9, 12]): encrypt each n_i as $E(n_i)$, send all $E(n_i)$ to some designated P_i (who does not have a decryption key), have P_i compute $S = \sum_i E(n_i)$ and send the result to the participants for decryption. Assuming that the encryption function E is *homomorphic*, i.e., that $\sum_i E(n_i) = E(\sum_i n_i)$, each party P_i can recover $\sum_i n_i$ upon decrypting S.

This scheme requires not just a one-way function, but a one-way function with a trapdoor since both encryption and decryption are necessary to obtain the result.

What we suggest in this section is a protocol that does not require any one-way function, but involves secure communication between some of the P_i. So, *our assumption* here is that there are k secure channels of communication between the k parties P_i, arranged in a cycle. *Our result* is computing the sum of private elements n_i without revealing any individual n_i to any $P_j, j \neq i$. Clearly, this is only possible if the number of participants P_i is greater than 2. As for the number of secure channels between P_i, we will show that it cannot be less than k, by the number of parties.

2.1. The protocol (computing the sum)

1. P_1 initiates the process by sending $n_1 + n_{01}$ to P_2, where n_{01} is a random element ("noise").

2. Each P_i, $2 \leq i \leq k-1$, does the following. Upon receiving an element m from P_{i-1}, he adds his $n_i + n_{0i}$ to m (where n_{0i} is a random element) and sends the result to P_{i+1}.

3. P_k adds $n_k + n_{0k}$ to whatever he has received from P_{k-1} and sends the result to P_1.

4. P_1 subtracts n_{01} from what he got from P_k; the result now is the sum $S = \sum_{1 \le i \le k} n_i + \sum_{2 \le i \le k} n_{0i}$. Then P_1 publishes S.

5. Now all participants P_i, except P_1, broadcast their n_{0i}, possibly over insecure channels, and compute $\sum_{2 \le i \le k} n_{0i}$. Then they subtract the result from S to finally get $\sum_{1 \le i \le k} n_i$.

Thus, in this protocol we have used k (by the number of the parties P_i) secure channels of communication between the parties. If we visualize the arrangement as a graph with k vertices corresponding to the parties P_i and k edges corresponding to secure channels, then this graph will be a k-cycle. Other arrangements are possible, too; in particular, a union of disjoint cycles of length ≥ 3 would do. (In that case, the graph will still have k edges.) Two natural questions that one might now ask are: (1) is any arrangement with less than k secure channels possible? (2) with k secure channels, would this scheme work with any arrangement other than a union of disjoint cycles of length ≥ 3? The answer to both questions is "no". Indeed, if there is a vertex (corresponding to P_1, say) of degree 0, then any information sent out by P_1 will be available to everybody, so other participants will know n_1 unless P_1 uses a one-way function to conceal it. If there is a vertex (again, corresponding to P_1) of degree 1, this would mean that P_1 has a secure channel of communication with just one other participant, say P_2. Then any information sent out by P_1 will be available at least to P_2, so P_2 will know n_1 unless P_1 uses a one-way function to conceal it. Thus, every vertex in the graph should have degree at least 2, which implies that every vertex is included in a cycle. This immediately implies that the total number of edges is at least k. If now a graph Γ has k vertices and k edges, and every vertex of Γ is included in a cycle, then every vertex has degree exactly 2 since by the "handshaking lemma" the sum of the degrees of all vertices in any graph equals twice the number of edges. It follows that our graph is a union of disjoint cycles.

2.2. Secure computation of a product

Now we show how to use the general ideas of the protocol for computing the sum (see Section 2.1) to securely compute a product. Again, there are k parties $P_1, \ldots, P_k$; each P_i has a private (nonzero) element n_i of a fixed constructible ring R. The goal is to compute the product of all n_i without revealing any of the n_i to any party $P_j, j \ne i$. Requirements on the ring R are going to be somewhat more stringent here than they were in Section 2. Namely, we require that R does not have zero divisors and, if an element r of R is a product $a \cdot x$ with a known a and an unknown x, then x can be efficiently recovered from a and r. Examples of rings with these properties include the ring of integers and any constructible field.

The protocol (computing the product)

1. P_1 initiates the process by sending $n_1 \cdot n_{01}$ to P_2, where n_{01} is a random nonzero element ("noise").

2. Each P_i, $2 \leq i \leq k-1$, does the following. Upon receiving an element m from P_{i-1}, he multiplies m by $n_i \cdot n_{0i}$ (where n_{0i} is a random element) and sends the result to P_{i+1}.

3. P_k multiplies by $n_k \cdot n_{0k}$ whatever he has received from P_{k-1} and sends the result to P_1. This result is the product $P = \Pi_{1 \leq i \leq k}\, n_i \cdot \Pi_{2 \leq i \leq k}\, n_{0i}$.

4. P_1 divides what he got from P_k by his n_{01}; the result now is the product $P = \Pi_{1 \leq i \leq k}\, n_i \cdot \Pi_{2 \leq i \leq k}\, n_{0i}$. Then P_1 publishes P.

5. Now all participants P_i, except P_1, broadcast their n_{0i}, possibly over insecure channels, and compute $\Pi_{2 \leq i \leq k}\, n_{0i}$. Then they divide P by the result to finally get $\Pi_{1 \leq i \leq k}\, n_i$.

2.3. Effect of coalitions

Suppose now we have $k \geq 3$ parties with k secure channels of communication arranged in a cycle, and suppose 2 of the parties secretly form a coalition. Our assumption here is that, because of the circular arrangement of secure channels, a secret coalition is only possible between parties P_i and P_{i+1} for some i, where the indices are considered modulo k; otherwise, attempts to form a coalition (over insecure channels) will be detected. If two parties P_i and P_{i+1} exchanged information, they would, of course, know each other's elements n_i, but other than that, they would not get any advantage if $k \geq 4$. Indeed, we can just "glue these two parties together", i.e., consider them as one party, and then the protocol is essentially reduced to that with $k - 1 \geq 3$ parties. On the other hand, if $k = 3$, then, of course, two parties together have all the information about the third party's element.

Secure Multiparty Computation

For an arbitrary $k \geq 4$, if $n < k$ parties want to form a (secret) coalition to get information about some other party's element, all these n parties have to be connected by secure channels, which means there is a j such that these n parties are $P_j, P_{j+1}, \ldots, P_{j+n-1}$, where indices are considered modulo k. It is not hard to see then that only a coalition of $k-1$ parties $P_1, \ldots, P_{i-1}, P_{i+1}, \ldots, P_k$ can suffice to get information about the P_i's element.

2.4. Ramification: voting/rating over insecure channels

In this section, our scenario is as follows. There are k parties $P_1, \ldots, P_k$; each P_i has a private integer n_i. There is also a computing entity B (for Boss) who shall compute the sum of all n_i. The goal is to let B compute the sum of all n_i without revealing any of the n_i to him or to any party $P_j, j \neq i$.

The following example from real life is a motivation for this scenario.

Example 1. Suppose members of the board in a company have to vote for a project by submitting their numeric scores (say, from 1 to 10) to the president of the company. The project gets a green light if the total score is above some threshold value T. Members of the board can discuss the project between themselves and exchange information privately, but none of them wants his/her score to be known to either the president or any other member of the board.

In the protocol below, we are again assuming that there are k channels of communication between the parties, arranged in a cycle: $P_1 \to P_2 \to \cdots \to P_k \to P_1$. On the other hand, communication channels between B and any of the parties are not assumed to be secure.

2.5. The protocol (rating over insecure channels)

1. P_1 initiates the process by sending $n_1 + n_{01}$ to P_2, where n_{01} is a random number.

2. Each P_i, $2 \leq i \leq k - 1$, does the following. Upon receiving a number m from P_{i-1}, he adds his $n_i + n_{0i}$ to m (where n_{0i} is a random number) and sends the result to P_{i+1}.

3. P_k adds $n_k + n_{0k}$ to whatever he has received from P_{k-1} and sends the result to B.

4. P_k now starts the process of collecting the "adjustment" in the opposite direction. To that effect, he sends his n_{0k} to P_{k-1}.

5. P_{k-1} adds $n_{0(k-1)}$ and sends the result to P_{k-2}.

6. The process ends when P_1 gets a number from P_2, adds his n_{01}, and sends the result to B. This result is the sum of all n_{0i}.

7. B subtracts what he got from P_1 from what he got from P_k; the result now is the sum of all n_i, $1 \leq i \leq k$.

3. Application: the "two millionaires problem"

The protocol from Section 2, with some adjustments, can be used to provide an elegant and efficient solution to the "two millionaires problem" introduced in [19]: there are two numbers, n_1 and n_2, and the goal is to solve the inequality $n_1 \geq ? n_2$ without revealing the actual values of n_1 or n_2.

To that effect, we use a "dummy" as the third party. Our concept of a "dummy" is quite different from a well-known concept of a "trusted third party"; importantly, our "dummy" is not supposed to generate any randomness; it just does what it is told to. Basically, the only difference between our "dummy" and a usual calculator is that there are secure channels of communication between the "dummy" and either "real" party. One possible real-life interpretation of such a "dummy" would be an online calculator that can combine inputs from different users. Also note that in our scheme below *the "dummy" is unaware of the committed values* of n_1 or n_2, which is useful in case the two "real" parties do not want their private numbers to ever be revealed. This suggests yet another real-life interpretation of a "dummy", where he is a mediator between two parties negotiating a settlement.

Thus, let A (Alice) and B (Bob) be two "real" parties, and D (Dummy) the "dummy". Suppose A's number is n_1, and B's number is n_2.

3.1. The protocol (comparing two numbers)

1. A splits her number n_1 as a difference $n_1 = n_1^+ - n_1^-$. She then sends n_1^- to B.

2. B splits his number n_2 as a difference $n_2 = n_2^+ - n_2^-$. He then sends n_2^- to A.

3. A sends $n_1^+ + n_2^-$ to D.

4. B sends $n_2^+ + n_1^-$ to D.

5. D subtracts $(n_2^+ + n_1^-)$ from $(n_1^+ + n_2^-)$ to get $n_1 - n_2$, and announces whether this result is positive or negative.

Remark 1. Perhaps a point of some dissatisfaction in this protocol could be the fact that the "dummy" ends up knowing the actual difference $n_1 - n_2$, so if there is a leak of this information to either party, this party would recover the other's private number n_i. This can be avoided if n_1 and n_2 are represented in the binary form and compared one bit at a time, going left to right, until the difference between bits becomes nonzero. However, this method, too, has a disadvantage: the very moment the "dummy" pronounces the difference between bits nonzero would give an *estimate* of the difference $n_1 - n_2$ *to the real parties*, not just to the "dummy".

We note that the original solution of the "two millionaires problem" given in [19], although lacks the elegance of our scheme, does not involve a third party, whereas our solution does. On the other hand, the solution in [19] uses encryption, whereas our solution does not, which makes it by far more efficient. Finally, we mention that since our paper [13] was published, we have come up with several other solutions of the "two millionaires problem" without using either one-way functions or a dummy [14], [11]. Some of those solutions use simple laws of (classical) physics instead.

4. Secure computation of symmetric functions

In this section, we show how our method can be easily generalized to allow secure computation of any expression of the form $\sum_{i=1}^{k} n_i^r$, where n_i are parties' private numbers, k is the number of parties, and $r \geq 1$ an arbitrary integer. We simplify our method here by removing the "noise", to make the exposition more transparent. Otherwise, the protocol is the same as the protocol for secure computation of a sum in Section 2.

4.1. The protocol (computing the sum of powers)

1. P_1 initiates the process by sending a random element n_0 to P_2.

2. Each P_i, $2 \le i \le k-1$, does the following. Upon receiving an element m from P_{i-1}, he adds his n_i^r to m and sends the result to P_{i+1}.

3. P_k adds his n_k^r to whatever he has received from P_{k-1} and sends the result to P_1.

4. P_1 subtracts $(n_0 - n_1^r)$ from what he got from P_k; the result now is the sum of all n_i^r, $1 \le i \le k$.

Now that the parties can securely compute the sum of any powers of their n_i, they can also compute any symmetric function of n_i. However, in the course of computing a symmetric function from sums of different powers of n_i, at least some of the parties will possess several different polynomials in n_i, so chances are that at least some of the parties will be able to recover at least some of the n_i. On the other hand, because of the symmetry of all expressions involved, there is no way to tell which n_i belongs to which party.

4.2. Open problem

Now it is natural to ask:

Problem 1. What other functions (other than the sum and the product) can be securely computed without revealing intermediate results to any party?

To be more precise, we note that one intermediate result is inevitably revealed to the party who finishes computation, but this cannot be avoided in any scenario. For example, after the parties have computed the sum of their private numbers, each party also knows the sum of all numbers except his own. What we want is that no other intermediate results are ever revealed.

To give some insight into this problem, we consider a couple of examples of computing simple functions different from the sum and the product of the parties' private numbers.

Example 2. We show how to compute the function $f(n_1, n_2, n_3) = n_1 n_2 + n_2 n_3$ in the spirit of the present paper, without revealing (or even computing) any intermediate results, i.e., without computing $n_1 n_2$ or $n_2 n_3$.

1. P_2 initiates the process by sending a random element n_0 to P_3.

2. P_3 adds his n_3 to n_0 and sends $n_3 + n_0$ to P_1.

3. P_1 adds his n_1 to $n_0 + n_3$ and sends the result to P_2.

4. P_2 subtracts n_0 from $n_0 + n_3 + n_1$ and multiplies the result by n_2. This is now $n_1 n_2 + n_2 n_3$.

Example 3. The point of this example is to show that functions that can be computed by our method do not have to be homogeneous (in case the reader got this impression based on the previous examples).

The function that we compute here is $f(n_1, n_2, n_3) = n_1 n_2 + g(n_3)$, where g is any computable function.

1. P_1 initiates the process by sending a random element a_0 to P_2.

2. P_2 multiplies a_0 by his n_2 and sends the result to P_3.

3. P_3 multiplies $a_0 n_2$ by a random element c_0 and sends the result to P_1.

4. P_1 multiplies $a_0 n_2 c_0$ by his n_1, divides by a_0, and sends the result, which is $n_1 n_2 c_0$, back to P_3.

5. P_3 divides $n_1 n_2 c_0$ by c_0 and adds $g(n_3)$, to end up with $n_1 n_2 + g(n_3)$.

Note that in this example, the parties used more than just one loop of transmissions in the course of computation. Also, information here was sent "in both directions" in the circuit.

Remark 2. Another collection of examples of multiparty computation without revealing intermediate results can be obtained as follows. Suppose, without loss of generality, that some function $f(n_1, \ldots, n_k)$ can be computed by our method in such a way that the last step in the computation is performed by the party P_1, i.e., P_1 is the one who ends up with $f(n_1, \ldots, n_k)$ while no party knows any intermediate result $g(n_1, \ldots, n_k)$ of this computation. Then, obviously, P_1 can produce any function of the form $F(n_1, f(n_1, \ldots, n_k))$ (for a computable function F) as well. Examples include $n_1^r + n_1 n_2 \cdots n_k$ for any $r \geq 0$; $n_1^r + (n_1 n_2 + n_3)^s$ for any $r, s \geq 0$, etc.

5. Mental poker

"Mental poker" is the common name for a set of cryptographic problems that concerns playing a fair game over distance without the need for a trusted third party. One of the ways to describe the problem is: how can 2 players deal cards fairly over the phone? Several protocols for doing this have been suggested, including [18], [5], [9] and [1]. As with bit commitment, it is rather obvious that fair card dealing to two players over distance is impossible without a one-way function, or even a one-way function with trapdoor. However, it turns out to be possible if the number of players is at least 3, assuming, of course, that there are secure channels for communication between at least some of the players. In our proposal, we will be using k secure channels for $k \geq 3$ players $P_1, \ldots, P_k$, and these k channels will be arranged in a cycle: $P_1 \to P_2 \to \ldots \to P_k \to P_1$.

To begin with, suppose there are 3 players: P_1, P_2, and P_3 and 3 secure channels: $P_1 \to P_2 \to P_3 \to P_1$.

The first protocol, Protocol 1 below, is for distributing *all* integers from 1 to m to the players in such a way that each player gets about the same number of integers. (For example, if the deck that we want to deal has 52 cards, then two players should get 17 integers each, and one player should get 18 integers.) In other words, Protocol 1 allows one to randomly split a set of m integers into 3 disjoint sets.

The second protocol, Protocol 2, is for collectively generating random integers modulo a given integer M. This very simple but useful primitive can be used: **(i)** for collectively generating, uniformly at random, a permutation from the group S_m. This will allow us to assign cards from a deck of m cards to the m integers distributed by Protocol 1; **(ii)** introducing "dummy" players as well as for "playing" after dealing cards.

5.1. Protocol 1

For notational convenience, we are assuming below that we have to distribute integers from 1 to $r = 3s$ to 3 players.

To begin with, all players agree on a parameter N, which is a positive integer of a reasonable magnitude, say, 10.

1. each player P_i picks, uniformly at random, an integer (a "counter") c_i between 1 and N, and keeps it private.

Secure Multiparty Computation

2. P_1 starts with the "extra" integer 0 and sends it to P_2.

3. P_2 sends to P_3 either the integer m he got from P_1, or $m + 1$. More specifically, if P_2 gets from P_1 the same integer m less than or equal to c_2 times, then he sends m to P_3; otherwise, he sends $m + 1$ and keeps m (i.e., in the latter case m becomes one of "his" integers). Having sent out $m + 1$, he "resets his counter", i.e., selects, uniformly at random between 1 and N, a new c_2. He also resets his counter if he gets the number m for the first time, even if he does not keep it.

4. P_3 sends to P_1 either the integer m he got from P_2, or $m + 1$. More specifically, if P_3 gets from P_2 the same integer m less than or equal to c_3 times, then he sends m to P_1; otherwise, he sends $m + 1$ and keeps m. Having sent out $m + 1$, he selects a new counter c_3. He also resets his counter if he gets the number m for the first time, even if he does not keep it.

5. P_1 sends to P_2 either the integer m he got from P_3, or $m + 1$. More specifically, if P_1 gets from P_3 the same integer m less than or equal to c_1 times, then he sends m to P_2; otherwise, he sends $m + 1$ and keeps m. Having sent out $m + 1$, he selects a new counter c_1. He also resets his counter if he gets the number m for the first time, even if he does not keep it.

6. This procedure continues until one of the players gets s integers (not counting the "extra" integer 0). After that, a player who already has s integers just "passes along" any integer that comes his way, while other players keep following the above procedure until they, too, get s integers.

7. The protocol ends as follows. When all $3s$ integers, between 1 and $3s$, are distributed, the player who got the last integer, $3s$, keeps this fact to himself and passes this integer along as if he did not "take" it.

8. The process ends when the integer $3s$ makes $N + 1$ "full circles".

We note that the role of the "extra" integer 0 is to prevent P_3 from knowing that P_2 has got the integer 1 if it happens that $c_2 = 1$ in the beginning.

We also note that this protocol can be generalized to arbitrarily many players in the obvious way, if there are k secure channels for communication between k players, arranged in a cycle.

5.2. Protocol 2

Now we describe a protocol for generating random integers modulo some integer M collectively by 3 players. As in Protocol 1, we are assuming that there are secure channels for communication between the players, arranged in a cycle.

1. P_2 and P_3 uniformly at random and independently select private integers n_2 and n_3 (respectively) modulo M.
2. P_2 sends n_2 to P_1, and P_3 sends n_3 to P_1.
3. P_1 computes the sum $m = n_2 + n_3$ modulo M.

Note that neither P_2 nor P_3 can cheat by trying to make a "clever" selection of their n_i because the sum, modulo M, of any integer with an integer uniformly distributed between 0 and $M-1$, is an integer uniformly distributed between 0 and $M-1$.

Finally, P_1 cannot cheat simply because he does not really get a chance: if he miscalculates $n_2 + n_3$ modulo M, this will be revealed at the end of the game. (All players keep contemporaneous records of all transactions, so that at the end of the game, correctness could be verified.)

To generalize Protocol 2 to arbitrarily many players $P_1, \ldots, P_k$, $k \geq 3$, we can just engage 3 players at a time in running the above protocol. If, at the same time, we want to keep the same circular arrangement of secure channels between the players that we had in Protocol 1, i.e., $P_1 \to P_2 \to \ldots P_k \to P_1$, then 3 players would have to be P_{i+1}, P_i, P_{i+2}, where i would run from 1 to k, and the indices are considered modulo k.

Protocol 2 can now be used to collectively generate, uniformly at random, a permutation from the group S_m. This will allow us to assign cards from a deck of m cards to the m integers distributed by Protocol 1. Generating a random permutation from S_m can be done by taking a random integer between 1 and m (using Protocol 2) sequentially, ensuring that there is no repetition. This "brute-force" method will require occasional retries whenever the random integer picked is a repeat of an integer already selected. A simple algorithm to generate a permutation from S_m uniformly randomly without retries, known as the *Knuth shuffle*, is to start with the identity permutation or any other permutation, and then go through the positions 1 through $(m-1)$, and for each position i swap the element currently there with an arbitrarily chosen element from positions i through m, inclusive (again, Protocol 2 can be used here to produce a random integer between i and m). It is easy to verify that any permutation of m elements will be produced by this algorithm with probability exactly $\frac{1}{m!}$, thus yielding a uniform distribution over all such permutations.

After this is done, we have m cards distributed uniformly randomly to the players, i.e., we have:

Proposition 1. *If m cards are distributed to k players using Protocols 1 and 2, then the probability for any particular card to be distributed to any particular player is $\frac{1}{k}$.*

5.3. Using "dummy" players while dealing cards

We now show how a combination of Protocol 1 and Protocol 2 can be used to deal cards to just 2 players. If we have 2 players, they can use a "dummy" player (e.g. a computer), deal cards to 3 players as in Protocol 1, and then just ignore the "dummy"'s cards, i.e., "put his cards back in the deck". We note that the "dummy" in this scenario would not generate randomness; it will be generated for him by the other two players using Protocol 2. Namely, if we call the "dummy" P_3, then the player P_1 would randomly generate c_{31} between 1 and N and send it to P_3, and P_2 would randomly generate c_{32} between 1 and N and send it to P_3. Then P_3 would compute his random number as $c_3 = c_{31} + c_{32}$ modulo N.

Similarly, "dummy" players can help k "real" players each get a fixed number s of cards, because Protocol 1 alone is only good for distributing *all* cards in the deck to the players, dealing each player about the same number of cards. We can introduce m "dummy" players so that $(m + k) \cdot s$ is approximately equal to the number of cards in the deck, and position all the "dummy" players one after another as part of a circuit $P_1 \to P_2 \to \ldots P_{m+k} \to P_1$. Then we use Protocol 1 to distribute all cards in the deck to $(m + k)$ players taking care that each "real" player gets exactly s cards. As in the previous paragraph, "dummy" players have "real" ones generate randomness for them using Protocol 2.

After all cards in the deck are distributed to $(m + k)$ players, "dummy" players send all their cards to one of them; this "dummy" player now becomes a "dummy dealer", i.e., he will give out random cards from the deck to "real" players as needed in the course of a subsequent game, while randomness itself will be supplied to him by "real" players using Protocol 2.

6. Summary of the properties of our card dealing (Protocols 1 and 2)

Here we summarize the properties of our Protocols 1 and 2 and compare, where appropriate, our protocols to the card dealing protocol of [1].

1. Uniqueness of cards. Yes, by the very design of Protocol 1.

2. Uniform random distribution of cards. Yes, because of Protocol 2; see our Proposition 1 in Section 5.2.

3. Complete confidentiality of cards. Yes, by the design of Protocol 1.

4. Number of secure channels for communication between $k \geq 3$ players: k, arranged in a cycle.

By comparison, the card dealing protocol of [1] requires $3k$ secure channels.

5. Average number of transmissions between $k \geq 3$ players: $O(\frac{N}{2}mk)$, where m is the number of cards in the deck, and $N \approx 10$. This is because in Protocol 1, the number of circles (complete or incomplete) each integer makes is either 1 or the minimum of all the counters c_i at the moment when this integer completes the first circle. Since the average of c_i is at most $\frac{N}{2}$, we get the result because within one circle (complete or incomplete) there are at most k transmissions. We note that in fact, there is a precise formula for the average of the minimum of c_i in this situation: $\frac{\sum_{j=1}^{N} j^k}{N^k}$, which is less than $\frac{N}{2}$ if $k \geq 2$.

By comparison, in the protocol of [1] there are $O(mk^2)$ transmissions.

6. Total length of transmissions between $k \geq 3$ players: $\frac{N}{2}mk \cdot \log_2 m$ bits. This is just the average number of transmissions times the length of a single transmission, which is a positive integer between 1 and m.

By comparison, total length of transmissions in [1] is $O(mk^2 \log k)$.

7. Computational cost of Protocol 1: negligible (because computation amounts to selecting random integers from a small interval).

By comparison, the protocol of [1] requires computing products of up to k permutations from the group S_k to deal just one card; the total computational cost therefore is $O(mk^2 \log k)$.

Secure Multiparty Computation   79

7. Secret sharing

Secret sharing refers to method for distributing a secret amongst a group of participants, each of whom is allocated a share of the secret. The secret can be reconstructed only when a sufficient number of shares are combined together; individual shares are of no use on their own.

More formally, in a secret sharing scheme there is one dealer and k players. The dealer gives a secret to the players, but only when specific conditions are fulfilled. The dealer accomplishes this by giving each player a share in such a way that any group of t (for threshold) or more players can together reconstruct the secret but no group of fewer than t players can. Such a system is called a (t, k)-threshold scheme (sometimes written as a (k, t)-threshold scheme).

Secret sharing was invented by Shamir [17] and Blakley [2], independent of each other, in 1979. Both proposals assumed secure channels for communication between the dealer and each player. In our proposal here, the number of secure channels is equal to $2k$, where k is the number of players, because in addition to the secure channels between the dealer and each player, we have k secure channels for communication between the players, arranged in a cycle: $P_1 \to P_2 \to \ldots \to P_k \to P_1$.

The advantage of our scheme over Shamir's and other known secret sharing schemes is that nobody, including the dealer, ends up knowing the shares (of the secret) owned by any particular players. The disadvantage is that our scheme is a (k, k)-threshold scheme only.

We start by describing a subroutine for distributing shares by the players among themselves. More precisely, k players want to split a given number in a sum of k numbers, so that each summand is known to one player only, and each player knows one summand only.

7.1. The subroutine (distributing shares by the players among themselves)

Suppose a player P_i receives a number M that has to be split in a sum of k private numbers. In what follows, all indices are considered modulo k.

1. P_i initiates the process by sending $M - m_i$ to P_{i+1}, where m_i is a random number (could be positive or negative).

2. Each subsequent P_j does the following. Upon receiving a number m from P_{j-1}, he subtracts a random number m_j from m and sends the result to P_{j+1}. The number m_j is now P_j's secret summand.

3. When this process gets back to P_i, he adds m_i to whatever he got from P_{i-1}; the result is his secret summand.

Now we get to the actual secret sharing protocol.

7.2. The protocol (secret sharing (k, k)-threshold scheme)

The dealer D wants to distribute shares of a secret number N to k players P_i so that, if P_i gets a number s_i, then $\sum_{i=1}^{k} s_i = N$.

1. D arbitrarily splits N in a sum of k integers: $N = \sum_{i=1}^{k} n_i$.

2. The loop: at Step i of the loop, D sends n_i to P_i, and P_i initiates the above Subroutine to distribute shares n_{ij} of n_i among the players, so that $\sum_{j=1}^{k} n_{ij} = n_i$.

3. After all k steps of the loop are completed, each player P_i ends up with k numbers n_{ji} that sum up to $s_i = \sum_{j=1}^{k} n_{ji}$. It is obvious that $\sum_{i=1}^{k} s_i = N$.

Acknowledgement

Both authors are grateful to Max Planck Institut für Mathematik (MPI), Bonn for its hospitality during the work on this paper. The first author is also grateful to MCCME, Moscow for excellent working conditions and inspiring atmosphere.

References

[1] I. Bárány and Z. Füredi. Mental poker with three or more players. *Inform. and Control*, 59:84–93, 1983.

[2] G. R. Blakley. Safeguarding cryptographic keys. *Proceedings of the National Computer Conference*, 48:313–317, 1979.

[3] G. Brassard, C. Crépeau, and J.-M. Robert. *All-or-nothing disclosure of secrets*. Advances in Cryptology – CRYPTO '86, pages 234–238, Lecture Notes Comp. Sci. 263. Springer, 1986.

[4] D. Chaum, C. Crépeau, and I. Damgård. *Multiparty unconditionally secure protocols (extended abstract)*, In Proceedings of the Twentieth ACM Symposium on the Theory of Computing, pages 11–19, ACM, 1988.

[5] C. Crépeau. *A zero-knowledge poker protocol that achieves confidentiality of the players' strategy or how to achieve an electronic poker face*, Advances in Cryptology – CRYPTO '86, pages 239–247, Lecture Notes Comp. Sci. 263, Springer, 1986.

[6] I. Damgård, M. Geisler, and M. Kroigard. Homomorphic encryption and secure comparison. *Int. J. Appl. Cryptogr.*, 1:22–31, 2008.

[7] I. Damgård, Y. Ishai. *Scalable secure multiparty computation*, Advances in Cryptology – CRYPTO 2006, pages 501–520, Lecture Notes Comp. Sci. 4117, Springer, Berlin, 2006.

[8] O. Goldreich. *Foundations of Cryptography: Volume 1, Basic Tools*. Cambridge University Press, 2007.

[9] S. Goldwasser and S. Micali. *Probabilistic Encryption and How to Play Mental Poker Keeping Secret All Partial Information*, In Proceedings of the 14th Annual ACM symp. on Theory of computing, ACM-SIGACT, pages 365–377, May 1982.

[10] S. Goldwasser and S. Micali. Probabilistic encryption. *J. Comput. System Sci.*, 28:270–299 1984.

[11] D. Grigoriev, L. Kish, and V. Shpilrain. *Yao's millionaires' problem with computationally unbounded parties and public-key encryption without computational assumptions*, preprint.

[12] D. Grigoriev and I. Ponomarenko. Constructions in public-key cryptography over matrix groups. *Contemp. Math., Amer. Math. Soc.*, 418:103–119, 2006.

[13] D. Grigoriev and V. Shpilrain. Secrecy without one-way functions. *Groups, Complexity, and Cryptology*, 5:31–52, 2013.

[14] D. Grigoriev and V. Shpilrain. Yao's millionaires' problem and decoy-based public key encryption by classical physics. *J. Foundations Comp. Sci.*, 25:409–417, 2014.

[15] R. Impagliazzo and M. Luby. *One-way functions are essential for complexity based cryptography*. In FOCS'89, pages 230–235, IEEE Computer Society, 1989.

[16] A. Menezes, P. van Oorschot, and S. Vanstone. *Handbook of Applied Cryptography*. CRC-Press, 1996.

[17] A. Shamir. How to share a secret. *Comm. ACM* 22:612–613, 1979.

[18] A. Shamir, R. Rivest, and L. Adleman. *Mental poker*, Technical Report LCS/TR-125, Massachusetts Institute of Technology, April 1979.

[19] A. C. Yao. *Protocols for secure computations* (Extended Abstract). 23rd annual symposium on foundations of computer science (Chicago, Ill., 1982), pages 160–164, IEEE, New York, 1982.

Received: 28 May 2016

IF Logic and Linguistic Theory

Jaakko Hintikka

Department of Philosophy, Boston University
745 Commonwealth Avenue, Boston MA 02215, USA
and
Collegium for Advanced Studies, University of Helsinki
Fabianinkatu 24, 00014 University of Helsinki, Finland

Abstract: Originally, modern symbolic logic was supposed to be a disambiguated and streamlined version of the logic of natural language. It has nevertheless failed to provide a full account of several telltale semantical phenomena of ordinary language, including Peirce's paradox, "donkey sentences" and more generally conditionals and different kinds of anaphora. It is shown here by reference to examples how these phenomena can be treated by means of IF logic and its semantical basis, game-theoretical semantics. Furthermore, methodological questions like compositionality and logical form will be discussed.

1. Frege-gate

The relations between symbolic logic and linguistic theorizing have been (and still are) complicated, close and confused. Symbolic logic was first thought, typically if not universally, as a minor regimentation and smoothlining of ordinary language. In another direction, mathematicians were formulating much of their reasoning in terms of ordinary prose, not in terms of manipulation of equations or other complexes of symbols. In fact mathematicians like Cauchy or Weierstrass were using – as they had to do – an explicit but unformalized logic of quantifiers in the guise of the so-called epsilon-delta technique, expressed in such ordinary language terms as "given such-and-such a number", "one can find" etc. (See here and in the following Hintikka [3,5]).

But then a huge scientific scandal, a veritable Frege-gate, took place without anyone's noticing. Frege undertook to formalize our entire logic, to present a notation (a *Schrift*) for all our concepts. Yet he failed to understand his fellow mathematicians' quantifier logic, and instead gave his followers a flawed logic that is only a part of the full story. Subsequent logicians unfortunately followed Frege and used this defective logic as their basic working logic. This alone would not been serious, for Frege's logic of quantifiers (by which I mean what is nowadays called first-order logic) is correct as far as its expressive powers go. The catastrophic mistake the logicians made was to think in effect that it is the full logic of quantifiers. The first specific disaster this caused was the bunch of paradoxes of set theory, which prompted the entire crisis of the foundations of set theory. This in turn led to further catastrophes, such as Zermelo-Fraenkel first-order set theory and the wishful belief that such results as Gödel's, Tarski's or Paul Cohen's tell us something about the limitations of logic and axiomatization or about the continuum hypothesis (Hintikka [4]).

This "Frege-gate" scandal came to light only recently when it was pointed out that the logic that mathematicians were using already hundred years ago was not the received first-order logic, but the richer logic that had been meanwhile rediscovered and systematized under the title "independence-friendly logic" (IF logic). (see e.g. Mann et al. [7]). However, the Frege-gate scandal has not hit headlines yet even in logic journals.

2. IF logic and linguistics

In this paper, I will discuss one aspect of the new problem situation, viz., its impact on linguistic theorizing. That there must be such an impact is obvious. To mention only one indication, at one time Chomsky thought that his syntactical counterparts to logical forms, the LF's, were essentially like formulas of (the received) first-order logic (see e.g. Chomsky, [1, p. 197]; [2, p. 67]). If they are not adequate representations of logical and, *a fortiori*, semantical forms of ordinary language sentences, we do not only need a better logic, but also a better syntactical theory.

Now IF logic, at least in its simplest version, has been around for a while and has even become an established research area in logic. Hence there has in fact been some discussion of its role in natural language. Much ingenuity has been expended on the first examples of purportedly IF sentences in ordinary language. They have been mostly so-called branching quantifier sentences like

IF Logic and Linguistic Theory

(2.1)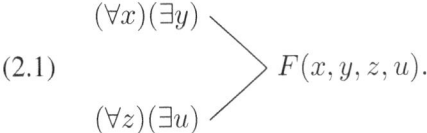

Its meaning can be expressed by the IF sentence

(2.2) $(\forall x)(\forall z)(\exists y /\forall z)(\exists u /\forall x) F(x, y, z, u)$.

This meaning cannot be expressed by a first-order quantifier sentence without the independence indication slash.

Examples from ordinary language were presented and discussed. An example was

(2.3) Every villager has a friend and every townsman has a relative who know each other.

Here choice of a friend is independent of the choice of a relative and *vice versa*.

Such examples are sufficiently complicated for confusing some philosophers. However, it has turned out that the examples are only the tip of an iceberg. Other examples look syntactically simple but still turn out to be semantically rather complex, e.g.

(2.4) Everybody has a different friend.

Its logical form can be seen to be

(2.5) $(\forall x_1)(\forall x_2)(\exists y_1 /\forall x_2)(\exists y_2 /\forall x_1)(((x_1 = x_2) \leftrightarrow (y_1 = y_2))\ \&\ F(y_1, x_1)\ \&\ F(y_2, x_2))$.

What was explained in these early linguistic applications of IF logic are particular examples, rather than general semantical or syntactical phenomena. In this paper, we concentrate on one particular relatively unexplored semantical phenomenon, viz., informational independence involving propositional connectives instead of (or in addition to) quantifiers.

3. Peirce's paradox

Ironically, the shortcomings of the usual ("Fregean") first-order logic were known already at the time of its formulation to Frege's co-inventor Charles S. Peirce (see Peirce [8, 4.546 and 4.580]). He pointed out a problem about the following pair of English sentences:

(3.1)   Someone is such that, if he fails in business, he commits suicide.

(3.2)   Someone is such that if everybody fails in business, he commits suicide.

Their respective logical forms seem to be

(3.3) $(\exists x)(F(x) \supset S(x))$,

(3.4) $(\exists x)((\forall y)F(y) \supset S(x))$.

Here (3.4) is equivalent to

(3.5) $(\exists x)((\exists y) \sim F(y) \vee S(x))$.

But something is paradoxical here. Formulas (3.1) and (3.2) obviously mean something different whereas, as Peirce pointed out, in the usual first-order logic (3.3) and (3.4) are logically equivalent.

Various *ad hoc* explications have been proposed, but they remain just that: adhockey. Yet game-theoretical semantics yields a diagnosis of the problem without any further assumptions or considerations. The problem is how the conditional (3.4) can be as strong as (3.3).

An answer is found by examining the meaning of (3.1) or (3.2) in game-theoretical terms. What (3.5) says is that it is true. That truth means in the existence of a winning strategy for the verifier ("myself") in the semantical game associated with (3.2). The first part of this strategy is a specification of the value c of x in $(\exists x)$. In order for it to be part of a winning strategy, there has to be a similar winning strategy in the game with

(3.6) $(\exists y) \sim F(y) \vee S(c)$.

The next step in a play of the game is the verifier's choice of one of the disjuncts. Whether or not this makes (3.6) true does depend on what the world is like.

If the world is such that everybody fails in business, the right choice of c is one of the people who commit suicide. But the world might be such that there are no such persons, so that the choice of $x = c$ must make the other disjunct true, in other words must satisfy $\sim F(x)$. This is guaranteed only if x satisfies $\sim F(x)$, in other words if it is a case that

(3.7) $\sim S(x) \supset \sim F(x)$.

In other words only

(3.8) $F(x) \supset S(x)$.

In that case, (3.2) can be true only if its antecedent is false, in other words only if not everybody fails in business. Hence the choice of x must provide a counter-example to everybody's failing in business. And the choice $x = c$ provides such an counter-example only if

(3.9) $\sim S(c) \supset \sim F(c)$.

The existence of such a counter-example means the truth of (3.3). Hence (the truth of) (3.5) implies the (the truth of) (3.3), which is Peirce's paradox.

In still other words, (3.3) is true only if there is an x such that if he fails in business, he commits suicide. Depending on what the world is like in (3.1) the verifier might have to choose $\sim F(c)$ or $S(c)$. In other words, c depends on the world. This means that the x in (3.5) or (3.4) is not the same individual independently of what the world is like. It is not really a choice of an "individual" as is required in (3.1) and (3.2).

4. Peirce's paradox and independence

This is clear interpretationally. But what does it mean in terms of the semantical games that convey our sentences their meaning? What is the right logic translation of (3.2)?

The analysis carried out above shows that the choice of a disjunct ("of a world") must be neutral with respect to the choice of objects. Hence the solution is to make $\vee$ independent of $(\exists x)$. Instead of (3.4) one should have

(4.1) $\quad (\exists x)((\forall y) F(y)(\vee/\exists x)\, S(x))$.

Thus the true representation of (3.2) is not (3.4) but (4.1). It cannot be formulated in IF logic in the usual narrow sense, but it can be formulated if this logic is amplified by allowing extra independencies between quantifiers and connectives. This opens up a new dimension of the entire hierarchy of different logics, besides further illustrating the inadequacy of Frege's logic.

5. Hierarchies of IF logics

In IF logic in the narrowest sense – which is the one in which it currently being used in the literature – the only extra kind of independence allowed is an independence of existential-force quantifiers of universal-force quantifiers within the formal scope of which they occur. (Quantifiers, whose scopes are not nested are automatically independent.) Only strong negations, $\sim$, are admitted. If we admit sentence-initial contradictory negations, $\neg$, we obtain richer and more satisfactory logic which is usually called extended independence-friendly logic (EIF) logic. It should perhaps be considered as the "real" basic IF logic. If we allow arbitrary extra independencies (existential quantifiers on existentials, universal quantifiers on universals, and universal quantifiers on existentials) we obtain a still much stronger logic that might be called generalized IF logic.

Here we are dealing with yet another way of enriching the basic or extended IF logic. This way is to allow extra independencies between quantifiers and propositional connectives. From the Peircean example and from others it is seen that this dimension of expressive enrichment is independent of quantifier independencies.

6. Simple donkey sentences

This new dimension also facilitates analysis of many interesting linguistic phenomena. One instructive example is constituted by the so-called donkey sentences. The interpretation of these sentences is a routine question discussed in the linguistic literature on definite and indefinite pronouns. The simplest example has the same form as the following sentence:

(6.1)   If Peter owns a donkey, he beats it.

This is *prima facie* of the following form

(6.2) $(\exists x)(D(x) \& O(p,x)) \supset B(p,x)$.

This would have to be equivalent with

(6.3) $(\forall x)(\sim D(x) \vee \sim O(p,x)) \vee B(p,x)$.

But (6.2) is ill-formed in that the last x is not bound to (is outside the scope of) $(\exists x)$. But the alternative

(6.4) $(\exists x)((D(x) \& O(p,x)) \supset B(p,x))$

says only that there is at least one animal such that if it is a donkey and is owned by Peter, he beats it. The true semantical form of (6.1) seems to be intuitively

(6.5) $(\forall x)((D(x) \& O(p,x)) \supset B(p,x))$.

But why? How come (6.1) should be translated as (6.5)? An indefinite article has the force of an existential quantifier. So why does it seem to have here the force of an universal one?

The answer can be obtained by analyzing the meaning of (6.1) the same way as the meaning of Peirce's paradoxical sentence (3.1) was analyzed earlier. The crucial point is that the choice of x must be independent of the choice between different relevant semantics codified in the second $\vee$ in (6.3). The solution is now to make the quantifier and the connective $\vee$ independent of each other. Here the covert logic translation of (6.1) will be

(6.6) $(\exists x)(D(x) \& O(p,x))(\supset /\exists x)B(p,x)$

which is equivalent with

(6.7) $(\forall x)(\sim D(x) \vee \sim O(p,x))(\vee /\forall x)B(p,x).$

When is there a winning strategy for the verifier in the game with (6.7), as (6.7) says? In that strategy, since $\vee$ is independent of $(\forall x)$, the falsifier chooses a value d for x. The resulting sentence

(6.8) $(\sim D(d) \vee \sim O(p,d)) \vee B(p,d)$

must be true, i.e. the verifier must be able to choose a true disjunct. Such a choice is possible for any d if it is the case that for any donkey d owned by Peter it is true that he beats it i.e. that $B(p,d)$ is true. But this is obviously just what (6.1) says.

7. Complex donkey sentences

This shows that the extensive literature designed to account for donkey sentences is, if not wrong, then at least redundant. Many purported explanations do not work for more complex donkey sentences like

(7.1) If you give each child a gift for Christmas, some child will open it to-day.

Here even a merely linguistic account of the role of the anaphoric phenomenon "it" is very tricky. No usual IF logic expression captures the meaning of (7.1) either. Yet its logic translation in terms of connective independence is possible.

The right translation is perhaps best seen if we first eliminate the existential quantifier in terms of its Skolem function and express (7.1) as

(7.2) $(\exists g)(\forall x)(G(x,g(x)) \supset (\exists z)O(z,g(z))).$

This is a second-order sigma one-one sentence. It is possible to translate such sentences to the corresponding IF first-order language, but not without independent connectives. Here is a translation:

(7.3) $(\forall x_1)(\forall x_2)(\exists y_1 /\forall x_2)(\exists y_2 /\forall x_1)((x_1 = x_2) \supset (y_1 = y_2))$ &

$G(x_1, y_1) \& G(x_2, y_2)(\supset /\forall x_1, \forall x_2)(\exists z)((z = x_1) \vee (z = x_2) \supset (O(z, y_1) \& O(z, y_2)))$.

This explains the meaning of (7.1).

8. Conditional reasoning

This is in explicit terms what the idea of "remembering" a strategy used in earlier subgame amounts to.

In general we have found an important distinction. It may be called a distinction between deductive reasoning and conditional reasoning. A deductive conclusion B from a premise A is a proposition that is true as soon as A is true. In the language of possible world semantics, B is true in each world in which A is true.

But the premise A does more than put forward a truth condition. It presents a situation, a fragment of one particular possible world, maybe a world in which Peter owns a donkey. We can then ask what else must be true in that particular world. This is a different question from asking what is true of all the worlds in which the premise A is true, for instance all worlds in which Peter owns some donkey or other. We are asking about the fate of that particular donkey postulated by the premise. Does Peter beat it?

What has been shown in this paper is how this question can be spelled out in sample cases by means of quantifiers independent of propositional connectives. These independencies are the gist of conditionality. It cannot be captured by ordinary "conditional" sentences of the form $(A \supset B)$ or by ordinary logical consequence relations. It is also the gist of the linguistic phenomenon of conditionality.

What is especially striking in all these examples is that the extra-connective-independence is not just one formally possible explanation of certain semantical phenomena, but the overwhelmingly natural one. This naturalness is easily converted into generality. When (in game-theoretical terms) a quantifier invites a player to choose an individual, the choice must not depend on what there may turn up later in the game. Thus the normal logic translation of disjunctive "or" appears to be, not $\vee$, but $(\vee /Q_1 x_1, Q_2 x_2, \ldots)$, where $(Q_i x_i)$ are the quantifiers within whose scope $\vee$ occurs in the translation.

9. Conditionality explained

These case studies illustrate *ipso facto* some of the explanatory possibilities in linguistic theorizing that are opened here. Consider, for example, the equivalence of (7.1) and (7.2). I have much earlier presented a semi-formal analysis of conditionals in a game-theoretical framework (Hintikka & Kulas [6]). It worked, but it was not purely logical. I had to resort to pre-formal ideas, e.g. the idea that a player in a semantical game could "remember" a strategy from another subgame. Such semiformal ideas can now be replaced by purely logical ones. For instance, look at (7.2). The Skolem function g there codifies (a part of a) strategy. This is used in a subgame with the antecedent of (7.1). From (7.2) one can see how it figures also as a strategy function (partial) in a game with the consequent.

In (7.2), this transfer of a strategy becomes the possibility of making use of the connection between x and y (subscripts do not matter) that was introduced in the antecedent also in the consequent. This is precisely what is made possible by the independence of $\vee$ of the quantifiers $(\forall x_1)$, $(\forall x_2)$.

This shows how by means of independences involving connectives we can capture the very conditionality of conditionals. This means that by means of such independences we can develop a viable general theory of conditionals.

10. Explaining anaphora

Even more generally much of any first-order logic can be thought as framework for a semantical representation of such phenomena as co-reference and anaphora. Not all such logics can be applied directly to the analysis of these phenomena in natural languages, mainly owing to the syntactical differences between them and natural language.

Certain general advantages of the kind of treatment of anaphora based on IF logic over some typical linguistic theories can presently be pointed out. Linguistic approaches to anaphora and co-reference often rely on the head-anaphora relation as one of their explanatory concepts. Of course linguists are aware that there are examples where there is no head to be found for a given anaphora or where the head and the anaphora cannot be said to be literally co-referential, that is, refer to one and the same entity. But such cases are typically considered somehow exceptional, not automatically explainable by the normal operation of anaphora.

We have already analyzed such an apparently anomalous case. In the complex donkey sentence (7.1), the obviously anaphoric pronoun "it" is not literally co-referential with any other phrase in the sentence. (It is not a "pronoun of laziness" either.) Yet (7.1) has an explicit logical form (7.2).

An explanation is implicit in what has been said earlier. We can interpret "it" because it is co-referential with an object that is functionally determined by other referring phrases in the same sentence or the same discourse. The functions that effect this determination are sometimes expressed in the sentence in question by a separate phrase. But they need not be. As we saw in our analysis of complex donkey sentences, existential quantifiers can introduce such dependencies through their Skolem functions. Sometimes the dependence is mediated by background information that the actual or hypothetical speaker if assumed to possess.

Hence a purely syntactical approach to the phenomena of anaphora and co-reference, such as Chomsky's government and binding theory, is bound to be incomplete account these phenomena.

11. Limits of compositionality

There is another general methodological moral in the story of this paper. The mode of operation of independent connectives illustrates a phenomenon that is as prevalent as it is important both in natural and formal languages. It is non-compositionality. (For a collection of articles on different aspects of compositionality, see Werning et al. [9].)[1]

Compositionality is rightly understood tantamount to semantical context-independence. Now we have seen in this paper how the logical force of a connective is different according to what quantifiers in its context it depends on. Of course a similar non-compositionality is obvious (though it was not to Frege) already in the dependence of quantifiers on other quantifiers. The main reason why this context dependence has not been emphasized more is that in

---

[1] I take this opportunity to correct a group of mistakes. On page 10 the authors say that Hodges has refuted "Hintikka's claim that Independence-Friendly logic is non-compositional". I have never made such a claim *simpliciter*, and on the contrary suggested a way in which any logic can in principle be given a compositional "semantics". What is the case (also according to Hodges) is that IF first-order logic cannot have a compositional semantics on the first order level.

the received first-order logic quantificational dependencies are expressed by the syntactical device of nesting scopes. But the only thing the necessity of so doing shows is the inadequacy of traditional first-order logic in semantic theorizing.

References

[1] N. Chomsky. *Essays on Form and Interpretation.* North-Holland, Amsterdam, 1977.

[2] N. Chomsky. *Knowledge of Language. Its Nature, Origin and Use.* Praeger, New York, 1986.

[3] J. Hintikka. What is the Significance of Incompleteness Results? Retrieved February 27, 2016 from http://people.bu.edu/hintikka/Papers_files/What_is_the_significance_of_the_incompleteness_%20results_JHintikka_0211_032211.pdf.

[4] J. Hintikka. IF Logic, Definitions and the Vicious Circle Principle. *Journal of Philosophical Logic*, Volume 41(Issue 2):505–517, April 2012.

[5] J. Hintikka. Which Mathematical Logic is the Logic of Mathematics? *Logica Universalis*, Volume 6(Special Issue 3 commemorating Jean van Heijenoort):459–475, December 2012.

[6] J. Hintikka and J. Kulas. *Anaphora and Definite Descriptions: Two Applications of Game-Theoretical Semantics.* Synthese language library; v. 17. D. Reidel, Dordrecht, Holland, 1985.

[7] A. Mann, G. Sandu, and M. Sevenster. *Independence-Friendly Logic: A Game-Theoretic Approach.* Cambridge University Press, Cambridge, 2011.

[8] C. S. Peirce. Exact Logic. The Simplest Mathematics. In Ch. Hartshorne and P. Weiss, editors, *Collected Papers of Charles Sanders Peirce*, volume 3 and 4. Harvard University Press, Cambridge MA, 1933.

[9] M. Werning, E. Machery, and G. Schurz, editors. *The compositionality of meaning and content. Volumes I & II.* Ontos verlag, Heusenstamm, 2005.

Received: 15 May 2012

Commentary on Jaakko Hintikka's "IF Logic and Linguistic Theory"

Gabriel Sandu

Department of Philosophy, History, Culture and Arts, University of Helsinki
Unioninkatu 40 A, 00014 Helsinki, Finland
sandu@mappi.helsinki.fi

In the paper Hintikka gives various arguments for the need of an extension of Independence-Friendly logic (IF-logic) with informationally independent disjunctions, i.e. connectives of the form

$$(\vee/\forall x)$$

that I will render more simply as $(\vee/x)$. Actually such an extension has been studied in Sandu and Väänänen [4], Hella and Sandu [2] and Mann, Sandu and Sevenster [3] but no application to natural language has been given. Thus I welcome Hintikka's endeavour. He introduces the case for informationally independent connectives by first offering a solution to what he calls Peirce's paradox which consists in the equivalence of

Hintikka compares

(3.1)   Someone is such that if he fails in business, he commits suicide.

with

(3.2)   Someone is such that if everybody fails in business, he commits suicide.

when they are represented in ordinary first-order logic as

$$\exists x(F(x) \to S(x))$$

and

$$\exists x(\forall y F(y) \to S(x))$$

respectively. (I will use '$\to$' instead of Hintikka's '$\supset$'). Hintikka analyzes the equivalence between these two sentences in game-theoretical semantics. This is a good idea, although I prefer a more straightforward game-theoretical argument than the one he offers. We establish the logical equivalence between

$$\exists x(\neg F(x) \lor S(x))$$

and

$$\exists x(\exists y \neg F(y) \lor S(x))$$

by showing that the Verifier has a winning strategy in one game if and only if she has a winning strategy in the other game (on any underlying model). As usual, these claims are established by a copy cat strategy argument. (Again a notational point: Hintikka makes a distinction between game-theoretical negation that he symbolizes by '$\sim$' and contradictory negation that he symbolizes by '$\neg$'. I will simply use the latter given that for ordinary first-order formulas the two are equivalent.)

Suppose there is a winning strategy for the Verifier in the first game. It consists of the choice of an individual, $x = a$ and the choice of a disjunct, left or right. Given that the strategy is winning, then, if left is chosen, a must satisfy $\neg F(x)$ and if right is chosen, then x must satisfy $S(x)$. Here is a winning strategy for Verifier in the second game. If in the first game Verifier chooses left, then in the second game she chooses $x = a$, then left, and then $y = a$. Given that a satisfies $\neg F(x)$ then this is a winning strategy. If in the first game Verifier chooses right, then in the second game Verifier chooses $x = a$ then right. Given that a satisfies $S(x)$, then this is a winning strategy in the second game.

For the converse, suppose the Verifier has a winning strategy in the second game. It is: choose $x = a$; then choose left or right. If left, choose $y = b$; if right, do nothing. Given that this is a winning strategy, then if right is chosen, x must satisfy $S(x)$. If left is chosen, then b must satisfy $\neg F(y)$. Here is a winning strategy for Verifier in the first game. If Verifier chooses right in the second game, then choose $x = a$ and then right in the first game. Then a satisfies $S(x)$ and thus this is a winning strategy. If Verifier chooses left and then $y = b$ in the second game, then in the first game she chooses $x = b$ and then left. Clearly given that b must satisfy $\neg F(y)$ this is a winning strategy.

Actually $\exists x(\exists y \neg F(y) \lor S(x))$ is logically equivalent with

$$(\exists y)\neg F(y) \lor \exists x S(x)$$

and thus "Peirce's paradox" is seen to be an instance of the more general law

$$\exists x(A(x) \vee B(x)) \equiv \exists x A(x) \vee \exists y B(y) \equiv \exists x A(x) \vee \exists x B(x).$$

Hintikka's suggestion in the paper is to block the paradox by blocking the above equivalence in this particular case, that is, by taking the logical form of (3.2) to be (there is a misprint in the text):

$$\exists x(\neg \forall y F(y) \ (\vee/x) \ S(x))$$

that is

$$\exists x(\exists y \neg F(y) \ (\vee/x) \ S(x))$$

where $(\vee/x)$ means that when Verifier chooses a disjunct, she does not know the value chosen earlier for x. Now apart from creating interpretational problems of its own, the proposal will not help him. Informally the proposal says that the choice of a disjunct should take place before the choice of a value of x takes place. But this renders the last sentence logically equivalent with

$$\exists y \neg F(y) \vee \exists x S(x)$$

which is, as pointed out above, logically equivalent with $\exists x(\neg F(x) \vee S(x))$. We are back to square one! I guess Hintikka has in mind another way to analyze the informational independence of Verifier of its own move than the one I proposed (games of imperfect information), that is, a proposal that does not render $\exists x(\exists y \neg F(y) \ (\vee/x) \ S(x))$ equivalent with $\exists x(\neg F(x) \vee S(x))$. I remember he once in conversation objected to the equivalence between $\exists x(\exists y/x)x = y$ with $\exists x \exists y x = y$ which holds in IF-logic. Fausto Barbero [1, forthcoming] has a notion of independence which does not render the two equivalent. It might be that Hintikka is relying in his proposal on a notion of independence on the basis of which $\exists x(A \ (\vee/x) \ B$ is not equivalent with $\exists x A \vee \exists x B$ but this is something for future work.

Based on his attempted solution to Peirce's paradox, Hintikka suggests also a new way to analyze simple donkey sentences like

(6.1)   If Peter owns a donkey, he beats it.

He takes the force of this sentence to be that of

(6.5) $\forall x(D(x) \wedge O(p, x)) \rightarrow B(p, x))$.

He asks: How do we get from (6.1) to (6.5)? One way to proceed is to take literally the surface structure of (6.1) where the indefinite is in the "scope" of the implication, and translate the indefinite "a donkey" by an existential quantifier, as standardly done. The result is, as Hintikka correctly points out:

(6.2) $\exists x(D(x) \wedge O(p, x)) \to B(p, x)$

which is equivalent, as he points out with

(6.3) $\forall x(\neg D(x) \vee \neg O(p, x)) \vee B(p, x)$.

But Hintikka is right to point out that (6.2) (and consequently (6.3)) is ill formed given that the last occurrence of the variable x is not bound. On the other side, if we try to bind the variable x by the existential quantifier, we get

(6.4) $\exists x(D(x) \wedge O(p, x)) \to B(p, x))$

which, as Hintikka correctly points out, says only that "there is at least one animal such that if it is a donkey and is owned by Peter, he beats it." So it seems we cannot obtained the true logical form of (6.1) which is (6.5).

Hintikka proposes an answer which is to go back to (6.2) and to take the implication to be independent of the existential quantifier

(6.6) $\exists x(D(x) \wedge O(p, x)) (\to /\exists x) B(p, x)$

or, if we operate instead on (6.3) which he takes to be equivalent to (6.2), he takes disjunction to be independent of the universal quantifier:

(6.7) $\forall x(\neg D(x) \vee \neg O(p, x)) (\vee /\forall x) B(p, x)$.

We are then told that the existence of a winning strategy for the Verifier in (6.7) means that for any choice d by the Falsifier, the sentence

(6.8) $(\neg D(d) \vee \neg O(p, d)) \vee B(p, d)$

must be true. And this yields (6.5).

Hintikka's analysis is ingenious but it does not get through, as it stands. I claim that the independence $(\to /\exists x)$ of implication from the existential quantifier in (6.6), or, equivalently the independence $(\vee /\forall x)$ of disjunction from the universal quantifier (6.7), does not make sense. The reason for this, focusing on the latter, is simply that in IF-logic as it currently stands, for a

Commentary on IF Logic and Linguistic Theory

move to be informationally independent from another, the first must be in the syntactical scope of the second. Or this is not the case in (6.7).

Finally Hitnikka motivates the use of informationally independent disjunctions by its role in the logical representation of complex donkey sentences like

(7.1) If you give each child a gift for Christmas, some child will open it to-day.

that Hintikka represents in second-order logic by

(7.2) $\exists g \forall x \, (G(x, g(x)) \to \exists z O(z, g(z)))$.

He then tells us that (7.2) can be represented on the first-order level by the IF sentence (7.3) which involves informationally independent disjunctions.

Hintikka's claim is not true. (7.3) is a second-order existential formula and as such known to be equivalent, by standard results of Walkoe [5], to an ordinary IF-formula which does not involve informationally independent disjunctions. Let me reproduce the procedure by which the IF-formula is obtained (I am grateful to Fausto Barbero here).

1. First in (7.2) we push the existential quantifier in front of the conditional and then Skolemize it:

$$\exists f \exists g \forall x (G(x, g(x)) \to O(f(x), g(f(x))))).$$

2. Next we eliminate the nesting of functions to obtain

$$\exists f \exists g \forall x \forall y (y = f(x) \to (G(x, g(x)) \to O(y, g(y))))).$$

3. Third we want each function to have a unique set of arguments (so we replace the second g with a new h):

$$\exists f \exists g \exists h \forall x \forall y (x = y \to g(x) = h(y) \land \\ \land [(y = f(x) \to (G(x, g(x)) \to O(y, h(y))))]).$$

4. Finally we replace each function by its appropriate pair of quantifiers and obtain the IF-formula which is the logical form of (7.1):

$$\forall x \forall y (\exists u/y)(\exists v/y, u)(\exists w/x, u, v)(x = y \to v = w \land \\ \land [(y = u \to (G(x, v) \to O(y, w)))]).$$

References

[1] F. Barbero. Cooperation in games and epistemic readings of Independence-Friendly sentences. Forthcoming in *Journal of Logic, Language, and Information*.

[2] L. Hella and G. Sandu. Partially Ordered Connectives and Finite Graphs. In M. Krynicki, M. Mostowski, and L. W. Szczerba, editors, *Quantifiers: Logics, Models and Computation. Volume Two: Contributions*, pages 79–88. Springer Netherlands, 1995.

[3] A. Mann, G. Sandu, and M. Sevenster. *Independence-Friendly Logic: A Game-theoretic Approach*. Cambridge University Press, Cambridge, 2011.

[4] G. Sandu and J. Väänänen. Partially ordered connectives. *Mathematical Logic Quarterly*, 38(1):361–372, 1992.

[5] J. Walkoe. Finite partially-ordered quantification. *J. Symbolic Logic*, 35(4):535–555, 1970.

Received: 30 April 2017

Foundations as Superstructure
(*Reflections of a practicing mathematician*)

Yuri I. Manin

Max–Planck–Institut für Mathematik,
Bonn, Germany

manin@mpim-bonn.mpg.de

Abstract: This talk presents foundations of mathematics as a historically variable set of principles appealing to various modes of human intuition and devoid of any prescriptive/prohibitive power. At each turn of history, foundations crystallize the accepted norms of interpersonal and intergenerational transfer and justification of mathematical knowledge.

Introduction

Foundations vs Metamathematics. In this talk, I will interpret the idea of Foundations in the wide sense. For me, Foundations at each turn of history embody currently recognized, but historically variable, principles of organization of mathematical knowledge and of the interpersonal/transgenerational transferral of this knowledge. When these principles are studied using the tools of mathematics itself, we get a new chapter of mathematics, *metamathematics*.

Modern philosophy of mathematics is often preoccupied with *informal interpretations of theorems, proved in metamathematics of the XX–th century*, of which the most influential was probably Gödel's incompleteness theorem that aroused considerable existential anxiety.

In metamathematics, Gödel's theorem is a discovery that a certain class of finitely generated structures (statements in a formal language) contains substructures that are not finitely generated (those statements that are true in a standard interpretation).

It is no big deal for an algebraist, but certainly interesting thanks to a new context.

© The Author(s) and College Publications 2017

Existential anxiety can be alleviated if one strips "Foundations" from their rigid *prescriptive/prohibitive*, or *normative* functions and considers various foundational matters simply from the viewpoint of their mathematical content and on the background of whatever historical period.

Then, say, the structures/categories controversy is seen in a much more realistic light: contemporary studies fuse (Bourbaki type) structures and categories freely, naturally and unavoidably.

For example, in the definition of *abelian categories* one starts with *structurizing sets of morphisms:* they become abelian groups. In the definition of 2–categories, the sets of morphisms are even *categorified:* they become objects of categories, whose morphisms become then the *morphisms of the second level* of initial category. Since in this way one often obtains vast mental images of complex combinatorial structure, one applies to them principles of *homotopy topology* (structural study of topological structures up to homotopy equivalence) in order to squeeze it down to size etc.

I want to add two more remarks to this personal credo.

First, the recognition of quite restrictive and historically changing normative function of Foundations makes this word somewhat too expressive for its content. In a figure of speech such as "Crisis of Foundations" it suggests a looming crash of the whole building (cf. similar concerns expressed by R. Hersh, [8]).

But, second, the first "Crisis of Foundations" occurred in a very interesting historical moment, when the images of formal mathematical reasoning and algorithmic computation became so precise and detailed that they could be, and were, described as *new mathematical structures*: formal languages and their interpretations, partial recursive functions. They could easily fit Bourbaki's universe, even if Bourbaki himself was too slow and awkward to really appreciate the new development.

At this juncture, contemporary *"foundations"* morphed into a *superstructure*, high level floor of mathematics building itself. This is the reason why I keep using the suggestive word "metamathematics" for it.

This event generated a stream of philosophical thought striving to recover the lost normative function. One of the reasons of my private mutiny against it (see e.g. [11]) was my incapability to find any of the philosophical arguments more convincing than even the simplest mathematical reasonings, whatever "forbidden" notions they might involve.

In particular, whatever doubts one might have about the scale of Cantorial cardinal and ordinal infinities, the basic idea of set embodied in Cantor's famous "definition", as a collection of definite, distinct objects of our thought, is as alive as ever. Thinking about a topological space, a category, a homotopy type, a language or a model, we start with imagining such a collection, or several ones, and continue by adding new types "of distinct objects of our thought", derivable from the previous ones or embodying fresh insights.

To summarize: good metamathematics is good mathematics rather than shackles on good mathematics.

Plan of the article. Whatever one's attitude to mathematical Platonism might be, it is indisputable that human minds constitute an important part of habitat of mathematics. In the first section, I will postulate three basic types of mathematical intuition and argue that one can recognize them at each scale of study: personal, interpersonal and historical ones.

The second section is concerned with historical development of the dichotomy *continuous/discrete* and evolving interrelations between its terms.

Finally, in the third section I briefly recall the discrete structures of linear languages studied in *classical metamathematics,* and then sketch the growing array of language–like non–discrete structures that gradually become the subject–matter of *contemporary metamathematics.*

1. Modes of mathematical intuition

1.1. Three modes. I will adopt here the viewpoint according to which *at the individual level* mathematical intuition, both primary and trained one, has three basic sources, that I will describe as *spatial, linguistic,* and *operational* ones.

The neurobiological correlates of the *spatial/linguistic* dichotomy were elaborated in the classical studies of *lateral asymmetry of brain.* When its mathematical content is objectivized, one often speaks about the opposition *continuous/discrete*.

The *linguistic/operational* dichotomy is observed in many experiments studying proto–mathematical abilities of animals. Animals, when they solve and communicate solutions of elementary problems related to counting, use not words but *actions*: cf. some expressive descriptions by Stanislas Dehaene in [6], Chapter 1: "Talented and gifted animals". Operational mode, when it is externalized and codified, becomes a powerful tool for social expansion of mathematics. Learning by rote of "multiplication table" became almost a symbol of democratic education.

The sweeping subdivision of mathematics into Geometry and Algebra, to which at the beginning of modern era was added Analysis (or Calculus) can be considered as a correlate on the scale of whole (Western) civilization of the trichotomy that we postulated above on the scale of an individual (cf. [2]).

It is less widely recognized that *even at the civilization scale*, at various historical periods, each of the spatial, linguistic and operational modes of mathematical intuition can dominate and govern the way that basic mathematical abstractions are perceived and treated.

I will consider as an example "natural" numbers. Most of us nowadays immediately associate them with their *names*: decimal notation $1, 2, 3, \ldots,$ $1984, \ldots,$ perhaps completed by less systemic signs such as 10^6 or XIX.

This was decidedly not always so as the following examples stretching over centuries and millennia show.

1.2. Euclid and his "Elements": spatial and operational *vs* linguistic.

For Euclid, a number was a "magnitude", a potential result of measurement. Measurement of a geometric figure A by a "unit", another geometric figure U, was conceived as a "physical activity in mental space": moving a segment of line inside another segment, step by step; paving a square by smaller squares etc. Inequality $A < B$ roughly speaking, meant that a figure A could be moved to fit inside B (eventually, after cutting A into several pieces and rearranging them in the interior of B).

In this sense, Euclidean geometry might be conceived as "physics of solid bodies in the dimensions one, two and three" (or more precisely, after Einstein, physics *in gravitational vacuum* of respective dimension). This pervasive identification of Euclidean space with our physical space probably influenced the history of Euclid's "fifth postulate". This history includes repeating attempts *to prove it*, that is, to deduce properties of space "at infinity" from observable ones at a finite distance, and then only reluctant acceptance of the Bólyai and Lobachevsky non–Euclidean spaces as "non–physical" ones.

As opposed to addition and subtraction, *the multiplication of numbers* naturally required passage into *a higher dimension*: multiplying two lengths, we get a surface. This was a great obstacle, but, I think, also opened for trained imagination the door to higher dimensions. At least, when Euclid has to speak

Foundations as Superstructure

about the product of an arbitrary large finite set of primes (as in his proof involving $p_1 \ldots p_n + 1$), he is careful to explain his general reasoning by the case of *three* factors, but without doubt, he had some mental images overcoming this restriction.

In fact, the strength of spatial and operational imagination required and achieved by modern mathematics can be glimpsed on a series of examples, starting, say with Morse theory and reaching Perelman's proof of Poincaré conjecture. Moreover, physicists could produce such wonders as Feynman's path integral and Witten's topological invariants, which mathematicians include in their more rigidly organized world only with considerable efforts.

At first sight, it might seem strange that the notion of *a prime number*, theorem about (potential) *infinity of primes*, and theorem about *unique decomposition* could have been stated and proved by Euclid in his geometric world, when no systematic notation for integers was accepted as yet, and no computational rules *dealing with such a notation* rather than numbers themselves were available.

But trying to rationalize this historical fact, one comes to a somewhat paradoxical realization that an efficient notation, such as Hindu–Arabic numerals, actually *does not help*, and even *hinders* the understanding of properties related to divisibility, primality etc. that is, all properties that refer to numbers themselves rather than their *names*.

In fact, the whole number theory could come into being only unencumbered by any efficient *notation* for numbers.

1.3. "Algorist and Abacist": linguistic *vs* operational. The dissemination of a positional number system in Europe after the appearance of Leonardo Fibonacci's *Liber Abaci* (1202) was, in essence, the beginning of the expansion of a universal, truly global language. Its final victory took quite some time.

The book by Gregorio Reisch, *Margarita Philosophica*, was published in Strasbourg in 1504. One engraving in this book shows a female figure symbolizing Arithmetics. She contemplates two men, sitting at two different tables, an *abacist* and an *algorist*.

The abacist is bent over his *abacus*. This primitive calculating device survived until the days of my youth: every cashier in any shop in Russia, having accepted a payment, would start calculating change clicking movable balls of her abacus.

The algorist is computing something, writing Hindu–Arabic numerals on his desk. The words "algorist" and modern "algorithm" are derived from the name of the great Al Khwarezmi (born in Khorezm c. 780).

In the context of this subsection, the abacus illustrates the operational mode whereas computations with numerals do the same for linguistic one (although in other contexts the operational side of such computations might dominate).

This engraving in the reception of contemporary readers was more politicised. It symbolized coming of a new epoch of democratic learning.

The Catholic Church supported the Roman tradition, usage of Roman numerals. They were fairly useless for practical commercial bookkeeping, calender computations such as dates of Easter and other moveable feasts etc. Here the abacus was of great help.

The competing tribe of algorists was able to compute things by writing strange signs on paper or sand, and their art was associated with dangerous, magical, secret Muslim knowledge. Al Khwarezmi teaching became their (and our) legacy.

Arithmetics blesses both practitioners.

1.4. John Napier and Alan Turing: operational. The nascent programming languages for centuries existed only as informal subdialects of a natural language. They had a very limited (but crucially important) sphere of applicability, and were addressed to human calculators, not electronic or mechanical ones. Even Alan Turing in the 20th century, when speaking of his universal formalization of computability, later called Turing machine, used the word "computer" to refer to *a person* who mechanically follows a finite list of instructions lying before him/her.

The ninety–page table of natural logarithms that John Napier published in his book *Mirifici Logarithmorum Canonis Descriptio* in 1614 was a paradoxical example of this type of activity that became a cultural and historical monument on a global scale. Napier, who computed the logarithms manually, digit by digit, combined in one person the role of creator of new mathematics and that of computer–clerk who followed his own instructions. His assistant Henry Briggs later performed this function.

Napier's tables were tables of (approximate values of) *natural logarithms*, with the base $e = 2,718281828....$ However, it seems that he neither referred to e explicitly, nor even recognized its existence. Roughly speaking, after having chosen the precision which he wanted to calculate logarithms, say with error $< 10^{-7}$, he dealt with integer powers of the number $1 + 10^{-8}$, whose 10^8 power was close to e.

This is one more example of the seemingly paradoxical fact, that an efficient and unified notation for objects of mathematical world can *hinder a theoretical understanding* of this world.

All the more amazing was the philosophical insight of Leibniz, who in his famous exhortation *Calculemus!* postulated that not only numerical manipulations, but any rigorous, logical sequence of thoughts that derives conclusions from initial axioms can be reduced to computation. It was the highest achievement of the great logicians of the 20th century (Hilbert, Church, Gödel, Tarski, Turing, Markov, Kolmogorov,...) to draw a precise map of the boundaries of the Leibnizian ideal world, in which

- *reasoning is equivalent to computation;*
- *truth can be formalized, but cannot always be verified formally;*
- *the "whole truth" even about the smallest infinite mathematical universe, natural numbers, exceeds potential of any finitely generated language to generate true theorems.*

The central concept of this program, *formal languages*, inherited the basic features of both natural languages (written form fixed by an alphabet) and the positional number systems of arithmetic. In particular, any classical formal language is one–dimensional (linear) and consists of discrete symbols that explicitly express the basic notions of logic.

Euclid found the remedy for the deficiencies of this linearity by strictly restricting role of natural language to the expression of *logic* of his proofs. The *content* of his mathematical imagination was transmitted by pictures.

2. Continuous or discrete? From Euclid to Cantor to homotopy theory

2.1. From continuous to discrete in "Elements". As we have seen, integers (and a restricted amount of other real numbers) for Euclid were results of (mental) measurement: *discrete came from continuous*. This was one–way road: continuous could not be produced from discrete. The idea that a line "consists" of points, so familiar to us today, does not seem to belong to Euclid's mental world and, in fact, to mental worlds of many subsequent generations of mathematicians until Georg Cantor. For Euclid, a point can be (a part of) the boundary of a (segment) of line, but such a segment cannot be scattered to a heap of points.

Geometric images are the source and embodiment not only of numbers, but of logical reasoning as well: in "Elements" at least a comparable part of its logic is encoded in figures rather than in words.

This was made very clear in the London publication of 1847, entitled

THE FIRST SIX BOOKS OF

THE ELEMENTS OF EUCLID

IN WHICH COLOURED DIAGRAMS AND SYMBOLS

ARE USED INSTEAD OF LETTERS FOR THE

GREATER EASE OF LEARNERS

whose author was Oliver Byrne, *"Surveyor of her Majesty's settlements in the Falklands Islands",* (see a recent republication [5]).

Byrne literally writes algebraic formulas whose main components are triangles, colored sectors of circle, segments of line etc. connected by more or less conventional algebraic signs.

2.2. From discrete to continuous: Cantor, Dedekind, Hausdorff, Bourbaki ...

This way is so familiar to my contemporaries that I do not have to spend much time to its description. The description of a mathematical structure, such as a group, or a topological space, according to Bourbaki starts with one or several unstructured sets, to which one adds elements of a these sets or derived sets satisfying restrictions formulated in terms of set theory.

Thus the twentieth century idea of "continuous" is based upon two parallel notions: that of *topological space* X (a set with the system of "open" subsets) and that of a "continuous map" $f: X \to Y$ between topological spaces. Further elaboration involving sheaves, topoi etc does not part with this basic intuition.

However, the set–theoretic point of departure helped enrich the geometric intuition by images that were totally out of reach earlier. The discovery of difference between *continuous* and *measurable* (from Lebesgue integral to Brownian motion to Feynman integral) was a radical departure from Euclidean universe.

Foundations as Superstructure

In a finite–dimensional context, one could now think about Cantor sets, Hausdorff dimension and fractals, curves filling a square, Banach–Tarski theorem. In infinite–dimensional contexts wide new horizons opened, starting with topologies of Hilbert and Banach linear spaces and widening in an immense universe of topology and measure theory of non–linear function spaces.

2.3. From continuous to discrete: homotopy theory. One of the most important development of topology was the discovery of main definitions and results of homotopy theory. Roughly speaking, a homotopy between two topological spaces X, Y is a continuous deformation producing Y from X, and similarly a homotopy between two continuous maps $f, g \colon X \to Y$ is a continuous deformation producing g from f. A *homotopy type* is the class of spaces that are homotopically equivalent pairwise. To see how drastically the homotopy can change a space, one can note that a ball, or a cube, of any dimension is *contractible*, that is, can be homotopically deformed to a point, so that dimension ceases to be invariant.

The basic discrete invariant of the homotopy type of X is the set of its connected components $\pi_0(X)$. To see, how this invariant gives rise to one of the basic structures of mathematics, ring of integers **Z**, consider a real plane P with a fixed orientation, a point x_0 on it, different from $(0,0)$, and the set of homotopy classes of loops (closed paths) in P, starting and ending at x_0 and avoiding $(0,0)$. This latter set can be canonically identified with **Z**: just count the number of times the loop goes around $(0,0)$. Each loop going in the direction of orientation counts as $+1$, where as the "counter–clockwise" loops count as -1.

On a very primitive level, this identification shows how the ideas of homotopy naturally introduce negative numbers. In the historically earlier periods when integers were measuring geometric figures (or counting real/mental objects) even idea of zero was very difficult and slowly gained ground in the symbolic framework of positional notation. Introduction of negative numbers required appellation to an extra–mathematical reality, such as *debt* in economics.

More generally, Voevodsky in his research project [14] introduces the following hierarchy of homotopy types graded by their h–levels. Zero level homotopy type consists of one point representing contractible spaces. If types of level n are already defined, types of level $n + 1$ consist of spaces such that the space of paths between any two points belongs to type of level n.

He further interprets types of level 1, represented by one point and empty sets, as *truth values*, and types of level 2 as sets. All sets in this universe are thus of the form $\pi_0(X)$.

Higher levels are connected with theory of categories, poly–categories etc, and we will return to them in the next section. At this point, we mention only that Voevodsky hierarchy does not *replace* sets but rather systematically embeds set–theoretical and categorical constructions and intuitions into a vaster universe where continuous and discrete are treated on an equal footing.

3. Language–like mathematical structures and metamathematics

3.1. Metamathematics: mathematical studies of formalized languages of mathematics.
Philosophy of mathematics in the XX–th century had to deal with lessons of *metamathematics,* especially of Gödel's incompleteness theorem.

As I have already said, I will consider metamathematics as *a special chapter of mathematics itself*, whose subject is the study of *formal languages and their interpretations.* On the foreground here were the first order formal languages, a formalization of Euclid's and Aristotle's legacy. Roughly speaking, to Euclid we owe the mathematics of spatial imagination (and/or kinematics of solid bodies), whereas Aristotle founded *the mathematics of logical deduction*, expressed in "Elements" by natural language *and* creative usage of drawings.

An important parallel development of formal languages involved languages formalizing *programs for and processes of computation,* of which chronologically first in the XX–th century was *Church's lambda calculus* [9].

An important feature of lambda calculus is the absence of formal distinctions between the language of programs and the language of input/output data (unlike Turing's machines, where a machine "is" the program, whereas input/output are represented by binary words). When, due to von Neumann's insights, this feature became implemented in hardware, lambda calculus was rediscovered and became in the 1960's the basis of development of programming languages.

These languages are *linear*, in the following sense: the set of all syntactically correct expressions in a formal language L could be described as a Bourbaki structure consisting of a certain *words,* – finite sequences of letters in a given *alphabet*, and finite sequences of such words, *expressions*. Words and ex-

Foundations as Superstructure

pressions must be *syntactically correct* (precise description of this is a part of definition of each concrete language). Letters of alphabet are subdivided into *types*: variables, connectives and quantifiers, symbols for operations, relations ... Syntactically correct expressions can be terms, formulas, ...

Such Bourbaki structures can be sufficiently rich to produce formal versions of real mathematical texts, existing and potential ones, and make them an object of study.

I will explain how the advent of category theory (and, to a lesser degree, theory of computability) required enriched languages, that after formalization become at first *non–linear,* and then *multidimensional.* Such languages require for their study *homotopy theory* and suggest a respective enrichment of the universe in which interpretations/models are supposed to live, from Sets to Homotopy Types, as in the Voevodsky's project (cf. above).

3.2. One–dimensional languages of diagrams and graphs. With the development of homological algebra and category theory in the second half of the XX–th century, the language of commutative diagrams began to penetrate ever wider realms of mathematics. It took some time for mathematicians to get used to "diagram-chasing." A simply looking algebraic identity $kg = hf$, when it expresses a property of four morphisms in a category, means that we are contemplating a simple *commutative diagram*, in which, besides morphisms f, g, k, h, also the objects A, B, C, D invisible in the formula $kg = hf$ play key roles:

$$\begin{array}{ccc} A & \xrightarrow{g} & B \\ f \downarrow & & \downarrow k \\ C & \xrightarrow{h} & D \end{array}$$

Although this square is not a "linear expression", one may argue that it, and its various generalizations of growing size (even the whole relevant category), are still "one–dimensional". This means that they can be encoded in a graph, whose vertices are labeled by (names of) objects of our category, whereas edges are labeled by pairs consisting of an orientation and a morphism between the relevant objects.

Similarly, a program written in a linear programming language can be encoded in a graph whose vertices are labeled by (names of) elementary operations that can be performed over the relevant data. To understand labeling of (oriented) edges, one must imagine that they encode channels, forwarding output data calculated by the operation at the start (input) of the edge to the its endpoint where they become input of the next operation (or the final output, if the relevant vertex is labeled respectively). Labels of edges might then include *types* of the relevant data.

3.3. From graphs to higher dimensions. Generally, a square of morphisms as above need not be commutative (i. e. it is possible that $kg \neq hf$). In order to distinguish these two cases graphically, we may decide to associate with *a commutative square the two–dimensional picture*, by glueing the interior part of the square to the relevant graph.

A well known generalization of this class of spaces are *cell complexes*, or, in more combinatorial and therefore more language–like version, *simplicial complexes*. Of course, we must allow labels of cells as additional structures.

In this way, we can get, for example, a geometric encoding of the category $\mathcal{C}$ by a simplicial complex, in which labeled $(n+1)$-complexes are sequences of morphisms

$$X_0 \xrightarrow{f_0} X_1 \xrightarrow{f_1} \ldots \xrightarrow{f_{n-1}} X_n$$

whereas the face map ∂^i omits one of the objects X_i and, if $1 \leq i \leq n-1$, replaces the pair of arrows around X_i by one arrow labeled by the composition of the relevant morphisms. The resulting simplicial space encodes the whole category in a simplicial complex that is called *the nerve* of the category. Clearly, not only objects and morphisms, but also all compositions of morphisms and relations between them can be read off it.

Thus the language of commutative diagrams becomes a chapter of algebraic topology, and when the study of functors is required, the chapter of homotopical topology.

3.4. Quillen's homotopical algebra and univalent foundations project. In his influential book [13], Quillen developed the idea that the natural language for homotopy theory should appeal *not* to the initial intuition of continuous deformation itself, but rather to a codified list of properties of category of topological spaces stressing those that are relevant for studying homotopy.

Quillen defined *a closed model category* as a category endowed with three special classes of morphisms: *fibrations, cofibrations*, and *weak equivalences*.

The list of axioms which these three classes of morphisms must satisfy is not long but structurally quite sophisticated. They can be easily defined in the category of topological spaces using homotopy intuition but remarkably admit translation into many other situations. An interesting new preprint [7] even suggests the definition of these classes in appropriate categories of discrete sets, contributing new insights to old Cantorian problems of the scale of infinities.

Closed model categories become in particular a language of preference for many contexts in which objects of study are quotients of "large" objects by "large" equivalence relations, such as homotopy.

It is thus only natural that the most recent Foundation/Superstructure, Voevodsky's Univalent Foundations Project (cf. [14] and [3]) is based on direct axiomatization of the world of homotopy types.

As a final touch of modernism, the metalanguage of this project is a version of typed lambda calculus, because its goal is to develop a tool for the computer assisted verification of programs and proofs. Thus computers become more and more involved in the interpersonal habitat of "theoretical" mathematics.

It remains to hope that humans will not be finally excluded from this habitat, as some aggressive proponents of databases replacing science suggest (cf. [1]).

Post Scriptum: Truth and Proof in Mathematics

As I have written in [12], the notion of "truth" in most philosophical contexts is a reification of a certain relationship between humans and *texts/utterances/statements*, the relationship that is called "belief", "conviction" or "faith".

Professor Blackburn in [4] in his keynote speech to the Balzan Symposium on "Truth" (where [12] was delivered) extensively discussed other relationships of humans to texts, such as *scepticism, conservatism, relativism, deflationism*. However, in the long range all of them are secondary in the practice of a researcher in mathematics.

I will only sketch here what must be said about texts, sources of conviction, and methods of conviction peculiar to mathematics.

Texts. Alfred North Whitehead said that all of Western philosophy was but a footnote to Plato.

The underlying metaphor of such a statement is: "Philosophy is a text", the sum total of all philosophic utterances.

Mathematics decidedly is *not* a text, at least not in the same sense as philosophy. There are no authoritative books or articles to which subsequent generations turn again and again for wisdom. Already in the XX–th century, researchers did not read Euclid, Newton, Leibniz or Hilbert in order to study geometry, calculus or mathematical logic. The life span of any contemporary mathematical paper or book can be years, in the best (and exceptional) case decades. Mathematical wisdom, if not forgotten, lives as an invariant of all its (re)presentations in a permanently self–renewing discourse.

Sources and methods of conviction. Mathematical truth is not revealed, and its acceptance is not imposed by any authority.

Ideally, the truth of a mathematical statement is ensured by *a proof*, and the ideal picture of a proof is a sequence of elementary arguments whose rules of formation are explicitly laid down before the proof even begins, and ideally are common for all proofs that have been devised and can be devised in future. The admissible starting points of proofs, "axioms", and terms in which they are formulated, should also be discussed and made explicit.

This ideal picture is so rigid that it became the subject of mathematical study in metamathematics.

But in the creative mathematics, the role of proof is in no way restricted to its function of carrier of conviction. Otherwise, there would be no need for Carl Friedrich Gauss to consider eight (!) different proofs the quadratic reciprocity law (cf. [10] for an extended bibliography; I am grateful to Prof. Yuri Tschinkel for this reference).

One metaphor of proof is a route, which might be a desert track boring and unimpressive until one finally reaches the oasis of one's destination, or a foot path in green hills, exciting and energizing, opening great vistas of unexplored lands and seductive offshoots, leading far away even after the initial destination point has been reached.

References

[1] Ch. Anderson. The End of Theory: The Data Deluge Makes the Scientific Method Obsolete. Retrieved Mar. 23, 2016, from http://www.wired.com/2008/06/pb-theory/, Jun. 23, 2008.

[2] M. Atiyah. Geometry and Physics of the 20th Century. In J. Kouneiher, D. Flament, Ph. Nabonnand, and J.-J. Szczeciniarz, editors, *Géométrie au XXe siècle. Histoire et horizons*, pages 4–9. Hermann, Paris, 2005.

[3] S. Awodey. Type theory and homotopy. Retrieved Mar. 23, 2016, from http://arxiv.org/abs/1010.1810, 2010.

[4] S. Blackburn. Truth and Ourselves: the Elusive Last Word. In N. Mout and W. Stauffacher, editors, *Truth in Science, the Humanities, and Religion. Balzan Symposium 2008*, pages 5–14. Springer, 2010.

[5] O. Byrne. *The first six books of the Elements of Euclid*. Taschen, 2013. Facsimile of the first edition published by Pickering in 1847.

[6] S. Dehaene. *The Number Sense: How the Mind Creates Mathematics*. Oxford University Press, 1st edition, 1999.

[7] M. Gavrilovich and A. Hasson. Exercices de style: a homotopy theory for set theory, I. Retrieved Mar. 23, 2016, from http://arxiv.org/abs/1102.5562, 2012.

[8] R. Hersh. Wings, not Foundations! In G. Sica, editor, *Essays on the Foundations of Mathematics and Logic*, volume 1, pages 155–164. Polimetrica Int. Sci. Publisher, Monza, Italy, 2005.

[9] A. Jung. A short introduction to the Lambda Calculus. Retrieved Mar. 23, 2016, from http://www.cs.bham.ac.uk/~axj/pub/papers/lambda-calculus.pdf, Mar. 18, 2004.

[10] F. Lemmermeyer. Proofs of the quadratic reciprocity law. Retrieved Mar. 23, 2016, from http://www.rzuser.uni-heidelberg.de/~hb3/rchrono.html.

[11] Yu. Manin. *A Course in Mathematical Logic for Mathematicians*. Graduate Texts in Mathematics 53. Springer, 2nd edition, 2010. With new Chapters written by Yu. Manin and B. Zilber.

[12] Yu. Manin. Truth as a Value and Duty: Lessons of Mathematics. In N. Mout and W. Stauffacher, editors, *Truth in Science, the Humanities, and Religion. Balzan Symposium 2008*, pages 37–45. Springer, 2010. Preprint arXiv:0805.4057v1 [math.GM].

[13] D. G. Quillen. *Homotopical Algebra*, volume 43 of *Lecture Notes in Mathematics*. Springer-Verlag, Berlin, 1st edition, 1967.

[14] V. Voevodsky. Univalent Foundations Project. Retrieved Mar. 23, 2016, from http://www.math.ias.edu/~vladimir/Site3/Univalent_Foundations_files/univalent_foundations_project.pdf, Oct. 1, 2010.

Received: 27 March 2016

Classical and Intuitionistic Geometric Logic

Grigori Mints*

Department of Philosophy, Stanford University
Stanford, Ca, 94305, USA

Abstract: Geometric sequents "A implies C" where all axioms A and conclusion C are universal closures of implications of positive formulas play distinguished role in several areas including category theory and (recently) logical analysis of Kant's theory of cognition. They are known to form a Glivenko class: existence of a classical proof implies existence of an intuitionistic proof. Existing effective proofs of this fact involve superexponential blow-up, but it is not known whether such increase in size is necessary. We show that any classical proof of such a sequent can be polynomially transformed into an intuitionistic geometric proof of (classically equivalent but intuitionistically) weaker geometric sequent.

Keywords: geometric formulas, Glivenko classes, intuitionistic logic.

Introduction

Geometric sequents (see definition below) play distinguished role in several areas including category theory [3]. This fragment of first order logic attracted new attention in the light of recent work by Theodora Achourioti and Michiel van Lambalgen [1] who propose a translation of the philosophical language of Kant's theory of judgements into the language of elementary logic and provide a convincing justification of their view.

Geometric sequents are known to form a Glivenko class: existence of a classical proof of a geometric sequent S implies existence of an intuitionistic proof. Existing proofs of this fact involve superexponential blow-up, but we do not know whether such increase in size is necessary. We show that any classical proof of S can be polynomially transformed into an intuitionistic geometric proof of (classically equivalent but intuitionistically) slightly weaker geometric sequent.

We consider formulas of first order logic.

---

*Research was supported by the Russian Foundation *Dynasty* grant.

Definition 1. *Positive formulas* are constructed from atomic formulas and the constant $\bot$ by $\&, \vee, \exists$.

Geometric implications are positive formulas, implications of positive formulas and results of prefixing universal quantifiers to such implications.

Geometric sequents are expressions of the form

$$I_1, \ldots I_n \Rightarrow I$$

where $I_1, \ldots I_n, I$ are geometric implications.

A *geometric derivation* is a derivation consisting of geometric sequents.

The second proof of Theorem 1 given below is non-effective, but the first one allows one to derive some complexity bound. The proof begins with construction of a cut-free derivation, therefore the only obvious bound is the same as for cut-elimination, that is hyperexponential one. This contrasts with the most prominent Glivenko class, namely that of negative formulas. When a classical derivation of a negative formula is given, its intuitionistic derivation is constructed by "negativizing" all formulas in the derivation plus local changes to reinstate the inferences that were destroyed by this transformation. These transformations are polynomial.

We show here a weaker result for geometric sequents. Any classical proof (with cut) of a geometric sequent $\Gamma \Rightarrow I$ can be polynomially transformed into an intuitionistic geometric proof of a geometric sequent $D, \Gamma \Rightarrow I$ where D is obtained by introducing abbreviations for some formulas. In fact $D, \Gamma \Rightarrow I$ is intuitionistically derivable iff $\Gamma \Rightarrow I$ is intuitionistically derivable, but on the surface the definitions in D are only classical.

In section 1 we give two proofs of the Glivenko property of geometric sequents.

Section 2 describes depth-reducing transformations we need for our proofs. As far as I know, this use of formulas (17-19) especially to achieve that the whole proof is new. It is inspired by similar use of (18) by V. Orevkov [5] in a different situation.

Section 3 contains the proof of the main result.

We use $\equiv$ for literal coincidence of syntactic objects and $\leftrightarrow$ for a logical equivalence connective.

LK, LJ are Gentzen's systems for classical and intuitionistic logic, both with cut.

$\vdash^c, \vdash^i$ denote derivability in classical or intuitionistic logic, that is in LK, LJ with cut.

A *formula translation* of a sequent $S \equiv A_1, \ldots A_n \Rightarrow B_1, \ldots B_m$ is a formula $S^f \equiv (A_1 \& \ldots \& A_n \to B_1 \vee \ldots \vee B_m)$. Many notions defined for formulas are generalized to sequents via the formula translation. For example $S \leftrightarrow T$ for sequents S, T means $S^f \leftrightarrow T^f$.

c-models are ordinary models for the classical predicate logic, i-models are Kripke models.

1. Geometric sequents have Glivenko property

The next theorem is well-known. The deductive proof given here is due to V. Orevkov [5] and can be traced back to the work of H. Curry [2].

Theorem 1. *A geometric sequent is derivable classically iff it is derivable intuitionistically.*

1. *A deductive proof.* Consider a cut-free proof of a geometric sequent

$$\Gamma \to I$$

in LK. Since the succedent rules for $\to, \forall$ are invertible in LK, we can analyze away initial universal quantifiers and implication in I, then assume that I is a positive formula. After that *the sequent $\Gamma \Rightarrow I$ contains only connective occurrences that give rise to rules*

$$\Rightarrow \&, \Rightarrow \vee, \Rightarrow \exists, \& \Rightarrow, \vee \Rightarrow, \exists \Rightarrow, \to \Rightarrow .$$

These rules are common for LK and LJm, hence our LK-derivation is already LJm-derivation, as required. ⊢

2. *A model-theoretic proof.* The idea here is rather similar, but I have not seen this proof in literature. Suppose a geometric sequent $\Gamma \Rightarrow I$ with positive formula I is underivable in LJm. Consider its proof search tree in LJm (see for example Mints [4]). This tree is not a derivation, and hence has a non-closed branch generating a Kripke countermodel for $\Gamma \Rightarrow I$. The rules for analysis of the connectives $\forall, \to$ in succedent are not applied in this tree. But these are exactly the rules that add new worlds to a model. Therefore the resulting model has just one world, and hence it is a classical model refuting our sequent. ⊢

2. Reducing formula depth

Familiar depth-reducing transformations by introduction of new predicate variables are modified here to preserve geometric sequents. There are subtle points noted below. Let's first recall well-known facts.

Let's define a relation between formulas (widely used in literature without a special name) which is weaker than provable equivalence but in some respects similar to it.

Write $F \trianglerighteq^s G$ where $s \in \{c, i\}$ if

$$G \equiv F' \to F \text{ and } \vdash^s F'[P_1/F_1, \ldots P_n/F_n]$$

where P_i/F_i are substitutions (performed in this order) for predicate variables $P_1, \ldots P_n$ not occurring in F.

Lemma 1. *Assume $F \trianglerighteq^s G$. Then*

1. $\vdash^s F$ iff $\vdash^s G$,

2. *s-models for G are expansions (with respect to $P_1, \ldots P_n$) of s-models for F.*

Proof. 1. $\vdash^s F \to G$ is obvious. If $\vdash^s G$ then since $G \equiv (F' \to F)$ the substitutions $P_1/F_1, \ldots P_n/F_n$ and modus ponens yield $\vdash^s F$.

2. Similarly to 1.

$\vdash$

Notation **x** below stands for $x_1, \ldots x_n$ with distinct variables $x_1, \ldots x_n$.

Lemma 2. *If **x** contains all free variables of formulas $A(\mathbf{x}), B(\mathbf{x})$ then*

$$LJ \vdash \forall \mathbf{x}(A(\mathbf{x}) \leftrightarrow B(\mathbf{x})) \to (F(A) \leftrightarrow F(B)).$$

Proof. Induction on F. $\vdash$

Lemma 3. *If P is a fresh n-ary predicate symbol, **x** contains all free variables of the formula $A(\mathbf{x})$ then for $L \in \{LJ, LK\}$*

$$L \vdash \Rightarrow F(A) \text{ iff } \quad L \vdash \forall \mathbf{x}(A(\mathbf{x}) \leftrightarrow P(\mathbf{x})) \Rightarrow F(P)$$

Proof. If $L \vdash F(A)$, apply the previous Lemma.

If $L \vdash \forall \mathbf{x}(A(\mathbf{x}) \leftrightarrow P(\mathbf{x})) \Rightarrow F(P)$, substitute A for P. The antecedent of the sequent becomes $\forall \mathbf{x}(A(\mathbf{x}) \leftrightarrow A(\mathbf{x}))$. ⊢

For a given formula F assume that for every non-atomic subformula G of F a fresh predicate symbol P_G is chosen with the same arity as the number of free variables of G. In particular P_F has free variables of F as arguments. Atomic subformula $P(t_1, \ldots t_n)$ is not changed.

Symbols P_G can be treated as pointers to subformulas of F. This informal observation can be formalized by assigning equivalences E_G to subformulas G in the following way:

If $G(\mathbf{x}) \equiv H(\mathbf{y}) \odot K(\mathbf{z})$ for $\odot \in \{\&, \vee, \rightarrow\}$ then

$$E_G \equiv \forall \mathbf{x}(P_G(\mathbf{x}) \leftrightarrow (P_H(\mathbf{y}) \odot P_K(\mathbf{z}))) \tag{1}$$

where $\mathbf{y}, \mathbf{z} \subseteq \mathbf{x}$.

If $G(\mathbf{x}) \equiv QyH(\mathbf{x}, y)$ for $Q \in \{\forall, \exists\}$ then

$$E_G \equiv \forall \mathbf{x}(P_G(\mathbf{x}) \leftrightarrow QyP_H(\mathbf{x}, y)). \tag{2}$$

Lemma 4. *Let $G, H \ldots F$ be all non-atomic subformulas of F. Then for $L \in \{LJ, LK\}$*

$$L \vdash F \leftrightarrow L \vdash E_G, E_H, \ldots E_F \Rightarrow P_F.$$

Proof. Apply previous Lemma successively to subformulas, beginning with the innermost ones. ⊢

Let's rewrite equivalences (1),(2) as pairs or triples of implications, transforming these implications in LJ-equivalent way.

$$\forall \mathbf{x}(P_{G\&H}(\mathbf{x}) \to P_G(\mathbf{y})), \quad (3)$$
$$\forall \mathbf{x}(P_{G\&H}(\mathbf{x}) \to P_H(\mathbf{z})), \quad (4)$$
$$\forall \mathbf{x}(P_G(\mathbf{y})\&P_H(\mathbf{z}) \to P_{G\&H}(\mathbf{x})); \quad (5)$$
$$\forall \mathbf{x}(P_G(\mathbf{y}) \to P_{G\vee H}(\mathbf{x})), \quad (6)$$
$$\forall \mathbf{x}(P_H(\mathbf{z}) \to P_{G\vee H}(\mathbf{x})), \quad (7)$$
$$\forall \mathbf{x}(P_{G\vee H}(\mathbf{x}) \to (P_G(\mathbf{y}) \vee P_H(\mathbf{z})); \quad (8)$$
$$\forall \mathbf{x}(P_{\exists y P_G}(\mathbf{x}) \to \exists y P_G(\mathbf{x},y)) \quad (9)$$
$$\forall \mathbf{x}\forall y(P_G(\mathbf{x},y) \to P_{\exists y P_G}(\mathbf{x})) \quad (10)$$
$$\forall \mathbf{x}\forall y(P_{\forall y P_G}(\mathbf{x}) \to P_G(\mathbf{x},y)); \quad (11)$$
$$* \ \forall \mathbf{x}(\forall y P_G(\mathbf{x},y) \to P_{\forall y P_G}(\mathbf{x})) \quad (12)$$
$$* \ \forall \mathbf{x}(\neg P_G(\mathbf{x}) \to P_{\neg G}(\mathbf{x})) \quad (13)$$
$$\forall \mathbf{x}(P_G(\mathbf{x})\&P_{\neg G}(\mathbf{x}) \to \bot) \quad (14)$$
$$\forall \mathbf{x}(P_{G\to H}(\mathbf{x})\&P_G(\mathbf{y}) \to P_H(\mathbf{z})) \quad (15)$$
$$* \ \forall \mathbf{x}((P_G(\mathbf{y}) \to P_H(\mathbf{z})) \to P_{G\to H}(\mathbf{x})) \quad (16)$$

All these universally quantified implications are geometric except the three marked by a *. Let's replace them by classically equivalent geometric implications.

$$\forall \mathbf{x}\exists y(P_G(\mathbf{x},y) \to P_{\forall y P_G}(\mathbf{x})) \quad (17)$$
$$\forall \mathbf{y}(P_G(\mathbf{y}) \vee P_{\neg G}(\mathbf{y})) \quad (18)$$
$$\forall \mathbf{x}((P_H(\mathbf{z}) \to P_{G\to H}(\mathbf{x})) \ \& \ (P_G(\mathbf{y}) \vee P_{G\to H}(\mathbf{x}))) \quad (19)$$

Denote the resulting set of geometric implications (3-11), (14,15) and (17,18,19) for subformulas of a set **F** of formulas by DEF$_\mathbf{F}$.

3. Transformation of classical derivations

In this section we mean by intuitionistic predicate calculus a multiple-succedent formulation LJm (cf. Mints [4]) which differs from LK only in the requirement that the list Δ is empty in the succedent rules for $\to, \neg, \forall$:

$$\frac{A, \Gamma \Rightarrow \Delta}{\Gamma \Rightarrow \Delta, \neg A} \qquad \frac{A, \Gamma \Rightarrow \Delta, B}{\Gamma \Rightarrow \Delta, A \to B} \qquad \frac{\Gamma \Rightarrow \Delta, A(b)}{\Gamma \Rightarrow \Delta, \forall x A(x)}$$

Definition 2. Formulas $\neg A, A \to B, \forall x A$ introduced by these rules in an LK-derivation are called below *special* formulas when Δ is non-empty.

Let d be a derivation of a geometric sequent S in LK. Then $\mathbf{f}(d)$ denotes the set of all cut formulas in d and DEF_d denotes $\text{DEF}_{\mathbf{f}(d)}$.

Theorem 2.

1. *Let d be a derivation of a geometric sequent $\Pi \Rightarrow \Phi$ in LK. Then it can be* polynomially *transformed into a geometric derivation in LJm of the sequent*
$$\text{DEF}_d, \Pi \to \Phi$$
consisting of geometric sequents.

2. $\text{DEF}_d, \Pi \Rightarrow \Phi \trianglerighteq^c \Pi \Rightarrow \Phi$.

3. $\vdash^c \text{DEF}_d, \Pi \Rightarrow \Phi$ *iff* $\quad \vdash^i \text{DEF}_d, \Pi \Rightarrow \Phi$ *iff* $\quad \vdash^i \Pi \Rightarrow \Phi$

Proof. We assume that all axioms $A, \Gamma \to \Delta, A$ have atomic A. Using if needed inversion transformations we assume that Φ consists of positive formulas. Then every special formula F is traceable to a cut formula. More precisely, $F \equiv F'(\mathbf{t})$ where $F'(\mathbf{x})$ is a subformula of some cut formula. Formula F' has a "representative" $P_{F'}(\mathbf{x})$ in DEF_d where $\mathbf{x}$ are free variables of F'. In this sense any occurrence of a formula F traceable to a cut formula has a representative which we write as $P_F(\mathbf{t})$.

Denote by d^+ the result of replacing every such occurrence of $F(\mathbf{t})$ as a separate formula in a sequent in d by $P_F(\mathbf{t})$.

This replacement destroys inferences having such $F(t)$ as principal formulas. Consider these inferences in turn to show they can be repaired using DEF_d.

Axioms are assumed to be atomic, therefore they are preserved. The cut inferences become cuts on atomic formulas.

Antecedent inferences are repaired using geometric implications in Def_d. For example $\to$-antecedent inference

$$\frac{\Gamma \Rightarrow \Delta, G(\mathbf{t}_1) \quad H(\mathbf{t}_2), \Gamma \Rightarrow \Delta}{G(\mathbf{t}_1) \to H(\mathbf{t}_2), \Gamma \Rightarrow \Delta}$$

goes into the figure

$$\frac{\Gamma \Rightarrow \Delta, P_G(\mathbf{t}_1) \quad P_H(\mathbf{t}_2), \Gamma \Rightarrow \Delta}{P_{G \to H}(\mathbf{t}), \Gamma \Rightarrow \Delta}$$

which is transformed using the formula $P_{G \to H}(\mathbf{t}) \& P_G(\mathbf{t}_1) \to P_H(\mathbf{t}_2)$ denoted below by I which is an instance of a formula (15) in DEF_d.

$$\dfrac{\dfrac{\dfrac{P_{G \to H}(\mathbf{t}) \Rightarrow P_{G \to H}(\mathbf{t}) \quad \Gamma \Rightarrow \Delta, P_G(\mathbf{t}_1)}{P_{G \to H}(\mathbf{t}), \Gamma \Rightarrow \Delta, P_{G \to H}(\mathbf{t}) \& P_G(\mathbf{t}_1)} \text{axiom} \quad P_H(\mathbf{t}_2), \Gamma \Rightarrow \Delta}{I, P_{G \to H}(\mathbf{t}), \Gamma \Rightarrow \Delta}}{\mathrm{DEF}_d, P_{G \to H}(\mathbf{t}), \Gamma \Rightarrow \Delta} \forall \Rightarrow$$

Other antecedent rules and succedent rules common to LK and LJm are treated similarly. Of the remaining rules consider $\neg, \to$ and $\forall$ in succedent. Given derivations are transformed as follows. The derivation

$$\dfrac{G, \Gamma \Rightarrow \Delta}{\Gamma \Rightarrow \Delta, \neg G}$$

goes to

$$\dfrac{\dfrac{P_G(\mathbf{t}), \Gamma \Rightarrow \Delta \quad P_{\neg G}(\mathbf{t}) \Rightarrow P_{\neg G}(\mathbf{t})}{P_G(\mathbf{t}) \vee P_{\neg G}(\mathbf{t}), \Gamma \Rightarrow \Delta, P_{\neg G}(\mathbf{t})}}{\mathrm{DEF}_d, \Gamma \Rightarrow \Delta, P_{\neg G}(\mathbf{t})} \vee \Rightarrow$$

The derivation

$$\dfrac{G, \Gamma \Rightarrow \Delta, H}{\Gamma \Rightarrow \Delta, G \to H}$$

goes to

$$\dfrac{\dfrac{P_G(\mathbf{t}_1), \Gamma \Rightarrow \Delta, P_H(\mathbf{t}_2) \quad \text{axioms}}{P_H(\mathbf{t}_2) \to P_{G \to H}(\mathbf{t}), P_G(\mathbf{t}_1) \vee P_{G \to H}(\mathbf{t}), \Gamma \Rightarrow \Delta, P_{G \to H}(\mathbf{t})}}{\mathrm{DEF}_d, \Gamma \Rightarrow \Delta, P_{G \to H}(\mathbf{t})} \vee \Rightarrow, \to \Rightarrow$$

The derivation

$$\dfrac{\Gamma \to \Delta, G(b)}{\Gamma \to \Delta, \forall y G(y)}$$

goes to

$$\dfrac{\dfrac{\dfrac{\Gamma \to \Delta, P_G(\mathbf{t}, b) \quad P_{\forall y G(y)}(\mathbf{t}) \to P_{\forall y G(y)}(\mathbf{t})}{P_G(\mathbf{t}, b) \to P_{\forall y G(y)}(\mathbf{t}), \Gamma \to \Delta, P_{\forall y G(y)}(\mathbf{t})}}{\exists y (P_G(y, \mathbf{t}) \to P_{\forall y G(y)}(\mathbf{t})), \Gamma \to \Delta, P_{\forall y G(y)}(b, \mathbf{t})}}{\mathrm{DEF}_d, \Gamma \to \Delta, P_{\forall y G(y)}(\mathbf{t})} \begin{array}{l} \to \Rightarrow \\ \exists \to \end{array}$$

This completes the proof of the first part of the theorem.

The second part follows from *classical* derivability of the results of substitution P_G/G into formulas in DEF_d.

For the third part, if $\vdash^c \text{DEF}_d, \Pi \to \Phi$ then substitution P_G/G for $G \in \mathbf{f}(d)$ yields $\vdash^c \Pi \Rightarrow \Phi$, then (by Theorem 1) $\vdash^i \Pi \Rightarrow \Phi$ and hence $\vdash^c \text{DEF}_d, \Pi \Rightarrow \Phi$ completing the chain of equivalences. As pointed out in the Introduction, the transformation in Theorem 1 is not polynomial. ⊢

References

[1] T. Achourioti and M. van Lambalgen. A Formalization of Kant's Transcendental Logic. *The Review of Symbolic Logic*, 4:254–289, 2011.

[2] H. B. Curry. *Foundations of Mathematical Logic*. Dover Publications, New York, 1977.

[3] R. Goldblatt. *Topoi. The Categorial Analysis of Logic*. Elsevier Science Publishers, Amsterdam, 1984.

[4] G. Mints. *A Short Introduction to Intuitionistic Logic*. Kluwer Academic Publishers, New York, 2000.

[5] V. P. Orevkov. Glivenko's sequence classes. In V. P. Orevkov, editor, *Trudy Mat. Inst. Steklov: Vol. 98. Logical and logical-mathematical calculus. Part 1*, pages 131–154. Nauka, Leningrad, 1968.

Received: 14 May 2012

Commentary on Grigori Mints' "Classical and Intuitionistic Geometric Logic"

Roy Dyckhoff

University of St Andrews, Scotland
roy.dyckhoff@gmail.com

Sara Negri

University of Helsinki, Finland
sara.negri@helsinki.fi

This paper studies a *Glivenko sequent class*, i.e. a class of sequents where classical derivability entails intuitionistic derivability; more specifically, the paper is about "geometric sequents". The main old result in this topic is a direct consequence [11] of Barr's theorem.[1] As background, Mints sketches an old deductive proof (from [7]) and an old model-theoretic proof, as in Exercise 2.6.14 of [10]; but, his interest being in complexity of proof transformations, he gives a third proof, of a result both more and less general.

A modern reconstruction [6] of Orevkov's proof [7, Theorem 4.1, part (1)] relies on what we would now call the "cut-free **G3c** calculus" [9], in which *Cut* and other structural rules are admissible and all the logical rules are invertible (indeed, height-preserving invertible). His result is that the list (or "σ-class") $[\to^+, \neg^+, \forall^+]$ is a "completely Glivenko class"; in other words, he shows that if a sequent with a single succedent has no positive occurrences of $\to$, $\neg$ or $\forall$ then its classical derivability implies its intuitionistic derivability. In modern terminology, this means just that if a sequent $\Gamma \Rightarrow A$ (where Γ consists of geometric implications and A is a positive formula) is derivable in cut-free **G3c**, then it is already derivable in the intuitionistic calculus **m-G3i** (also from [9]). The proof method actually shows the stronger result, that the cut-free

---

[1] "Let $\mathcal{E}$ be a Grothendieck topos. Then there is a complete Boolean algebra $\mathcal{B}$ and an exact cotripleable functor $\mathcal{E} \to \mathcal{FB}$", $\mathcal{FB}$ being the topos of sheaves over $\mathcal{B}$ [1].

© The Author(s) and College Publications 2017

G3c derivation is already a **m-G3i** derivation. The weaker result extends to the case where A is a geometric implication by using the invertibility in cut-free **G3c** of the succedent rules for the three mentioned connectives. Other work, such as [4], related to the deductive proof of this result, is cited in the bibliographies of [5] and [6]. The usefulness of cut-free **G3** calculi in the study of Glivenko classes has been further demonstrated in [6], with direct proofs of generalisations of results in [7].

Mints' interest, however, in this paper is in derivations in **G3c** with *Cut*. One can apply standard cut-elimination transformations, and then those corresponding to the inversions; but this leads to a "super-exponential blow-up", as can be seen in a similar context in [9, Section 5.2]. How can this be avoided? One solution is just to start with a cut-free derivation. One can go even further, using the cut-free calculi introduced in [5], where the axioms Γ are replaced by inference rules: this avoids proof transformations entirely (since, in such calculi, classical proofs of a geometric implication A are already intuitionistic proofs). But, Mints would insist that **G3c** with *Cut* is a traditional (i.e. respectable) starting point.

The question then arises: can the transformation be changed so that there is an at most polynomial expansion of the derivation? Clearly it should not begin with cut elimination, so a trick is needed to handle instances of the *Cut* rule rather than eliminating them. The trick is attributed to Orevkov [7]; one might also attribute it to Skolem, who pioneered in [8] the use of what [2] should have called "relational Skolemisation", i.e. the replacement, by introduction of new relation symbols, of complex formulae by atomic formulae. When this is sufficiently thorough to ensure that every formula is equivalent to an atomic formula, it is called "atomisation" or "Morleyisation"; this paper doesn't go so far.

The novel result of this paper is now the result (both weaker and stronger) that, if d is a classical proof of a geometric sequent, then it can be polynomially transformed into an intuitionistic proof of the sequent conservatively extended by extra antecedent formulae that are geometric implications. These extra implications are generated by relational Skolemisation of the subformulae of the cut formulae in d. The result is weaker by virtue of having these extra implications; it is stronger by virtue of the complexity reduction.

There are the following points at which the paper is incorrect:

Commentary on "Classical and Intuitionistic Geometric Logic" 129

1. Mints' (9) should be $\forall \mathbf{x}(P_{\exists yG}(\mathbf{x}) \to \exists y P_G(\mathbf{x}, y))$ rather than $\forall \mathbf{x}(P_{\exists y P_G}(\mathbf{x}) \to \exists y P_G(\mathbf{x}, y))$;

2. His (10) should be $\forall \mathbf{x} \forall y (P_G(\mathbf{x}, y) \to P_{\exists yG}(\mathbf{x}))$ rather than $\forall \mathbf{x} \forall y (P_G(\mathbf{x}, y) \to P_{\exists y P_G}(\mathbf{x}))$;

3. His (11) should be $\forall \mathbf{x} \forall y (P_{\forall yG}(\mathbf{x}) \to P_G(\mathbf{x}, y))$ rather than $\forall \mathbf{x} \forall y (P_{\forall y P_G}(\mathbf{x}) \to P_G(\mathbf{x}, y))$;

4. His (12) should be $\forall \mathbf{x}(\forall y P_G(\mathbf{x}, y) \to P_{\forall yG}(\mathbf{x}))$ rather than $\forall \mathbf{x}(\forall y P_G(\mathbf{x}, y) \to P_{\forall y P_G}(\mathbf{x}))$;

5. His (19) (replacing (16)) is not a geometric implication;

6. His (17) (replacing (12)) is not a geometric implication.

The first four of these problems are minor: note that in Mints' (9) the suffix $\exists y P_G$ is not a subformula of one of the cut-formulae, and similarly for (10), (11) and (12). The penultimate problem can be fixed by distributing $\forall \mathbf{x}$ across the conjunction, thus obtaining two geometric implications: $\forall \mathbf{x}(P_H(\mathbf{z}) \to P_{G \to H}(\mathbf{x}))$ and $\forall \mathbf{x}(P_G(\mathbf{y}) \lor P_{G \to H}(\mathbf{x}))$. [It has already been made clear that **y** and **z** are subsets of the set **x** of variables.]

The final problem is not so easily fixed: the paper wrongly claims that the formula $\forall \mathbf{x} \exists y (P_G(\mathbf{x}, y) \to P_{\forall y P_G}(\mathbf{x}))$ is a geometric implication. This is not fixed by changing (12) (as proposed above) to $\forall \mathbf{x}(\forall y P_G(\mathbf{x}, y) \to P_{\forall yG}(\mathbf{x}))$ and then obtaining $\forall \mathbf{x} \exists y (P_G(\mathbf{x}, y) \to P_{\forall yG}(\mathbf{x}))$; this is still not geometric, because of the implication within the scope of the existential quantifier.

A partial solution may be had by changing this formula to the geometric implication

$$\forall \mathbf{x}(\exists y P_{\neg G}(\mathbf{x}, y) \lor P_{\forall yG}(\mathbf{x})) \tag{17}$$

but this introduces a new relation symbol $P_{\neg G}$, where $\neg G$ may not be a subformula of one of the cut formulae. To fix this problem, the relational Skolemisation needs to be applied not just to all such subformulae but also to **all** their negations.

With these changes, the application of the extra formulae (i.e. members of DEF_d) to deal with the special formulae of the derivation is unchanged for implication. We show (for example) the effects of improving (9) on the treatment of an antecedent $\exists$-inference and of correcting the treatment of universal quantification.

The improved version of (9) is $\forall \mathbf{x}(P_{\exists yG}(\mathbf{x}) \to \exists y P_G(\mathbf{x}, y))$. The step

$$\frac{G(b), \Gamma \Rightarrow \Delta}{\exists y G(y), \Gamma \Rightarrow \Delta}$$

is transformed to

$$\frac{\dfrac{P_{G(y)}(\mathbf{t}, b), \mathrm{DEF}_d, \Gamma \Rightarrow \Delta}{\exists y P_{G(y)}(\mathbf{t}, y)), \mathrm{DEF}_d, \Gamma \Rightarrow \Delta} \, \exists\Rightarrow}{\mathrm{DEF}_d, P_{\exists y G(y)}(\mathbf{t}), \Gamma \Rightarrow \Delta}.$$

Using the improved version of (17), the step

$$\frac{\Gamma \Rightarrow \Delta, G(\mathbf{t}, b)}{\Gamma \Rightarrow \Delta, \forall y G(\mathbf{t}, y)}$$

is transformed (with some implicit weakenings to save space and aid readability) to

$$\frac{\dfrac{\dfrac{\dfrac{\dfrac{\dfrac{\dfrac{\mathrm{DEF}_d, \Gamma \Rightarrow \Delta, P_{G(\mathbf{x},y)}(\mathbf{t}, b)}{P_{\neg G(\mathbf{x},y)}(\mathbf{t}, b), \mathrm{DEF}_d, \Gamma \Rightarrow \Delta, P_{G(\mathbf{x},y)}(\mathbf{t}, b)} \, Wkn \quad \Rightarrow\wedge, \text{axiom}}{P_{\neg G(\mathbf{x},y)}(\mathbf{t}, b), \mathrm{DEF}_d, \Gamma \Rightarrow \Delta, P_{\neg G(\mathbf{x},y)}(\mathbf{t}, b) \wedge P_{G(\mathbf{x},y)}(\mathbf{t}, b)}}{P_{\neg G(\mathbf{x},y)}(\mathbf{t}, b), \neg(P_{\neg G(\mathbf{x},y)}(\mathbf{t}, b) \wedge P_{G(\mathbf{x},y)}(\mathbf{t}, b)), \mathrm{DEF}_d, \Gamma \Rightarrow \Delta} \, \neg\Rightarrow}{P_{\neg G(\mathbf{x},y)}(\mathbf{t}, b), \mathrm{DEF}_d, \Gamma \Rightarrow \Delta}}{\dfrac{\exists y P_{\neg G(\mathbf{x},y)}(\mathbf{t}, y), \mathrm{DEF}_d, \Gamma \Rightarrow \Delta}{\exists y P_{\neg G(\mathbf{x},y)}(\mathbf{t}, y) \vee P_{\forall y G(\mathbf{x},y)}(\mathbf{t}), \mathrm{DEF}_d, \Gamma \Rightarrow \Delta, P_{\forall y G(\mathbf{x},y)}(\mathbf{t})} \, \exists\Rightarrow \quad \dfrac{}{P_{\forall y G(\mathbf{x},y)}(\mathbf{t}), \Gamma \Rightarrow \Delta, P_{\forall y G(\mathbf{x},y)}(\mathbf{t})} \, \text{axiom}}}{\mathrm{DEF}_d, \Gamma \Rightarrow \Delta, P_{\forall y G(\mathbf{x},y)}(\mathbf{t})} \, \vee\Rightarrow$$

Note the importance of having $P_{\forall y G(\mathbf{x},y)}(\mathbf{t})$ (rather than, from the succedent of the old (17), Mints' $P_{\forall y P_G}(\mathbf{t})$) in the antecedent of the lowest axiom step. It is not the case that $\forall y P_{G(\mathbf{x},y)}$ (i.e. Mints' $\forall y P_G$) is a subformula of one of the cut formulae; the presence of the fresh predicate symbol $P_{G(\mathbf{x},y)}$ forbids this.

Note also the use of the Weakening rule *Wkn*; either this rule should be included in the **m-G3i** calculus or the admissibility of the rule exploited once the derivation has been fully transformed.

References

[1] M. Barr. Toposes without points. *J. Pure and Applied Algebra*, 5:265–280, 1974.
[2] R. Dyckhoff and S. Negri. Geometrisation of first-order logic. *Bulletin of Symbolic Logic*, 21:123–163, 2015.

[3] J. E. Fenstad, editor. *T. Skolem, Selected Works in Logic*. Universitetsforlaget, Oslo, 1970.

[4] G. Nadathur. Correspondence between classical, intuitionistic and uniform provability. *Theoretical Computer Science*, 232:273–298, 2000.

[5] S. Negri. Contraction-free sequent calculi for geometric theories, with an application to Barr's theorem. *Archive for Mathematical Logic*, 42:389–401, 2003.

[6] S. Negri. Glivenko sequent classes in the light of structural proof theory. *Archive for Mathematical Logic*, 55:461–473, 2016.

[7] V. P. Orevkov. Glivenko's sequence classes. In V. P. Orevkov, editor, *Trudy Mat. Inst. Steklov: Vol. 98. Logical and logical-mathematical calculus. Part 1*, pages 131–154. Nauka, Leningrad, 1968.

[8] T. Skolem. Logisch-kombinatorische Untersuchungen über die Erfüllbarkeit oder Beweisbarkeit mathematischer Sätze nebst einem Theoreme über dichte Mengen. In *Skrifter utgit av Videnskapsselskapet i Kristiania. I, Matematisknaturvidenskabelig Klasse*, volume 1920 bd. 1, No 4. Kristiania : I Kommission hos J. Dybwad, 1920. http://www.biodiversitylibrary.org/item/52015#page/111/mode/1up. Also reprinted in [3, pages 103–136].

[9] A. S. Troelstra and H. Schwichtenberg. *Basic Proof Theory*. CUP, 2nd edition, 2000.

[10] A. S. Troelstra and D. van Dalen. *Constructivism in Mathematics (vol 1)*. North Holland, 1988.

[11] G. Wraith. Intuitionistic algebra: some recent developments in topos theory. In *Proceedings of International Congress of Mathematics*, pages 331–337, Helsinki, 1978.

Received: 6 October 2016

The Historical Role of Kant's Views on Logic

Andrei Patkul

Department of Ontology and Epistemology, St. Petersburg State University
7/9, nab. Universitetskaya, St. Petersburg 199034, Russia

a.patkul@spbu.ru

Abstract: The article represents the outcomes of the reconstruction of Kant's classification of kinds of logic. It demonstrates that it would be impossible to form the traditional understanding of so-called formal logic, as well as of today's symbolic logic, without Kant's notion of pure general logic. It was formed by Kant within the framework of his critique of reason. Critique has changed our understanding of logic from seeing it as *organon* to an understanding of it as a *canon* of finite cognition. In conclusion we pose the question of the status of Kant's transcendental logic with regard to its connection to classification of types of logic given by him and possibility of its formalization in account of Kant's idea of pure general logic.

Keywords: critique of reason, general use of understanding, particular use of understanding, pure general logic, applied logic, transcendental logic, epistemology of logic.

1. Introduction

Currently, at a casual glance, it might appear that logic, of the shape it was created in by Aristotle, is a science of forms of thought. There have been many speculations presuming that logic of such type deals only with the formal thought-structures, independent of any content. Therefore, logic, understood in a such way, could be entitled *formal logic*, i.e. a pure formal science. At the same time, this kind of logic is often treated as wrong, obsolete or, at least, as insufficient one. Such critique generally goes from the point of contemporary logic, which overcame the notion of a thought-form by reducing of object-matter of logic, to the pure symbols and the rules of their combinations. The representatives of so-called symbolic logic understand it as the direct opposite of the traditional formal one. Herein, the question of what the proper *quid juris* of their criticism of traditional logic actually is, arises. In the fact, there is the following reason for such a question to arise. Symbolic logic deals with forms as well, but contemporary logicians understand them quite otherwise. Moreover, being self-evident but not distinct enough the notion of formality is presupposed in both cases. It is obvious that symbols and rules of their combinations are formal as the thought-forms *pari passu*. Namely, both forms of thinking and signs are independent of their possible content. Hence, we have to ask whether symbolic logic is a later descendant of formal logic which has forgotten its own roots.

In other words, is it permissible to consider the definition, given currently both to ontological and epistemological statuses of such symbols, symbolic structures and rules of their combinations, as strict and correct? It is a well-known fact that pure forms of such a kind play a significant role in the process of so-called formalization as one of the basic methods of the contemporary scientific knowledge, but what kind of formalization would allow them to become forms empty of any content? If we tried to formalize the things themselves, which we can accomplish any formalization by, it would lead us into *regressus ad infinitum*. Hence, the formality of logical symbols remains problematic both for ontology and epistemology.

Further, we can inquire into the following matter. Which understanding of the essence of logic and its object domain does play the role of the basis of the differentiation of the mentioned kinds of logic at all? Moreover, it raises another question. Which understanding of the essence of logic and its objects domain does make the basis of the notion of the formality of logical forms in each case?

Indeed, the necessity of the separation of both forms of thought and symbols from their specific content is not self-evident. Moreover, it is perfectly possible that the differentiation existing between them is the extremely later term-division which was being made through the abstractive work of "pure reason" during the history of philosophy. Therefore, this differentiation, as such, has to be justified both ontologically and epistemologically. It is to our regret, that it is impossible to accomplish such a justification in the systematical regard here, but we could undertake a reconstruction of an indicative example of interpretation of logic from the history of philosophy which would emphasize the problematic character of the logical formality. In this way, the validity of the logical formalism could be justified not by a formal deduction of its possibility, which, as it was said, would lead to *regressus ad infinitum*, but by detection of the transcendental genesis of the separation of form and content and, hence, of the notion of logic as of a pure formal discipline in any sense. To be more precise, the matter concerns the transformation of the understanding of the essence of logic made by Immanuel Kant.

It is doubtless that Kant's revolutionary conception of logic is one of the most important points in the historical path of the separation of the notions of form and content which remains, even latently, a live issue nowadays. Hence, it is a matter of great importance for the genuine conception of conditions of possibility and limits of applicability of logic as a formal science. We believe that today it could induce our philosophical disputations on the nature of logical forms and structures.

In this context, on one hand, the main aim of the article is to demonstrate the conditions of shaping of notion of form in its logical sense, by reconstructing Kant's logical views. On the other hand, we should remember that Kant showed of all others that logic in its transcendental shape could have some content. Its content can be only the *a priori* one. Thereby, the correlation between logic as a discipline of the pure universal forms and transcendental logic has to remain controversial. However, we presuppose that the transcendental conditions of Kant's interpretation of logic are historically responsible for the genesis of symbolic logic of nowadays. And therefore, it might be justified theoretically only by an ontological and epistemological justification of Kant's logical views.

We believe that the transformation of the understanding of the essence of logic made by Kant cannot be attacked from the standpoint of today's logic and semantics which finally have been derived from Kant's position. Moreover, there is no need in justifying Kant's logic through them.

In this context, in the course of this consideration, we referred to the following paper by Achourioti and van Lambalgen [1]. It is devoted to a justification of the Kant's idea of logic from the perspective of today's symbolic logic. As it was stated, our thesis is the opposite one. One ought to verify today's logical approaches through the reconstruction of the transcendental and historical genesis of the ontological and epistemological conditions of logic, shaping it as a discipline of formal symbols and structures from the transformation of the essence of logic made by Kant.

2. Traditional logic

Nevertheless, it remains very questionable whether logic could be described as a formal discipline starting right from its origin in Aristotle. It is well-known fact that Aristotle treated form (μορφή) as a shape (εἶδος) or even a prototype (παράδειγμα). For instance, he speaks about form in the sense of *causa formalis*, "The form and template, which is the account of the what-it-was-to-be-that-thing. Also the kinds of form are causes in this way. [...] Also the intrinsic parts of the account."[1] [3, p. 115]. Hence, to be more exact one ought to say that form as εἶδος is connected with *logos* as a meaningful definition. Certainly, εἶδος is one of the possible senses of the term of form. At the same time, it is the preferential one. Anyway, the form, treated in Aristotelian way, is something always rich in content. In this case, no form can be separated from its content. It remains questionable, whether it could be possible to differentiate form from formless content in εἶδος within the framework of Aristotle's views on logic. We believe that no distinction of form and content can exist inside εἶδος at all. Any form is always form of certain content, they correspond in an absolute way to each other in Aristotle's opinion. Any εἶδος, itself, is form (μορφή) as such which holds its own certain content as its "What" in itself. Therefore, one cannot understand logic only as a discipline of pure forms of thinking, independent of any content. Farther, it is doubtful whether logic (the title which Aristotle himself hasn't used) meant ἐπιστήμη, i.e. a science in the strict sense for him. We have to remember that any science, insofar it is a science, should have its own object domain which would have its own ontological status (being in things or only in our mind and so on) and which should be distinct strictly from the object domains of all other sciences in accordance with Aristotle.

Yet we could not find something alike logic in his set of sciences, both theoretical and practical. For instance, logic is absent in Aristotle's set of theoretical sciences, which *philosophia prima*, physics and mathematics belong to. It indicates that logic, from its creator's standpoint, does not have its own object, possessing necessary and invariable principles. Therefore, there is no specific domain of beings (including the domain of mathematical objects

---

[1]In his own words Aristotle even says, "...ἄλλον δὲ τὸ εἶδος καὶ τὸ παράδειγμα, τοῦτο δ' ἐστὶν ὁ λόγος τοῦ τί ἦν εἶναι καὶ τὰ τούτου γένη [...] καὶ τὰ μέρη τὰ ἐν τῷ λόγῳ"(*Metaph.*, V, 1013a, 25). Here ἄλλον δὲ means "one can speak about something as cause". See English translation in [2].

which are immovable and dependent upon our mind) which could be the object domain of logical investigations as such. Hence, any specific realm of pure logical forms exists neither *per se* nor *in rebus* nor *in mentis*. Therefore, logic cannot be understood as a true science.

Rather, logic is a kind of τέχνη which deals only with the rules of any correct cognition. Thus, we should learn it before we start to cognize any object domain.

3. The misinterpretation of traditional logic

In opposition, there is an existent conviction, that logic even in its Aristotelian version, is a science in the most rigorous sense. One can compare its status, with regard to exactness, only to the status of mathematics. (Assuming, that the question of propinquity or heterogeneity of both logic and mathematics remains disputable). Thereby, this conviction has an essential presupposition and, in accordance with it, logic, consistent with the general notion of a science, should have a specific object domain which ought to be cognized by it. This presupposition is generally missed. This domain is considered a region of empty thinking-forms or, in today's version, sign-forms. On the other hand, in this regard it might even appear that today's understanding of logic, in the constructive version, recovers original treating of it as τέχνη in Aristotelian sense. For instance, logic could be understood as τέχνη in a sense of creation of logical formulas by the combination of signs which corresponds to given rules. But what kind of ontological or epistemological status can these rules, as such, have?

Indeed, various types of today's logic do not need to possess any kind of object domain to be considered as an exact and rigorous discipline. But, as it was said, this circumstance does not exclude logic dealing with empty forms in the way which they attribute to Aristotelian *Analytics*. However, such forms mustn't be treated as forms of thinking. As it was stated, the status of such forms is ontologically problematic. In this regard, contemporary logic remains just a specification of an idea of a science of pure forms but technically it is more sophisticated. The truth is that Aristotle himself did not consider his analytic as a science of pure forms.

4. Kant's modification of logic

Now, the following question should be posed. What served as the origin of as well as the reason for transformation, the idea of logic undergone, from its Aristotelian understanding as τέχνη to treating it as a science of pure forms both of thought and signs? Perhaps, such transformation began a great while ago. We could even propose that it began a long while ago before Kant. However, Kant's treating of logic is one of the most notable examples of a reinterpretation of the essence of logical knowledge which could cast the light on the problem of the genesis of the notion of logic as a science which deals only with pure forms apart from their actual content. Nevertheless, Zinkstok emphasizes, "The first thing we should note is that Kant calls logic a *science*. This is, in fact, a break with most of the traditional views ... " [13, p. 39]. The possible answer to this question is the following. The origin of transformation of the understanding of the logic's essence could be found in Immanuel Kant's *Critique of Pure Reason*.

4.1. Logic, *noumena* and *phenomena*

It is a well-known fact that the transformation of idea of logic made by Kant is connected with the change in treating of this science as ὄργανον of knowledge (as it was for Aristotle) to treating it as κανών of any possible cognition. From Kant's standpoint, previously it was thought, that logic as ὄργανον is not only necessary condition for any knowledge but also the sufficient one. For instance, before Kant it was considered that logical criteria are absolutely sufficient for rational cognition, i.e. for such kind of cognition which does not refer to any possible experience.

Kant has turned the tide. In accordance with his standpoint, logic can serve as the necessary but not the sufficient condition of cognition. Any knowledge has to correspond to logic, i.e. it must not come into a contradiction with logical laws, in the first instance, to first analytical principle of *tertium non datur*. In other words, any cognition has to be free from contradiction within itself. At the same time, according to Kant, logic (without its connection with a possible experience) is not sufficient for acquiring new knowledge. Hence, it is impossible to obtain new knowledge in the pure rational disciplines with such object domains which cannot be given in any possible experience (soul as simple substance, world as a whole, God). These objects are just ideas of reason in Kant's terminology.

Kant justifies this new conception by the thesis that logic is applicable only to the things as phenomena, but not to the things as such or to entities as entities (*ens qua ens*) in the terms of Aristotle. Thus, logic corresponds to something that does not have objective (in the sense of being *in rebus* themselves) but belongs only to the field of subjectivity. The meaning of subjectivity in this and following expressions does not refer to subject as to a singular person. It is related only to a subject in general or, in other words, to a structures of subjectivity as such. Thereby, these structures should necessarily have a relation to the mode which something what exists in itself can be given to us as subjects in a transcendental sense of subjectivity in. As it was said, it implies that any cognition has to be measured by logic, but logic as such cannot give any new knowledge. Insofar, it is isolated from experience (i.e. from the way which the phenomena could appear in) logic can have only subjective but, at the same time, a general and necessary value. In such a way, logic acquires the meaning of *sine qua non* of any knowledge but not of actual cognition. Hence, it is unacceptable to treat Kant's view of logic without regard for the division made by him between *phenomena* and *noumena*.

At the same time, a science of logic acquires the meaning of a science of pure forms of thought which are originally on the subjective side and, hence, independent of the concrete content of *phenomena* given to us through experience. Such acquiring of the subjective character by logic is a necessary condition for the shaping of the notion of pure form. Nevertheless, such acquiring, insofar it is based on the transcendental character of the subjectivity, cannot relativize logic. By the possible relativization of logic we mean such kind of its treating, which implies that the common value of the logical forms is relative to an empirical subject or, in other words, to a singular person who uses these forms. Kant denied any empirical relativity of logic in this sense. Namely, the matter of experience appears in an accidental mode but forms of thought are general and necessary according to him. They have to be present in us *a priori* to make appearance of a matter through experience possible. However, they justify their objective value only by applying themselves to the things as *phenomena*, i.e. to content of our experience.

On one hand, one can understand the famous division of types of logic made by Kant in his *Critique of Pure Reason* only in the context of the described transformation of the role of logic for human cognition. On the other hand, this division can brilliantly demonstrate the reduction of logic to a science of pure forms of thought made by Kant. In order to do so we have to make an attempt to reconstruct an architectonic of logical disciplines given by Kant, in broad outline. For further information on the matter see, for example, [13].

4.2. The general and the particular use of understanding

Thus, Kant has primarily divided the general notion of logic into (i) logic of the general and (ii) logic of the particular use of our understanding. He stated,

> Now, logic in its turn may be considered as twofold, – namely, as logic of the general [universal], or of the particular use of understanding. (A52, B77) [9, pp. 46–47]

The type of logic, last mentioned, deals with a particular object domain in each case as well as with main rules of its cognition. The logic of the particular use of understanding always refers to a matter of one of object domains. In a manner of speaking, it should follow a content of this domain. Hence, as it depends on concrete content of an object domain, logic of such type cannot be detected as a pure or formal science.

Thus, it is quite noteworthy that Kant considers logic, which would refer to some matter, being possible only as a particular but not as an universal discipline. One can suppose that this circumstance goes back up to Aristotle's fundamental thesis, "That which is is spoken of in many ways" [3, p. 167],[2] or – in scholastic formula – to *analogia entis*. In fact, Kant does not refer to these formulas. It is unlikely at all that he actually knew this Aristotelian conception. For German-speaking philosophy it was rediscovered later, thanks to the efforts of Brentano.

It is more important that Kant accepted that logic can be non-universal in any case. The possibility of the non-universality of logic stems from the fact that it is connected with its content, i.e. with a matter of a certain object domain. Therefore, following the thesis τὸ 'ὸν λέγεται πολλαχῶς there is no universal object domain. Moreover, no universal object domain is ontologically possible. Hence, logic can be universal on the assumption of an abstraction from any particular object domain. Any type of universal logic or other formal calculus can, it seems, be only objectless. Here, the objectlessness implies that universal

---

[2]In Aristotle's own words: (τὸ 'ὸν λέγεται πολλαχῶς) (*Metaph.*, VII, 1, 1028 a, 10). See [2].

logic does not refer to any object, neither to universal (which is ontologically impossible) nor to particular (which are non-universal) ones. In this sense, it refers only to its own structures which has no objective character but only the (transcendental) subjective one.

On the other hand, as may be supposed, universal logic could not be objectless but it relates to all possible objects by the abstraction from differences of individual entities, of types of objects and so on. Indeed, it is acceptable to interpret Kant's notion of universal logic in this way. In such case, logic would be directed toward an object. But its object would be non-particular. We are opposed to such treating of Kant's view on universal logic. Here, one ought to emphasize two reasons for doing so. (i) We adhere to the above-mentioned Aristotelian thesis which we consider an ontological principle, any understanding of logic has to be founded on. Τὸ ’όν λέγεται πολλαχῶς. Therefore, as it was said, no universal object (even an empty and indifferent one) is ontologically possible. Any object should have its own essence as well as a way-of-being. "Object in general" is a *flatus vocis*. The source of the general validity of logic is quite different from any objectivity. It is subjective in a transcendental sense of subjectivity. (ii) Since Kant made a division between *noumena* and *phenomena* we cannot tell if logical forms could be applied to *noumena* which belong to "object in general". For instance, we do not know whether thinking of God has to be yielded to the principle of *tertium non datur*. After all, the mystics of all time have been telling us that God is being and non-being at the same moment.

Anyway, we assent to an opinion of MacFarlane who has stated the following,

> Kant's claim that logic is purely Formal – that it abstracts entirely from the objective content of thought – is in fact a radical innovation. [11, pp. 44–45]

MacFarlane demonstrated that this "radical innovation" was bounded to Kant's rejection of neo-Leibnizian views on logic, implying that logic is general but not objectless. It has its own most general content. In contraposition to them, Kant started to understand logic as a discipline which deals only with rules of thinking, i.e. which has only subjective sense.

One way or another, Kant himself described logic of the particular use of understanding in the following words,

> The logic of the particular use of the understanding contains the laws of correct thinking upon a particular class of objects. (A52, B77) [9, p. 47]

As opposed to logic of the general use of understanding, logic of its particular use may be *organon* of cognition of a specific object domain in accordance with Kant. He stated,

> The former[3] may be called elemental logic, – the latter, the organon of this or that particular science. The latter is for the most part employed in the schools, as a propaedeutic to the sciences, although, indeed, according to the course of human reason, it is the last thing we arrive at, when the science is already matured, and needs only in finishing touches toward its correction and completion; for our knowledge of the objects of our attempted science must be tolerably extensive and complete before we can indicate the laws by which a science of these objects can be established. (A52, B77) [9, p. 47]

We may presuppose that the logic of the particular use of our understanding could be identified with methodology of a particular science in contemporary word usage. It deals with the rules of cognition of a specified object domain but it can appear only after the maturity of one or another particular science.

However, we think that Kant's idea of the particular logic of understanding remains relevant at the present day. Namely, we believe that an attempt of comparison of Kant's logic of the particular use of understanding and the notion of the regional ontology in the phenomenological branch of today's philosophy could be productive in various methodological perspectives.[4] Though, the notion of the regional ontology is not derived directly from Kant's notion of particular logic, this Kant's term usage could clarify the proper meaning of the term "logic", for instance, in Heidegger's word-combination "productive logic of science" [6, p. 4] which was understood as a regional ontology by him.

Nevertheless, we have to dismiss this analogy between logic of the particular use of understanding and the regional ontology and revert to the above-mentioned division between the logic of the general use of understanding and the logic of its particular use accomplished by Kant. Now we should inquire into first part of this division.

4.2.1. Back to the division made by Kant

Here, we have to recall that Kant defined logic of the general use of understanding as a discipline which deals with general rules of thought regardless of any matter of applying of this thought. As it was said, in contrast to logic of the particular use of understanding, logic of its general use cannot be general

---

[3] "The former" means here the logic of the general use of understanding.
[4] See, for instance, [8].

organon of our finite cognition. It can be only its *canon*. Then, Kant called it "elemental logic".

Namely, he stated that logic of the general use of understanding

> [...] contains the absolutely necessary laws of thought, no use of understanding at all could be possible without, and therefore gave laws to understanding, regardless of the difference of objects, which it may be applied to. (A52, B77) [9, p. 47]

It is obvious that such distinction was a good step forward in the direction of shaping of the notion of logic as a science of pure forms abstracted from any content. Namely, Kant started to consider a logical form on the base of the notion of the law of thought. The laws of thought are of functional character. So, a form of thought is a function which prescribes the one and only mode which it could act in, regardless of its content. This functional character is based on the spontaneity of thinking as such. The condition, necessary for it, is the universality of a law of thought, i.e. its independence of a concrete matter or content. MacFarlane emphasized the normative character of general logic in this regard,

> The generality of logic, for Frege as for Kant, is a normative generality: logic is general in the sense that it provides constitutive norms for thought as such, regardless of its subject matter. [11, p. 35]

Only logic of such kind can be universal. In other words, it can be used indifferently to the peculiarity of an object domain. In particular, it has to be noted that such understanding of general logic leads to very productive and, yet, very disputable idea of formal ontology which reckons as its object not just subjective "laws of thought" (as it considered by Kant) but also the universal and the only formal definitions of something in general, or of "quasi-region" (Husserl). Nevertheless, here one has to dismiss the reason for a turn from pure formal logic (from the *mathesis universalis* in the widest sense) to formal ontology without prejudice.

In any case, the mentioned step is still insufficient for the ultimate formation of the notion of logic as of a science of pure forms in Kant's interpretation. We just have to point out here, that the distinction between logic of general and logic of the particular use of understanding is based on the quantitative principle of the difference between generality and particularity. Thereby this qualitative principle is aligned with the difference between dependence upon content and its independence stated by Kant. Only the independence of any content of logic ensures its quantitative generality.

4.3. Pure and applied universal logic

As distinct from the notion of logic of the particular use of understanding, the notion of logic of its general use is divisible further from Kant's standpoint. He believed that logic of the general use of understanding can be divided in two parts. Namely, they are, on one hand, pure logic and, on the other hand, applied logic. The common condition for both types of logic is their generality.

The first of the mentioned types of logic, as of a general discipline, deals only with the rules of thought regardless of the concrete conditions of its implementation by an empirical subject. The laws of thought are equally independent of the situation which someone applies them in. Nevertheless, they belong not to objectivity but to the subjective field only, they possess an ideal identity within themselves. The universal and transcendental character of subjectivity in accordance with Kant is the guarantee for such identity. Hence, it is indifferent for the formality of the laws of thought who, where, when and how applies them. In each case, they will remain the same. In this sense, it has to be said, that the laws of logic have "objective value" but it does not mean that the ground of this value lies in objects. It doesn't mean that their source lies in objectivity in its opposition to empirical subjects. This characteristic of the laws of logic refers only to "objectivity" in the sense of the ideal "universal validity".

On the contrary, general but applied logic takes into account such empirical conditions of thinking. This difference could be well-clarified by a few statement made by Kant in his *Critique of Pure Reason*,

> General logic is again either pure or applied. In the former, we abstract all the empirical conditions under which the understanding is exercised; for example, the influence of the senses, the play of the fantasy or imagination, the laws of the memory, the force of habit, of inclination etc, consequently also, the sources of prejudice, – in a word, we abstract all causes from which particular cognitions arise, because these causes regard the understanding under certain circumstances of its application, and, to the knowledge of them experience is required. (A52-3, B77) [9, pp. 47–48]

Hence, we could detect that Kant's term "applied logic" is equal to psychology of logical knowledge in the contemporary usage of terms. The author of *Critique of Pure Reason* has especially emphasized that he used the term "applied" with regard to logic in quite a different sense than it is commonly used. According to Kant's interpretation of applied logic, it doesn't belong to any kind of τέχνη. It is not a technical discipline or a kind of skill which teaches how to apply the logical rules and laws correctly. Applied logic is exactly general logic, not a particular one. Since it is one of subsections of general logic, one cannot use it as ὄργανον of cognition. In any event, it does not consider the logical laws in a strict sense. It treats the domain, which should be already yielded to the rules of the general use of understanding. Hence, it, in fact, remains questionable whether Kant's characteristic of this discipline as of logic is correct. At the same time, it cannot be understood as κανών of knowledge, in contrast to other subsection of logic of the general use of understanding.

The thinker stated the following, concerning general applied logic,

> General logic is called applied, when it is directed to the laws of the use of the understanding, under the subjective empirical conditions which psychology teaches us. It has therefore empirical principles, although, at the same time, it is in so far general, that it applies to the exercise of the understanding, without regard to difference of objects. On this account, moreover, it is neither a canon of the understanding in general, nor an organon of a particular science, but merely a cathartic of the human understanding. (A53, B77-78) [9, p. 48]

Since the domain, which applied general logic inquires into, is already yielded to the general rules of understanding, pure general logic has no need to follow such empirical conditions. It its origin lies not in our actual but in contingent experience,

> Pure general logic has to do, therefore, merely with pure *a priori* principles, and is a canon of understanding and reason, but only in respect of the formal part of their use, be the content what it may, empirical or transcendental. (A53, B77) [9, p. 48]

Therefore, the laws of use of understanding, which pure applied logic discovers, should correspond to the "pure a priori principles". For, in its turn, applied logic as such has to be commensurate to the common κανών of knowledge, i.e. to pure general logic. Hence, general logic is not deducible from pure applied logic. In other words, it is impossible to derive logical forms as such from the modes, we use and apply them in, in our empirical circumstances. As it was said, the logical forms should be already present in a way. They can be applicable in this case only. The usage and the application of logical forms are already yielded to these forms.

In this regard, Kant concluded,

> In general logic, therefore, that part which constitutes pure logic must be carefully distinguished from that which constitutes applied (though still general) logic. The former alone is properly science, although short and dry, as the methodical exposition of an elemental doctrine of the understanding ought to be. (A53-4, B78) [9, p. 48]

It is our belief that in this way Kant derived the notion of logic, very close to "formal logic" in a contemporary sense. In Kant's words, formal logic is the pure general one, i.e. it is logic, independent as well of concrete content given by the experience as of concrete conditions of accomplishing of thinking by an empirical subject.

In summary of this subsection, we would state that principle of the differentiation of pure and applied logic within the framework of logic of the general use of understanding lies in the difference of the notions of the transcendental and the empirical fields. Hence, it is a qualitative or, to be more precise, essential, principle as distinct from the basis of the differentiation of general and particular logics. As it becomes apparent, Kant pointed out two requirements for such kind of logic or conditions it could be formed in:

(i) One ought to differentiate form and content in a thought disregarding to the differences between objects which could be thought by a logical form. Then, one ought to expound form as that what belongs to subjective field. Content has to be considered as that what derives from objects. In that way, we can differentiate form and content finally and, then, get the notion of the form of thought, which would be independent of its content.

(ii) One ought to exclude "empirical principles" of usage of logical forms and, hence, to show that psychological conditions of application of logic have nothing to do with the laws of logic as such. In this way, one can justify why the relativity of the logical forms in their application does not follow from their subjective status. Namely, we show that the universality of these forms does not contradict to their subjectivity because of their subjectivity is not the empirical one. See, for instance, [11, p. 48].

Hence, formality of logical forms is defined in a privative way through the independence (i) of objective content and (ii) circumstances of accomplishing of thinking. As it was said, MacFarlane believed that Kant's understanding of logic as of a formal discipline, hence, the peculiarity of his concept of logical form, became a real "innovation". The path of shaping of the notion of logical form goes through abstraction from content both particular and general.

The Historical Role of Kant's Views on Logic   147

4.3.1. Pure logic and antipsychologism

One especially ought to emphasize Kant's second requirement for pure general logic. In fact, this thinker formed some conditions for arising of so-called antipsychologism in treating of the essence of logic. He made it by rigorous distinguishing between pure formal principles and subjective empirical conditions of thinking. It is a well-known fact that detailed critique of the reduction of logical laws as well as of mathematical objects and structures to the phenomena of the psychical life was elaborated subsequent to Kant in two very different schools of philosophy, namely in analytical philosophy (Frege) and phenomenology (Husserl). The members of the mentioned schools didn't accept Kant's way of treating nature and status of logic and, especially, of its relation to mathematics and their objects. For instance, Frege has elaborated a program of logical justification of the mathematics, conflicting with Kant's understanding of the essence of the mathematics (see, in particular, [4]). In this regard, we have to remember that logic as a science of pure forms of thought and mathematics as knowledge based on pure forms of sensibility have transcendentally different origin and nature form Kant's standpoint. Hence, he can be acknowledged as the forerunner of the mathematical intuitionism. On the contrary, it would be wrong to speak about the elements of intuitionism in Kant's treating of logic. Thereby, we can find the same situation with regard to a status of the universal validity of logic in all three cases. One can reach the mentioned only by separation of logic forms not only from content delivered from a side of objects but also from private or empirical-subjective conditions of the validity of logical statements.

Nevertheless, all the above-mentioned philosophers have quite different understandings of the logic. It is our belief that one ought to bear in view such kind of difference in presupposing of the possibility in the interpretation of Kant's logical views, for instance, from the point of view of Frege's philosophy. Namely, one should pose a question whether Frege's notion of logic could be justified on the base of the same conception of the (transcendental) subjectivity as Kant's treating of logic. Does Frege's idea of logic require any conception of subjectivity at all?

Conversely, it was demonstrated that Kant's notion of pure general logic is impossible without the admission of a distinction between pure and the applied logic and, thus, without the admission of subjectivity and a subjective character of such kind of logic at all. For instance, MacFarlane has shown that a possible problem in interpretations of the nature of logic made by Kant and Frege lies in the circumstance that both philosophers had very different understanding of a function of logic. MacFarlane stated that Kant's "[...] picture of logic is evidently incompatible with Frege view that logic can supply us with substantive knowledge about objects" [11, p. 29].

On the other hand, Husserl has agreed that one has to consider the logic as well as the *mathesis universalis*, as a whole, and as an effect of the constitutive activity of the transcendental subjectivity. Hence, the idea of transcendental subjectivity is presupposed in Husserl's case of treating of logic as well. However, it is very doubtful that it is the same activity of subjectivity which constitutes logical forms within Kant's and Husserl's understanding.

However, it is very doubtful that this is the same activity of subjectivity which constitutes logical forms in Kant and Husserl.

5. The idea of transcendental logic

Yet, the given reconstruction of Kant's view on logic remains insufficient. It was shown that Kant has created a notion of pure general logic as a science of pure forms of thought. Nonetheless, he laid down demands for creating of a very peculiar type of universal logic which, none the less, would have certain content. Namely, he introduced the notion of transcendental logic beyond his taxonomy of types of logic. It is clear that his notion of pure general logic, as logic in a current sense of formal science, belongs to this taxonomy. But a new type of logic also claims to deal with the universal and necessary knowledge.

In his later*Lectures on Logic*, Kant distinguished between these two types of logic in the following way,

> Now as propaedeutic to all use of the understanding in general, universal logic is distinct also on another side from *transcendental logic*, in which the object itself is represented as an object of the mere understanding; universal logic, on the contrary, deals with all objects in general. [10, p. 530]

Then, we have to ask again. How is it possible for logic to have universal validity and certain content at the same time? Does it not reduce the Kant's breakthrough with regard to the justification of the logic's purity and its universality to absurdity?

There could be an exact following answer to this question. Transcendental logic in Kant's term-use deals with very specific content, namely, with the relations of our cognitions with their objects. These relations are also of very peculiar kind. Namely, the transcendental logic should inquire into the origin of our cognitions of objects, insofar, it cannot be contained directly within these objects.

It deals with the transcendental origin of the cognitions *a priori*, therefore, the nature of transcendental logic is not analytic, but synthetic. Thereby, it differs principally from pure general logic which deals only with analytical forms. In opposition to it, transcendental logic is non-analytic and intends to explicate the grounds of a connection of logical forms with objects. Hence, transcendental logic cannot be explained through any formal analysis. It only could be justified through the transcendental synthesis. Since it has synthetic essence, transcendental logic deals with the transcendental genesis of both knowledge and objectivity.

Kant himself stated with concern to his idea of transcendental logic,

> In this case, there would exist a kind of logic, in which we should not make abstraction of all content of cognition; for that logic which should comprise merely the laws of pure thought (of an object), would of course exclude all those cognitions which were of empirical content. This kind of logic would also examine the origin of our cognitions of objects, so far as that origin cannot be ascribed to the objects themselves; while, on the contrary, general logic has nothing to do with the origin of our cognitions [...] (A55, B81) [9, p. 48]

It might appear at the first sight that Kant, himself, has destroyed his own idea of pure logic with the introduction of the notion of logic which would deal not only with pure forms, as such, but also with their origin *a fortiori*.

Yet, Kant's own opinion was the opposite one. The point is, that transcendental logic is not connected with the origin of cognition of all types of objects but only of the objects, which could be known *a priori*, exclusively. However, in Kant's opinion *a priori* possesses only a formal character. To be more exact, one also ought to limit the notion of a priori in the current context, for it could have a regard to the transcendental use of logic. This logic deals not with all *a priori* cognitions but with the cognitions of such kind, which allow us to know that some concepts are present *a priori* and can be applied only *a priori*. This type of logic clarifies how it can be possible at all. But it excludes from the consideration sensitive *a priori*, i.e. forms of sensibility as well as their relations to objects. Consideration of these forms belongs to transcendental aesthetics. Moreover, transcendental logic does not consider the very notion of the pure understanding, as such, but it treats relations of the mentioned to the objects only. As it was said, its subject is the origin and the limits of their applicability. To be more precise, from Kant's standpoint, transcendental logic deals with the possibility of relation of forms of thought (categories etc.) to objects as phenomena. Therefore, it takes a part in "substantive knowledge about objects".

Hence, we can speak about content of transcendental logic in some peculiar aspect. But if we try to analyze this content we will notice that such kind of content, in its turn, is, in some sense, the formal one. Namely, transcendental logic deals with synthetic formality of thought. For instance, Kant always thought that the categories of the pure understanding are pure forms of understanding. Their relations to objects, in the same measure, are the formal ones. In other words, transcendental logic presupposes abstraction of content too. But this abstraction is not total. One could describe abstraction of such kind as reduction of content to formal relations.

It is very significant that such forms and their formal relations to objects represent the proper content of this discipline. Hence, form, in a way, is content in case of transcendental logic. Thereby, transcendental logic is still pure logic. Since transcendental logical forms have a relation only to objects as *phenomena*, namely, to their realm as whole, insofar, it even is constituted by such forms. Transcendental logic, in a way, is universal. For further details, see [12].

Dealing with the genesis of *a priori* concepts, transcendental logic, in the way it was treated by Kant, has a productive moment within itself, but not absolutely. Namely, Kant searched after the subjective conditions *a priori* of a possible relation of our knowledge to objects. Hence, his understanding of transcendental logic remains subjective (but not empirical). Still, Kant's transcendental logic is not the logic of any, so to say, objective content. Kant's treatment of transcendental logic as of a discipline which has some content rooted in his doctrine of transcendental subjectivity and its structures *a priori*.

Hence, we can conclude that Kant's idea of transcendental logic meets the conditions of pure universal logic mentioned above. On one hand, although it does indeed relate to some specific kind of objects excludes all sensitive objects as well as the rules of the empirical thought from consideration, (i) it can obtain universal validity in a certain sense. Since it inquires only into *a priori* forms of thinking of objects, its content is also only form or, so to

The Historical Role of Kant's Views on Logic   151

say, it is empty of empirical content. On the other hand, since it treats only *a priori* structures of transcendental subjectivity, (ii) transcendental logic does not depend upon the empirical conditions of cognition and therefore upon an empirical accomplishing of our cognition. Therefore, it can have the universal validity in the domain of objects which are also constituted by pure notions of understanding. Yet, from Kant's standpoint, this domain is exclusively inside the realm of *phenomena*.

More to the point, Kant's idea of transcendental logic remains currently relevant. It has been existing in the phenomenological branch of today's philosophy, at least, since Husserl's *Formal and Transcendental Logic*.[5] Phenomenological treating of these two kinds of logic has indicated the following questions. Which correlation between the formal and transcendental kinds of logic is proper? Should formal logic be grounded by transcendental logic? On the contrary, should transcendental logic be understood as widening of formal logic which would underlie to it? Sadly, we have to shelve these questions here. But one ought to emphasize here that the relation of this new type of logic to reconstructed Kant's taxonomy of types of logic is initially ambiguous. Please, find the description of the possible modes of this relation, as well as various reconstructions of Kant's taxonomy of types of logic here: [13, pp. 1–10].

6. Exclusion of "speculative logic"

Now, we have to reject one noteworthy solution which, in our opinion, followed from Kant's dividing form of thought from content as its matter (as well as from division he made between pure general and transcendental logic). We think this solution intends to resolve the fundamental dualism of form and content in logic, which has arisen due to the mentioned divisions made by Kant. The solution we have in mind seems to be very effective and grandiose but, at the same time, very controversial.

Namely, we would like to make an exclusion of so-called "speculative logic", which has originated from Hegel's philosophy of absolute idealism and its materialistic reinterpretation made in the ideology of Marxism (so-called "dialectical logic") from our discussion. Nevertheless, it should be noted that this quasi-logical solution is not absolutely unusual in the history of logic. Moreover, it could shed the light on the problem of the possibility of formalization

---
[5]In this context, see [7].

of logic in some negative way. Namely, it can show possible limits of this formalization.

Thus, this kind of philosophical thinking makes a claim to elaborate the so-called logic of content. It seems to be initially incoherent, being undertaken after a sufficient separation of form from content in logic made by Kant. Yet, we would like just to emphasize the proper way which it was planned to be done in. Sadly, this way often remains undetected by logicians and historians of logic. For instance, Hegel referred not just to the possibility of explication of concrete content of concepts from their implicative condition but also to the possibility of producing and generating of such content through combining of pure logical forms. From his standpoint, the idea of logic, as of a formal discipline, roots in the abstract mode it is considered in. In fact, according to Hegel, logic itself can generate its content and provide a matter of thought to itself. He stated in the Introduction to his *Science of Logic*,

> More to the point is that the emptiness of the logical forms lies rather solely in the manner in which they are considered and dealt with. Scattered in fixed determinations and thus nor held together in organic unity, they are dead forms [. . .] Therefore they lack proper content [. . .] But logical reason is itself the substantial or real factor which, within itself, holds together all the abstract determinations and constitutes their proper, absolutely concrete, unity. [5, pp. 27–28]

Nevertheless, it is permissible to notice that the idea of "logic of content", based on Hegel's ontological premise of speculative identity of logic and ontology, should not be identified with Aristotelian understanding of form as of something rich in content. Therefore, "speculative logic" is situated outside of the main path of elaboration of logic as of a science of empty forms, whereas the Aristotelian shape of logic lies in the initial point of this path. Hegel and other "dialectical" logicians tried to get over a chasm between form and content in logic through logical (in the sense which they generally understood the logic in) tools. Hence, they aimed at unifying form and content through the quasi-logical combinations in the situation of historically already-actualized separation between form and content. Therefore, it could not be confused with the initial Aristotelian notion of form as something, rich in content within itself. So, these dialectical ideas don't belong to the mainstream of elaboration of logic starting from the Aristotelian treating of form to an idea of it as of a science of pure forms and form-combinations.

7. Summary and conclusion

In this way, Kant has discovered the possibility of logic which describes the correlation of the pure notions with objects. Thereby he has made room in his architectonic of logic for the kind of logic which would be general and pure, in a sense, yet it could not be independent of content. At the same time, its content doesn't have an empirical source. On this ground, this logic could be titled as properly "philosophical logic" which deals with the origin of our cognitions and their possible relation to objects unlike all the other types of logic which do not have proper philosophical sense. Since this logic considers conditions of our cognitions of objects, we would also call it epistemological logic. Since it discovers condition of relation to objects, we, as well, could define it as ontological logic.

Appositely, one ought to add that our earlier hypothesis implying that the possible conceptual origin of an idea of a regional ontology in phenomenology should lie not just in Kant's idea of logic of the particular use of understanding but also in his concept of transcendental logic. Indeed, logic of the particular use of understanding can, however, play an exclusively methodological role for a particular positive science but the regional ontology should ground one or another particular positive science on the basis of categories and their relation to the subject matter of this science.

Nevertheless, transcendental logic is quite different from pure logic, which has only a formal sense that does not contradict with formal characteristics which are present in both logics, philosophical and non-philosophical (pure formal logical) at the same time. However, the final questions arise. Is it possible to formalize this philosophical logic, which Kant's doctrine of kinds of judgment and categories belongs to? Does logic in interpretation given to it by Kant need any formalization or, at least, allow it? There is the following reason for such kind of questions. As it has been demonstrated already, Kant's transcendental logic is pure and formal (in the sense that its content is the pure formal relations) as well as independent of singular empirical conditions. Hence, we have to ask whether it is possible to formalize this type of logic which has long been logic of forms.

One ought to say that there have been some attempts made recently in order to rehabilitate Kant's transcendental logic with regard to today's semantics through the formalization with the tools of today's symbolic logic. They are of high interest and sophisticated. For instance, we could refer to the paper by

Achourioti and van Lambalgen [1] which was mentioned above. In particular, these authors speak about "typical modern dismissal of Kant's formal logic" [1, p. 254]. (They refers to Frege's and Strawson's works in this regard). Regarding the contemporary evaluation of his transcendental logic, they have stated, "Worse, Kant's transcendental logic does not seem to be a logic in the modern sense at all: no syntax, no semantics, inferences" [1, p. 254]. Achourioti and van Lambalgen think magnanimously that they will save Kant's transcendental logic by the demonstrating that "a logical system very much like Kant's formal logic is a distinguished fragment of first-order logic, namely, geometric logic" [1, p. 254]. And we do not think that it is a "hopeless enterprise." [1, p. 254].

Yet, from our standpoint, the following question arises in this regard. Is it still necessary to justify Kant's logics both pure general and transcendental from the point of view of symbolic logic or semantics? It is a problem (i) because, as was shown, the mentioned disciplines are possible in dimension which was cleared away only by the transformation of the understanding of logic with regard to the notion of a pure form made by Kant. Today's logic has just exchanged thought-forms to sign-forms but transcendental-philosophical conditions of the formality which were recognized by Kant remain the same. Moreover, (ii) Kant's logic (even transcendental one) does not need to be, as well as, it and cannot be formalized because, in a way, it has always possessed the formal status, as it is. Therefore, a question posed here should be not of how to justify Kant's views of logic from the perspective of symbolic logic and semantics but of existence of a possibility, as such, for justifying Kant's understanding of logic. It is a fact that both symbolic logic and semantics do not pose the question of their own *quid juris* unlike Kant did with regard to logic. Hence, it is still unclear where an epistemological source of the contemporary fetishism of "syntax, semantics and inferences" is. Maybe we could find it in Kant's philosophy itself. Therefore, it remains disputable whether it is possible or needed to formalize the Kantian conception of logic, even following the semantic character of contemporary logicism which was emphasized by MacFarlane.

References

[1] T. Achourioti and M. van Lambalgen. A Formalization of Kant's Transcendental Logic. *The Review of Symbolic Logic*, 4(2):254–289, 2011.

[2] Aristotle. *Aristotle's Metaphysics*. Clarendon Press, Oxford, 1924.
[3] Aristotle. *Metaphysics*. Penguin Books, London, 1998.
[4] G. Frege. *Foundations of Arithmetic*. Northwestern University Press, Evanston, 1980.
[5] G. W. F. Hegel. *The Science of Logic*. Cambridge University Press, Cambridge, 2010.
[6] M. Heidegger. *History of Concept of Time*. Indiana University Press, Bloomington, 1985.
[7] E. Husserl. *Formal and Transcendental Logic*. Martinus Nijhof, The Hague, 1969.
[8] E. Husserl. *Ideas Pertaining to a Pure Phenomenology and to a Phenomenological Philosophy: First Book: General Introduction to a Pure Phenomenology*. Kluwer Academic Publishing, Dordrecht, 1989.
[9] I. Kant. *Critique of Pure Reason*. Henry G. Bohn, London, 1855.
[10] I. Kant. *Lectures on Logic*. Cambridge University Press, Cambridge, 1992.
[11] J. MacFarlane. Frege, Kant and the Logic in Logicism. *The Philosophical Review*, 11(1):25–65, 2002.
[12] C. Tolley. The Generality of Kant's Transcendental Logic. *Journal of the History of Philosophy*, 50(3):417–446, 2012.
[13] J. H. Zinkstok. *Kant's Anatomy of Logic. Method and Logic in the Critical Philosophy*. Proefschrift, Groningen, 2013.

Received: 25 May 2016

The Logic for Metaphysical Conceptions of Vagueness

Francis Jeffry Pelletier

Department of Philosophy, University of Alberta
Edmonton, Alberta, Canada
Jeff.Pelletier@ualberta.ca

Abstract: Vagueness is a phenomenon whose manifestation occurs most clearly in linguistic contexts. And some scholars believe that the underlying cause of vagueness is to be traced to features of language. Such scholars typically look to formal techniques that are themselves embedded within language, such as supervaluation theory and semantic features of contexts of evaluation. However, when a theorist thinks that the ultimate cause of the linguistic vagueness is due to something other than language – for instance, due to a lack of knowledge or due to the world's being itself vague – then the formal techniques can no longer be restricted to those that look only at within-language phenomena. If, for example a theorist wonders whether the world itself might be vague, it is most natural to think of employing many-valued logics as the appropriate formal representation theory. I investigate whether the ontological presuppositions of metaphysical vagueness can accurately be represented by (finitely) many-valued logics, reaching a mixed bag of results.

Keywords: vagueness, many-valued logic, Evans-argument.

Introduction

Even though people sometimes point to vague memories (e.g., of that very first date you had) or vague objects (like the cloud above me as I write, or the mist that covered St. Petersburg a few nights ago), it is in language where vagueness most clearly manifests itself, and where most theorists focus their attention. The reasons for this are not hard to fathom:

- The majority of our linguistic terms admit borderline cases;

This published version has benefited greatly from the comments of a large number of anonymous referees.

- We are unable to resolve the application vs. non-application of many scalar predicates;

- We sometimes may not be able to determine what proposition (if any) is asserted when using certain vague terms.

But even if it is in the realm of language where we find vagueness manifested, there is still the question What is the "ultimate" cause of the vagueness? Is it perhaps a matter of lack of knowledge? Perhaps lack of knowledge of some relevant features of the world? Or perhaps lack of knowledge of the relevant context? Or is it instead that the precise language is correctly representing a vague reality? Or is it merely that language itself does not completely and precisely represent (the non-vague, precise) reality?

It is traditional to divide viewpoints concerning the ultimate cause of vagueness into three sorts: (a) Epistemological Vagueness, where vagueness is claimed to be due to a lack of knowledge – an inability to tell whether some statement is true or false, even though it might correctly represent reality or represent it incorrectly; (b) Linguistic Vagueness, where vagueness is claimed to be due to a shortcoming in the language itself – the language is not adequate to correctly or fully represent the detailed features of the world; and (c) Metaphysical (or Ontological) Vagueness, where vagueness is claimed to be inherent in reality – our language correctly represents reality, but these items are themselves vague. We will look briefly at each in turn, before we focus on the use (and motivation for the use) of many-valued logic.

Most accounts of vagueness, of all these different types, focus on *properties* that can manifest vagueness, particularly properties that characterize a "scale" – such as tallness, or being a heap, or intelligence, or .... Less time has been spent on the possibility of vague *objects*.[1] (Of course, some scholars think that one way to have a vague object is for it to manifest one of the vague properties in a vague manner.) In this paper we investigate the vague objects more closely than vague properties, although of necessity we talk also about vague properties.

---

[1]Compare the differences in focus and detail of the papers [1] and [20].

1. Epistemological vagueness

The natural way to understand this viewpoint on vagueness is that the "world" is precise, determinate, definite, and so on, but our apprehension of these precise facts is limited in one way or another by our finite epistemological powers. In the "world" there are objects that have precise boundaries, and all properties have sharp cut-off points (or maybe: all objects are in the clear extension or anti-extension of all the properties).

However, because we lack knowledge, vagueness is introduced: For example, old Prof. Worthington, the ancient don at Wembley College, only vaguely/indeterminately/fuzzily remembers who was present at his Doctoral Viva. He can *almost* remember someone with white or maybe blond hair. But he can't recall clearly whether it was the long-dead Coppleston or the equally dead Millingston.

That was a case of "individual vagueness" on Worthington's part. But there can be wider and wider cases of vagueness: *no one* quite remembers what the priest looked like at old Dr. Benoit's baptism. And maybe it is even more pervasive: a feature of the way the world has developed (all relevant people have died, and no one left any unambiguous memoirs) – for instance, did Galileo *actually* drop balls of different weights from on high? And did he also tether together different weighted balls in order to determine how fast the composite object fell? These are events that actually happened or didn't happen – totally and completely – in the actual world. But since there is now no evidence of any sort to decide which way the world actually went, we say it is vague whether Galileo dropped balls of different weights from a height. One might even go so far as to say that this is the category of "verifiable in principle but not actually verifiable".

And it could be more radical than this: For example, the Epistemological Vagueness position holds that in reality there is *in fact* a particular number of grains of sand that would make this pile of sand be a heap (say, m grains). However, we *can't know* that m grains of sand make a heap because all the evidence that we (or anyone) have available is the same for adding one grain of sand to an $(m-2)$-grains pile as it is for adding one grain of sand to an $(m-1)$-grains pile. (Since by hypothesis we can't discern a change when only one grain is added). Yet in the former case we don't know that a pile has become a heap (because by hypothesis it hasn't). So in the latter case we can't know either (even though it *has* become a heap).[2]

What would be an appropriate representational medium for this conception of vagueness? Well, since the view holds that

- *in the world* there is no indeterminacy...every factual sentence either is true or is false, every object is unique, distinct, and separate from all others, and

- vagueness comes from a lack of positive or negative knowledge of these facts, including lack of knowledge as to what proposition is being asserted,

it seems to follow that some sort of epistemic logic is called for. Thus the epistemic interpretation really involves two logics: classical, two-valued logic for "the world" and the just-mentioned epistemic logic to accommodate the state of knowledge of people. Vagueness seems then just to be identified with conceptual indeterminacy on the part of a speaker. Such an epistemic logic would employ a modal operator that means "is vague", but of course, in this conception, being vague is interpreted as being epistemically indeterminate, and so something can be *non*-vague by being definitely (in the epistemic sense) false, as well as by being definitely true (again, in the epistemic). Thus, if something is vague (epistemically), then so is its negation, under this conception. Using ∇ to represent this indeterminacy (and $\triangle$ for determinacy), typical postulates of such a logic include, among others

$$\text{if } \vDash \varphi, \text{ then } \triangle \varphi$$

$$\triangle \varphi \leftrightarrow \triangle \neg \varphi$$

$$\triangle \varphi \leftrightarrow \neg \nabla \varphi$$

So the required modal logic couldn't be a Kripke-normal modal logic. In [12], I proposed a class of logics of epistemic vagueness (or epistemic indeterminacy): every statement is in fact either true or false (at a world), but when inside the epistemic vagueness operator, we are to evaluate what is going on at a certain class of related worlds. But as I mentioned, this class is not determined in a classical Kripke-manner, but rather in terms of "neighbourhood semantics".

---

[2]The epistemic conception of vagueness is most famously championed by [21,22,29].

2. Linguistic vagueness

Linguistic Vagueness posits the same ontology as Epistemological Vagueness, namely that the "world" is precise, determinate, definite, and so on. But it differs from the epistemological version by saying that our *description* of these precise facts is limited in one way or another, rather than our knowledge of the precise facts. It holds that in the "world" there are objects that have precise boundaries, and all properties have sharp cut-off points. Vagueness in this conception is a matter of a kind of mismatch between language and "the world" and not a matter of a mismatch between people's knowledge and "the world", as it is in the epistemological conception. (Of course, different versions of Linguistic Vagueness will have differing accounts of what specific parts of language exhibit the mismatch.)

One version of this mismatch might hold, for example, that when Allen says that George is tall, the name 'George' picks out some specific individual in the world (namely, George) who has some specific height such as 180 cm. But it might hold that there is no such *primitive* property in the world as being tall, for only the specific heights count as primitive properties. In this view, either the property TALLNESS doesn't exist, or if it does, then at least it is not a "basic" property[3] but is instead defined, in one way or another, in terms of the more basic, specific properties and (perhaps) "contexts of use" (as in some of the "contextual theories of vagueness", [8, 16, 19]).

It is a shortcoming of our language, according to some (but not all) of the believers in Linguistic Vagueness, that it has developed with these sorts of predicate-terms. Some also hold it to be a shortcoming of our language that *the denotation relation* is not precise: the name 'Mt. Everest' does not unproblematically designate a specific region of the Earth; so when people use this linguistic term they are not accurately identifying what is the case "in the world". When a person says "This rock is a part of Mt. Everest", the imprecision of the denotation relation forces the sentence as a whole to be vague.[4]

---

[3] I use 'basic' and "primitive' in an intuitive manner, allowing that the relevant theories will be obliged to provide a detailed analysis of these notions.

[4] This view of vagueness – although without the feeling that it is a shortcoming – is expressed in [10]:

> The only intelligible account of vagueness locates it in our thought and language. The reason it's vague where the outback begins is not that there's

Advocates of the explication of vagueness in terms of a linguistic mismatch have formed the largest group of philosophers, at least starting with Frege. Some were dismayed by the fact that natural language had vague predicates, and saw the ideal language as remedying this:[5]

> We have to throw aside concept-words that do not have a *Bedeutung*. These are... such as have vague boundaries. It must be determinate for every object whether it falls under a concept or not; a concept word which does not satisfy this condition on its *Bedeutung* is *bedeutungslos*. [7, p. 178]

Some others who also thought that vagueness was linguistic believed instead that it was a *good* thing in natural language:

> ... a vague belief has a much better chance of being true than a precise one, because there are more possible facts that would verify it. ... Precision diminishes the likelihood of truth." [18, p. 91]

> Vagueness is a natural consequence of the basic mechanism of word learning. The penumbral objects of a vague term are the objects whose similarity to ones for which the verbal response has been rewarded is relatively slight. ... Good purposes are often served by not tampering with vagueness. Vagueness is not incompatible with precision. [15, pp. 113–115]

> There are contexts in which we are much better off using a term that is vague in a certain respect than using terms that lack this kind of vagueness. One such context is diplomacy. [2, pp. 85–86]

(For example, "We will take strenuous measures to block unwanted aggression whenever and wherever it occurs" allows for a wide course of actions, whereas any non-vague statement would not allow such freedom.)

What would be an appropriate representational medium for this conception of vagueness? Well, since the view holds that

- *in the world* there is no indeterminacy... every factual sentence that uses only the *basic* predicates and the correct denotation relation either is true or is false, and

> this thing, the outback with imprecise borders; rather, there are many things, with different borders, and nobody's been fool enough to try to enforce a choice of one of them as the official referent of the word 'outback'. (p. 212)

A similar view is expressed in [27].

[5]Actually, it is very difficult to find any theorist of vagueness – of whatever sort – who thinks that vagueness is a shortcoming in language as a whole. What is more problematic, they would say, is the use of some vague term or phrase in a context where more precision, accuracy, or definiteness is desired and is available for use but just not chosen.

- vagueness comes from the use of non-basic predicates (and "ambiguously denoting" singular terms) where there is no relevantly determined method of stating how they are related to the basic predicates,

it seems to follow that some semantic technique is needed for displaying the various types of results that might hold between the non-basic predicates used in some linguistic expression and the basic predicates that describe "the world".

For example, one might decide that one class of non-basic predicates actually are abbreviations of some (ordered) range of the basic predicates, and that it is "context" that determines which part of this ordered range is relevant to evaluating the truth value of the expression. (Supervaluations and maybe some other semantic techniques, as introduced by [23, 24], and developed by [3, 28], are plausible candidates for this sort of evaluation, as are some of the theories that employ context, like [8, 16, 19]).

Unlike the Epistemic conception of vagueness in which every (declarative) sentence either *is* true or *is* false (but in some cases we may not know which, so that vagueness is a type of epistemic shortfall), in the Linguistic conception only *some* sentences are true and only *some* are false. Among the ones that are true or are false are those composed with basic predicates (and no funny stuff with the denotation relation). Many of the sentences containing non-basic predicates will be given the value 'vague' (i.e., 'neither true nor false'). But not all of these latter type of sentence will be vague, as for example when the specific object of a predication *clearly* satisfies the predicate. For instance, when 200 cm. in height LeBron James is said to be tall, this is true despite the vagueness inherent in 'tall'.

And there can even be true (also false) sentences about the tallness of middle-height people... and similarly for other non-basic terms. For example, supervaluation theory allows that classical logical truths and contradictions are true/-false. And perhaps different semantic techniques, such as contextual theories, could generate other examples.

3. Ontological vagueness

Ontological/Metaphysical/Realistic Vagueness locates vagueness "in the world". So, as opposed to being unclear as to whether a situation actually obtains or not (Epistemic Vagueness), and as opposed to being vaguely described by a language that contains non-basic predicates (Linguistic Vagueness), Realistic Vagueness claims that certain objects in the world just plain are vague. (The intent here, which I will in general follow, is to target physical objects with this characterization, although it might also apply to abstracta, events, relations, and so on.) Few writers have explained it, but [18], who is an advocate of Linguistic Vagueness, assures us that it used to be a common view: "... it is a case of the fallacy of verbalism – the fallacy that consists in mistaking the properties of words for the properties of things."

One might also point to fictional entities as neither having nor lacking certain properties: Hamlet neither has nor lacks a 5 mm wart on his left shoulder. Even though this example is from the realm of fiction, Realistic Vagueness might claim that for an vague actual object, there is some property which it neither has nor lacks.

As is well-known, [5] claims that all views advocating ontological vagueness must invoke the claim that, for certain names a and b, the sentence $a = b$ is neither definitely true nor definitely false. (That is, Ontological Vagueness predicts that there are vague objects in the world, and when we have vague objects, then whether they are or are not the same object can also be indeterminate, at least according to some advocates. This gives at least a sufficient condition for metaphysical vagueness.) [25, 26] also proposed that the crucial test would be a situation in which the question 'In talking about x and y, how many things are we talking about?' has the features that 'none' is definitely a wrong answer; 'three', 'four', etc., are all definitely wrong answers; and neither 'one' nor 'two' is either definitely a right or definitely a wrong answer to it. I call this viewpoint about what is the underlying feature of Ontological Vagueness the Evans/van Inwagen criterion, or, when discussing specific argumentation that turns on exactly how this criterion is to be represented formally, the Evans assumption (or the van Inwagen assumption), and when considering the argumentation that makes use of the Evans assumption I call it Evans' argument and sometimes *an* Evans argument (to emphasize that, while it is not exactly Evans' argument, it is an argument inspired by Evans' argument).

What would be an appropriate representational medium for this conception of vagueness? Well, since the view holds that

- *in the world* there is indeterminacy... vague objects actually have the *real* property: being neither red nor not-red, for example, and

- for any object/property pair, either the object has the property (definitely), or the object lacks the property (definitely), or else it neither has nor lacks it – and this last fact is, in its own way, just as definite as the former two,

it seems that the appropriate representation of this conception will employ a many-valued logic. If a has property F, then Fa is true; if a lacks F, then Fa is false; so in the case where a neither has nor lacks F, Fa must take on some other truth value (counting 'neither True nor False' as a truth value).

Employing a modal logic would not accurately capture Realistic Vagueness, for a modal logic presumes that in each world, every sentence either *is* true or *is* false. Employing unusual semantic techniques also does not adequately capture Realistic Vagueness, for the Realist insists that *all* the properties under discussion *are* in fact "real" and "basic". Only a many-valued logic could capture the Realist's attitude toward vagueness.

And it is to many-valued logics that I now turn.

4. A 3-valued logic embodying vagueness

There are three values: intuitively, TRUE, FALSE, INDETERMINATE.[6] These are taken to describe *three different ways the actual world might relate to a sentence describing it*. That is, the portion of the actual world that is under discussion is *actually* one of: definitely the way being described, definitely not the way being described, or correctly described as indeterminate.

We would like our language to be able to express the facts that sentence φ is TRUE, FALSE or INDETERMINATE (calling these semantic values T, F, I). So let us invent sentence operators ("parametric operators") that do that: D_t, D_f, V. They are ordinary, extensional logic operators, having the following truth tables.

φ	$D_t\varphi$	$D_f\varphi$	$V\varphi$
T	T	F	F
F	F	T	F
I	F	F	T

We use standard 3-valued (Łukasiewicz) interpretations of negation, and, or. (And use the convention that the truth values are ordered: $T > I > F$).

---

[6]We turn later to logics with more than three values, when we discuss the possibility of describing "degrees of vagueness" as an account of "higher-order vagueness".

φ	$\neg\varphi$
T	F
I	I
F	T

$[\varphi \wedge \psi] = \min([\varphi],[\psi])$
$[\varphi \vee \psi] = \max([\varphi],[\psi])$

I am going to steer clear of the intricacies involved in the interpretation of the conditional and biconditional, other than to advocate on behalf of these principles:

$[D_t\text{-AXIOM}]: \quad \vDash D_t\varphi \to \varphi$

$[EQ\text{-RULES}]: \text{If } \vDash (\varphi \leftrightarrow \psi), \text{then infer } \vDash (D_t\varphi \leftrightarrow D_t\psi)$
$\qquad\qquad\qquad \text{If } \vDash (\varphi \leftrightarrow \psi), \text{then infer } \vDash (D_f\varphi \leftrightarrow D_f\psi)$
$\qquad\qquad\qquad \text{If } \vDash (\varphi \leftrightarrow \psi), \text{then infer } \vDash (V\varphi \leftrightarrow V\psi)$

Although neither $(\varphi \wedge \neg\varphi)$ nor $(\neg D_t\varphi \wedge \neg D_f\varphi)$ is a contradiction in a three-valued logic, contradictions *can* be described by insisting on the Uniqueness of Semantic Value in 3-valued logic.[7]

[USV$_3$]: Every sentence takes exactly one of the three values:
$(D_t\varphi \vee D_f\varphi \vee V\varphi) \wedge \neg(D_t\varphi \wedge D_f\varphi) \wedge \neg(D_t\varphi \wedge V\varphi) \wedge \neg(D_f\varphi \wedge V\varphi).$

Lemma. *If the main operator of formula* Φ *is* D_t, D_f, *or* V, *then* $[\![V\Phi]\!] = F$.

Proof. If the main connective of Φ is one of the three parametric operators, then (as can be seen from their truth tables) the value of Φ is either T or F. But then $V\Phi$ will be F. □

Corollary. *If all sentential parts of formula* Φ *are in the scope of any of* D_t, D_f, V, *then* $[\![V\Phi]\!] = F$, *or equivalently,* $[\![\neg V\Phi]\!] = T$ *or equivalently,* $[\![D_f V\Phi]\!] = T$.

With regards to using a three-valued interpretation in the predicate logic (with identity), I do not give a full characterization, but only three principles:

$[\text{V-}\forall]: \qquad \vDash V[(\forall x)F(x)] \to \neg(\exists x)D_f[F(x)]$

(i.e., if it is vague that everything is F, then there cannot be anything of which it is definitely FALSE that it is F)

$[\text{ref}_=]: \qquad \vDash D_t[\alpha = \alpha]$

---
[7] Although in Graham Priest's logic LP [14], the third value is claimed to be *both T and F simultaneously*, and "not really" a different third value.

(i.e., self-identities are definitely TRUE).

[LL]: $\vDash a = b \leftrightarrow (\forall F)(Fa \leftrightarrow Fb)$

(Leibniz's Law, as this is usually called: Two things are identical if and only if they share all properties.[8])

Now, while some scholars might find some of these principles questionable (I mention Graham Priest in footnote 9 below), the holders of Ontological Vagueness have pretty much uniformly taken them on.

5. An argument against vague objects in this logic

The argument

An Evans argument proceeds by assuming the Evans/van Inwagen criterion of what the believers in Ontological Vagueness hold: that it can be vague whether there is one or two objects before a person; and it continues, using the principles mentioned above about many-valued logic. The following version is given in [13].

a.	$\mathbf{V}[a = b]$	the Evans assumption
b.	$a = b \leftrightarrow (\forall F)(Fa \leftrightarrow Fb)$	**LL**
c.	$\mathbf{V}[a = b] \leftrightarrow \mathbf{V}(\forall F)(Fa \leftrightarrow Fb)$	(b) and [**EQ**-RULE]
d.	$\mathbf{V}(\forall F)(Fa \leftrightarrow Fb)$	(a), (c), $\leftrightarrow$-elim
e.	$\neg(\exists F)\mathbf{D}_f(Fa \leftrightarrow Fb)$	(d) and [**V**-$\forall$]
f.	$(\forall F)[\mathbf{D}_t(Fa \leftrightarrow Fb) \vee \mathbf{V}(Fa \leftrightarrow Fb)]$	(e) and [**USV**$_3$]
g.	$\mathbf{D}_t[\mathbf{D}_t[a = a] \leftrightarrow \mathbf{D}_t[a = b]] \vee$ $\mathbf{V}[\mathbf{D}_t[a = a] \leftrightarrow \mathbf{D}_t[a = b]]$	(f), instantiate ($\forall F$) to $\lambda x \mathbf{D}_t[a = x]$ and λ-convert
h.	$\mathbf{D}_t[\mathbf{D}_t[a = a] \leftrightarrow \mathbf{D}_t[a = b]]$	(g), [Lemma], disjunctive syllogism
i.	$\mathbf{D}_t[a = a] \leftrightarrow \mathbf{D}_t[a = b]$	(h) and [$\mathbf{D}_t$-AXIOM]
j.	$\mathbf{D}_t[a = b]$	(i) and [**ref**$_=$], $\leftrightarrow$-elim
k.	$\neg\mathbf{V}[a = b]$	(j) and [**USV**$_3$]

Although not every pair of formulas of the form φ and $\neg\varphi$ contradict one another in a three-valued logic, (a) and (k) do really contradict each other. (a) is either T, F, or I (by [**USV**$_3$]); by the truth-table for **V** it cannot be I; so it is either T or F. But this argument shows that if (a) is T then it is F, but if (a) is

---

[8]As both [6,11] remark, Leibniz himself only took pains to argue for the right-to-left aspect of [LL], and that with a restriction on the types of properties that F designate. Presumably everybody finds the left-to-right direction of [LL] undeniable. As is often noted, there is a peculiarity with this verbalization of the formula, since the formula gives a condition for there being just one thing under consideration, not two, and says that any property this one thing has is a property it has.

F then it is T (à la (k) and the truth table for ¬). But [**USV**$_3$] claims that no formula can be both T and F.[9]

Comments & defense of the argument

Was there any "cheating" going on in this proof, or with the postulates? Is there something "funny" about the **V**-operator? Might one question the λ-abstract: is it a "real" property? Might one have concerns about one of the principles used: [**D**$_t$-AXIOM], [**EQ**-RULES], [**USV**$_3$], [**LL**], [**V**-∀], [**ref**$_=$]?

The argument I presented proceeds by λ-abstraction, using the predicate: 'being definitely TRUE of x that it is identical to a'. Does that predicate correspond to a *real* property? If not, then this is not a legitimate case of λ-abstraction, by the standards of Ontological Vagueness.

For the advocate of Metaphysical Vagueness, the answer must be 'yes, it is a genuine property'. For, it is a feature of this position that *in the world* there is vagueness, and its contrary, definiteness. These are *real, actual* properties that are designated by these predicates. And unless the advocates of this position want there to be some sort of "ineffability" when it comes to their postulated properties-in-the-world, they have to admit that such expressions *do* designate such properties. The language is entirely extensional – there is no "funny business" going on about 'opaque contexts' or rigid vs. non-rigid names or . . . . The λ-abstraction picks out what the believer in Ontological Vagueness must acknowledge is a legitimate property.

But let's look again at this presumed notion of vagueness and definiteness. It cannot be the modal notion of the epistemicists, since that characterizes one's epistemic states rather than reality. *That* kind of (in)definiteness does *not* characterize an item "in the world" but rather cognizer's *apprehension* of objects. Any indefiniteness operator of this variety will endorse a principle like $\nabla \varphi \leftrightarrow \nabla \neg \varphi$, as I mentioned above, and such a principle does not characterize the usual ontological vagueness theorists' view.[10]

The view of Heck in [9] that ∇ should in fact obey this principle shows that his argumentation is not really directed against nor in favour of metaphysical

---

[9]This shows that if we were to interpret the middle value **V** as it is in Priest's [14] – as being both **T** and **F** – we would have to rephrase the interpretation of [**USV**$_3$]. And in fact, Priest (p.c.) says that he *denies* [**USV**$_3$], believing that there are but *two* truth values, but that some formulas can take both. Therefore, I propose this argument only against those who do not think that vagueness leads to true contradictions.

[10]Well, except perhaps for dialetheic views like that of Priest [14].

vagueness, but rather at or in favour of some hybrid view of metaphysical and epistemic vagueness.

I say again: only the many-valued logic viewpoint accurately captures the ontological vagueness theorist's view.

So far as I am aware, no one has faulted the following principles that are used in the proof (well, so long as they are willing to allow a 3-valued logic in the first place, and as long as they see the extra values as being truly distinct from TRUE and FALSE, contrary to Priest's viewpoint expressed in footnote 9):

- **USV$_3$**: $\vDash (D_t\varphi \vee D_f\varphi \vee V\varphi) \wedge \neg(D_t\varphi \wedge D_f\varphi) \wedge \neg(D_t\varphi \wedge V\varphi) \wedge \neg(D_f\varphi \wedge V\varphi)$;

- **D_t-AXIOM**: $\vDash D_t\varphi \rightarrow \varphi$;

- **ref$_=$**: $\vDash D_t[\alpha = \alpha]$;

- **V-$\forall$**: $\vDash V[(\forall x)\Phi(x)] \rightarrow \neg(\exists x)D_f[\Phi(x)]$.

In the case of those believers in ontological vagueness who hold there to be more "degrees of metaphysical vagueness" than just the three we have been assuming, a strictly analogous proof to the very same conclusion can be crafted, as discussed in [13]. One changes the [**USV**] axiom to accommodate the further truth values, and generalizes the [**V-$\forall$**] axiom for the extra truth values.

We will return to a discussion of the argument after a brief excursion into higher-order vagueness.

6. Higher-order vagueness

The topic of higher-order vagueness concerns the issue of whether it can be vague that something is vague (and even further iterations, such as being definite that it is vague that it is vague). For the believer in Ontological Vagueness, the first iteration amounts to wondering whether it can be vague that some aspect of reality is vague? It is not clear to me that a proponent of Vagueness-in-Reality will wish to accept this as a part of their doctrine concerning *Reality*. They might instead prefer to view it as a mixture of different types of vagueness: "We don't know whether it is true or false that such-and-so is metaphysically vague", and would thereby prefer some mixture of a many-valued logic with an epistemic logic of vagueness (like that of [12]) added on. Certainly, if they *do* wish to have metaphysical higher-order vagueness, they wouldn't represent it by iterating the V-operator (nor with iterated mixtures of any of the V-, D_t, and D_f-operators). The previously-mentioned Lemma precludes this.

Instead they would increase the number of truth values... and with them, the number of truth-operators in the language. For this purpose, it is common in discussions of many-valued logic to take the truth-values to be integers, with **1** being "most true" and (for an n-valued logic) to make **n** be the "most false" value. And then it is common to introduce the so-called J_i-operators [17]. Such an operator is a generalization of the ideas behind our D_t, D_f and V operators – like our operators, the **J**-operators have a formula as an argument, and are semantically valued as being "completely true" (that is, take the value **1**) if the formula-argument takes the value indicated in the subscript of the **J**-operator, and "completely false" (that is, take the value **n**) otherwise. Semantically this is to say, for any value i of an n-valued logic ($1 \leq i \leq n$)

$$[\![J_i(\varphi)]\!] = 1 \text{ if } [\![\varphi]\!] = i$$
$$= \mathbf{n} \text{ otherwise}$$

For example, a five-valued logic would have J_1, J_2, J_3, J_4, J_5 as **J**-operators), with truth tables:

φ	$J_1\varphi$	$J_2\varphi$	$J_3\varphi$	$J_4\varphi$	$J_5\varphi$
1	1	5	5	5	5
2	5	1	5	5	5
3	5	5	1	5	5
4	5	5	5	1	5
5	5	5	5	5	1

and this account might say that $J_3\varphi$ makes the claim that φ is completely vague, while $J_2\varphi$ asserts that it is vague whether φ is completely vague or is true; and that $J_4\varphi$ claims that it is vague whether φ is completely vague or false.

With suitable additions of the number of truth-values, this seems as plausible a way to represent higher-order metaphysical vagueness as it is in modal logic to represent higher order epistemological vagueness by the iteration of a modal operator $\triangledown\triangledown\Phi$. However: a version of the Argument can be made using *any* (finitely-) many valued logic (with suitable emendations to the various principles). I don't rehearse the proof of that fact here; details can be found in [13].

I think the ability to represent higher-order vagueness (of any finite number of iterations) shows that it is not that many-valued logics are incapable of giving some sense to higher-order vagueness, but that the argument in [13] demonstrates that there is some other, perhaps deeper, incoherency with Metaphysical Vagueness.

Although the ploy of increasing the number of truth values shows that many-valued logics are in fact capable of giving a plausible account of higher-order vagueness for any finite number of iterations, there remains still the issue just mentioned: no matter how (finitely) many truth-values our ontological vagueness proponent wishes to invoke as a way of handling higher-order vagueness for a finite number of truth values, there is a generalization of the Argument that can be turned against it. A question naturally arises then concerning the interaction of higher-order vagueness with the number of truth values. We've just seen that to have one iteration of higher-order vague, we would increase the number of truth values from three to five. If we had another iteration, and wanted all possible combinations to be represented, we would need many more. And if we thought it possible to have any level of iterated higher-order vagueness, then the conclusion would be that we need an infinite-valued logic to accommodate this. Infinite-valued logics come with their own share of unusual characteristics, such as that a quantified formula can be assigned true ($[\![J_i(\exists x Fx)]\!]$=1) even without there being any object a in the domain such that $[\![J_i(Fa)]\!]$=1. I think that most vagueness-in-reality theorists either hold that higher-order vagueness of any sort is impossible (as various authors have claimed, even independently of whether they believed in metaphysical vagueness), or else that it is bounded by some finite number of iterations. (It is not clear how this latter possibility might be argued for by our ontological metaphysicians. Most arguments to this conclusion come from the point of view of it being *cognitively impossible* to have infinite iterations... and that's not very relevant to the ontological conception of vagueness.) Anyway, I'm not going to consider it further.

7. Returning to the argument

In [4], Cowles and White object to the statement of Leibniz's Law in the form given in [LL] above, namely
$\vDash a = b \leftrightarrow (\forall F)(Fa \leftrightarrow Fb)$,
and prefer to see it as (what they call "Classical LL"):
$\boldsymbol{D}_t[a = b] \leftrightarrow \boldsymbol{D}_t[\forall F(Fa \leftrightarrow Fb)]$.

They also deny the full force of the [**EQ-RULES**]: They claim that just having
$$\vDash \varphi \leftrightarrow \psi$$
does not justify
$$\vDash \boldsymbol{V}[\varphi] \leftrightarrow \boldsymbol{V}[\psi],$$
nor
$$\vDash \boldsymbol{D}_f[\varphi] \leftrightarrow \boldsymbol{D}_f[\psi].$$
As they show, their position has the effect of denying both:

- $\boldsymbol{V}[a=b] \leftrightarrow \boldsymbol{V}[\forall F(Fa \leftrightarrow Fb)]$
- $\boldsymbol{D}_f[\forall F(Fa \leftrightarrow Fb)] \to \boldsymbol{D}_f[a=b]$

(although it does validate $\boldsymbol{D}_f[a=b] \to \boldsymbol{D}_f[\forall F(Fa \leftrightarrow Fb)]$).

Just how plausible are these denials? Not very, it seems to me. Is it really plausible to claim that when we have a *logical* truth that two formulas are equivalent, we cannot conclude that one of them is vague just in case the other one is? Nor that one of them is definitely false just in case the other one is? How plausible is it to claim that even when it is *definitely* false that two objects share all properties, it might yet not be definitely false that these are the same object?

On the other hand, I should admit that because of the plausibility of the [**EQ-RULES**], as well as the other rules, I had originally – when I wrote [13] – thought that the Argument showed the complete *im*plausibility of the conception of Metaphysical Vagueness. However, I hadn't internalized these facts (or maybe I hadn't even noticed them):

1. Although the proof given was framed as showing that a contradiction followed from the assumption of $\boldsymbol{V}[a=b]$, *it equally is a proof of*
   $$\boldsymbol{V}[a=b] \to \neg \boldsymbol{V}[(\forall F(Fa \leftrightarrow Fb)]$$
   i.e., even if it is vague that $a=b$, it can't be vague that they share all the same properties.

2. And of course: $\boldsymbol{D}_t[a=b] \to \neg \boldsymbol{V}[(\forall F(Fa \leftrightarrow Fb)]$
   i.e., if it is definitely TRUE that $a=b$, then it isn't vague that they share all the same properties: intuitively, it is definitely TRUE that they *do* share all the same properties.

3. Furthermore, clearly: $D_f[a = b] \to \neg V[\forall F(Fa \leftrightarrow Fb)]$

   i.e., given that it is definitely FALSE that $a = b$, it can't be vague that they share all the same properties: intuitively, it has to be definitely TRUE that they differ on at least one property.

But [USV] asserts that one of the three cases must hold, so we can conclude

$\vDash \neg V[\forall F(Fa \leftrightarrow Fb)]$.

That is, it is *never* the case that it is vague that two(?) objects have all properties in common. (Or, that it is *never* vague that an object has all the properties it has).

In conclusion

I used to think that the original argumentation showed:

- The conception of Metaphysical Vagueness is committed to representing its doctrines with a many-valued logic.

- The conception was committed to various logical principles (listed above), as a consequence of its metaphysics.

- Part of Metaphysical Vagueness was a commitment to the Evans/van Inwagen criterion.

- The Argument showed that any many-valued logic which embodied those principles led to a contradiction.

- And I concluded that Metaphysical Vagueness – Vagueness in Reality – was an incoherent notion.

But given that the Argument proves $\vDash \neg V[\forall F(Fa \leftrightarrow Fb)]$, (which by one of the [EQ-RULES] shows $\neg V[a = b]$), perhaps we should instead follow a different route:

- Continue to hold to the requirement of a many-valued logic with the specified logical principles to describe the view, *BUT*

- Deny the background assumption given to us by Evans [5] and van Inwagen [25, 26] that Metaphysical Vagueness is committed to instances of $V[a = b]$.

- And similarly deny the [25] version of Evans' assumption to the effect that one cannot count vague objects – because it is *never* TRUE or FALSE that such a thing is one object and it is *never* TRUE or FALSE that it is two objects.

So, by this line of thought the Argument does *not* show Metaphysical Vagueness to be incoherent, it shows instead that *the Evans/van Inwagen criterion of metaphysical vagueness is incorrect*. So believers in Vagueness-in-Reality should turn their attention to finding a different way of stating their basic metaphysical position, and not allow their opponents to define the field for them. Since I am not an advocate of metaphysical vagueness, I cannot therefore offer anything for them.

However, until they do that, Metaphysical Vagueness remains a "deeply dark and dank conception" that one should avoid.

References

[1] A. Abasnezhad and D. Husseini. Vagueness in the World. In K. Akiba and A. Abasnezhad, editors, *Vague Objects and Vague Identity*, pages 239–256. Springer, Dordrecht, 2014.

[2] W. Alston. *Introduction to Philosophy of Language*. Prentice-Hall, Englewood Cliffs, NJ, 1964.

[3] N. Asher, D. Dever, and C. Pappas. Supervaluations Debugged. *Mind*, 118:901–933, 2009.

[4] D. Cowles and M. White. Vague Objects for Those Who Want Them. *Philosophical Studies*, 63:203–216, 1991.

[5] G. Evans. Can There Be Vague Objects? *Analysis*, 38:208, 1978.

[6] P. Forrest. The Identity of Indiscernibles. In E. N. Zalta, editor, *The Stanford Encyclopedia of Philosophy*. Retrieved January 18, 2016, from http://plato.stanford.edu/entries/identity-indiscernible/, Winter 2012 edition, 2012.

[7] G. Frege. Comments on *Sinn* and *Bedeutung* (1892). In M. Beaney, editor, *The Frege Reader*, Wiley Blackwell Readers, pages 172–180. Wiley-Blackwell, Oxford, UK, 1997.

[8] D. Graff Fara. Shifting Sands: An Interest Relative Theory of Vagueness. *Philosophical Topics*, 28: 326–335. 2000.

[9] R. Heck. That there Might be Vague Objects (So Far as Concerns Logic). *The Monist*, 81:277–299, 1998.

[10] D. Lewis. *On the Plurality of Worlds*. Blackwell, Oxford, UK, 1986.

[11] B. C. Look. Gottfried Wilhelm Leibniz. In E. N. Zalta, editor, *The Stanford Encyclopedia of Philosophy*. Retrieved January 18, 2016, from http://plato.stanford.edu/entries/leibniz/, Spring 2014 edition, 2014.

[12] F. J. Pelletier. The Not-So-Strange Modal Logic of Indeterminacy. *Logique et Analyse*, 27:415–422, 1984.

[13] F. J. Pelletier. Another Argument Against Vague Objects. *Journal of Philosophy*, 86:481–492, 1989.

[14] G. Priest. *In Contradiction: A Study of the Transconsistent*. Martinus Nijhoff, Amsterdam, 1987.

[15] W. V. O. Quine. *Word and Object*. MIT Press, Cambridge, MA, 1960.

[16] D. Raffman. Vagueness and Context-Relativity. *Philosophical Studies*, 81: 175–192. 1996.

[17] J. B. Rosser and A. Turquette. *Many-Valued Logics*. North-Holland Press, Amsterdam, 1952.

[18] B. Russell. Vagueness. *Australasian Journal of Psychology and Philosophy*, 1:84–92, 1923.

[19] S. Shapiro. *Vagueness in Context*. Oxford University Press, Oxford, UK. 2006.

[20] S. Schiffer. Vague Properties. In R. Diaz and S. Monuzzi, editors. *Cuts and Clouds: Vagueness, its Nature and its Logic*, pages 109–130. Oxford UP, Oxford. 2010.

[21] R. Sorensen. *Blindspots*. Clarendon Press, Oxford, 1988.

[22] R. Sorensen. *Vagueness and Contradiction*. Oxford University Press, Oxford, 2001.

[23] B. van Fraassen. Singular Terms, Truth-Value Gaps, and Free Logic. *Journal of Philosophy*, 63:481–495, 1966.

[24] B. van Fraassen. Presuppositions, Supervaluations, and Free Logic. In Karel Lambert, editor, *The Logical Way of Doing Things*, pages 67–91. Yale University Press, New Haven, 1969.

[25] P. van Inwagen. How to Reason About Vague Objects, 1986. Paper read to the American Philosophical Association, Central Division, St. Louis.

[26] P. van Inwagen. *Material Beings*. Cornell Univ. Press, Ithaca, NY, 1990.

[27] A. Varzi. Vagueness in Geography. *Philosophy and Geography*, 4:49–65. 2001.

[28] A. Varzi. Supervaluationism and its Logics. *Mind*, 116:663–676, 2007.
[29] T. Williamson. *Vagueness*. Routledge, Oxford, 1994.

Received: 29 July 2016

A Note on the Axiom of Countability

Graham Priest

Departments of Philosophy, the University of Melbourne
and
the Graduate Center, City University of New York

`priest.graham@gmail.com`

> **Abstract:** The note discusses some considerations which speak to the plausibility of the axiom that all sets are countable. It then shows that there are contradictory but nontrivial theories of ZF set theory plus this axiom.

In this note, I will make a few comments on a principle concerning sets which I will call the Axiom of Countability. Like the Axiom of Choice, this comes in a weaker and a stronger form (local and global). The weaker form is a principle which says that every set is countable:

WAC $\forall z \exists f (f$ is a function with domain $\omega \wedge \forall x \in z \exists n \in \omega\, f(n) = x)$.

(The variables range over pure sets—including natural numbers. ω is the set of all natural numbers.) The stronger form is that the totality of all sets is countable:

SAC $\exists f (f$ is a function with domain $\omega \wedge \forall x \exists n \in \omega\, f(n) = x)$.

The stronger form implies the weaker. Any set, a, is a sub-totality of the totality of all sets. Hence, if the latter is countable, so is a. So I focus mainly on this.

Let us start by thinking about the so called Skolem Paradox. Take an axiomatization of set theory, say first-order classical ZF. This proves that some sets, and *a fortiori* the totality of all sets, are uncountable. Standard model theory assures us that there are models of this theory (in which '$\in$' really is the membership relation) where the domain of the model is countable. There *is* a function which enumerates the members of the domain. It is just one which has failed to get into the domain of the interpretation. Why should we not suppose, then, that the universe of sets really is countable? From the perspective of the metatheory, ZF^+ (ZF + 'There is a model of ZF'), the countable model is not the intended "interpretation". Our metatheory tells us that the domain of all sets is actually uncountable. But ZF^+ itself has a countable model, so the situation is exactly the same with this. We might suppose that the countable model of this tells us how things actually are. True, in the metatheory we are now working in, ZF^{++} (ZF^+ + 'There is a model of ZF^+'), that model will appear not to be the intended model. But we can reply in the exactly same way. Clearly, the situation repeats indefinitely. And at no stage are we *forced* to conclude that the universe of sets is really uncountable. We will always have a countable model at our disposal.

Indeed, it is not just the case that there is nothing that will force us to conclude that the universe of sets is really uncountable. There are certain conceptions of sethood which actually push us to that conclusion. Thus, suppose that one takes the not implausible view that sets are simply the extensions of predicates (or some predicates anyway).[1] Then, given that the language is countable, so it the universe of sets.

Now, imagine that the history of set theory had been slightly different. Suppose that set theory had been investigated for a few years before Cantor, and that those who investigated it took sets to be simply the extensions of predicates. Suppose also that the theory had actually been formalised, say by some mathematician, Zedeff. The (strong) Axiom of Countability, being an *a priori* truth about sets, was one of the axioms. Things were bubbling along nicely, until Cantor came along and showed that within the theory one could prove that some sets are uncountable. The theory was inconsistent. In this history, Cantor was playing Russell to Zedeff's Frege. We can imagine that the community was dismayed by this paradox, and started to try to amend the axiomatization in such a way as to avoid paradox. Perhaps, indeed, the hierarchy ZF, ZF^+, ZF^{++}, ... emerged—rather as the hierarchy of Tarski metalanguages emerged in our actual history.

In actual history, set theory was consistentized in response to Russell's paradox and related ones. However, as we now know, there is an alternative: maintain the naive comprehension schema—that is, the schema $\exists x \forall y (y \in x \leftrightarrow \psi)$, where ψ does not contain y—allow the paradoxes, and deploy a paraconsistent logic, which quarantines the paradoxes. The same was an option in our hypothetical history; maintain the Axiom of Countability, the paradoxes it generates, and deploy a paraconsistent logic.

---

[1] See [2, ch. 10]. See also [1].

A Note on the Axiom of Countability

Now back to reality. Is there such a theory? There is. Using the paraconsistent logic LP, we can show the existence of such a theory by applying a result called the *Collapsing Lemma*. Take a first-order language (without function symbols) for LP.[2] Let $M = \langle D, \delta \rangle$ be any interpretation for this. Let $\sim$ be any equivalence relation on D.[3] If $d \in D$, let $[d]$ be its equivalence class under $\sim$. We define a new interpretation (the collapsed interpretation), $M^\sim = \langle D^\sim, \delta^\sim \rangle$, as follows. $D^\sim = \{[d] : d \in D\}$. For any constant, c, $\delta^\sim(c) = [\delta(c)]$. For any n-place predicate, P, $\langle a_1, ..., a_n \rangle$ is in the extension of P in $M^\sim$ iff there are $d_1 \in a_1, ..., d_n \in a_n$, such that $\langle d_1, ..., d_n \rangle$ is in the extension of P in M. Similarly for the anti-extension of P. The collapse, in effect, simply identifies all the members of an equivalence class, producing an object with the properties of each of its members. The Collapsing Lemma tells us that any sentence in the language of M (i.e., the language augmented with a name for each member of D) which is true in M is true in $M^\sim$; and any sentence false in M is false in $M^\sim$.[4]

To apply this: let the language be the language of first-order ZF (without set abstracts). Take a (classical) interpretation of this, M, which is a model of ZF. Let k be any countable set in D. (Here, and in what follows, I mean countable—or uncountable—in the sense of M.) Consider the equivalence relation on D which identifies all uncountable sets with k, and otherwise leaves everything alone. That is, $x \sim y$ iff in M:

- x and y are uncountable
- or (x is uncountable and y is k)
- or (y is uncountable and x is k)
- or (x and y are both k)
- or (x and y are countable sets distinct from k, and $x = y$).

Now consider the collapsed model obtained with $\sim$. By the Collapsing Lemma, this is a model of ZF. But in $M^\sim$ every set is countable. For every constant, c, that denotes a countable set in M:

---

[2] For a presentation of the semantics of LP, see [2, sec. 16.3].
[3] If the language were to contain function symbols, $\sim$ would also have to be a congruence on their interpretations.
[4] For full details, including the proof, see [2, sec. 16.8].

- $\exists f(f \text{ is a function with domain } \omega \wedge \forall x \in c \exists n \in \omega \, f(n) = x)$

is true in M, and so by the Collapsing Lemma, in $M^\sim$. Since every member of $D^\sim$ has such a name in $M^\sim$, we have the WAC in $M^\sim$:

- $\forall z \exists f(f \text{ is a function with domain } \omega \wedge \forall x \in z \exists n \in \omega \, f(n) = x)$.

A slightly different equivalence relation delivers an interpretation which verifies SAC. Let k now be the object which is V_ω (the sets of rank ω) in M. Consider the equivalence relation which identifies all things of rank greater than ω with V_ω, and leaves everything else alone. That is, $x \sim y$ iff in M:

- $x, y \in V_\omega$ and $x = y$
- or $x, y \notin V_\omega$.

Again, this is a model of ZF. $k \cup \{k\}$ is countable in M. Let i be the name of the function that enumerates it, and let e be the name of any member of $k \cup \{k\}$. Then in M it is true that:

- i is a function with domain $\omega \wedge \exists n \in \omega \, i(n) = e$.

Hence this is true in $M^\sim$. But since every member of $D^\sim$ is named by some e of this kind, we have in $M^\sim$:

- i is a function with domain $\omega \wedge \forall x \exists n \in \omega \, i(n) = x$.

Hence we have the SAC in $M^\sim$:

- $\exists f(f \text{ is a function with domain } \omega \wedge \forall x \exists n \in \omega \, f(n) = x)$.

For good measure, $M^\sim$ is also a model of the naive comprehension schema, $\exists z \forall x(x \in z \equiv A)$, too.[5] If sets just are the extensions of predicates, one would expect this schema to hold. I note also that both of the models we have constructed are non-trivial. Thus, if c and d refer to two distinct objects in D that are not involved in the collapse, $c = d$ is not true in the collapsed model.

What we see, then, is that there are (non-trivial) theories that contain the (strong or weak) Axiom of Countability, plus ZF (plus, in one case, the naive comprehension schema). If T is the set of things true in either of the collapsed models we have constructed, T is one such theory. Within such a theory, every

---
[5]For details see [2, sec. 18.4].

set is countable; but, because of Cantor's Theorem, some sets are uncountable as well. It is Cantor's Theorem that generates the hierarchy of different sizes of infinity. And as seen from the perspective of one of these theories, the Theorem is recognizably paradoxical. The whole hierarchy of infinities is therefore a consequence of the paradox. The transfinite, then, is generated by the transconsistent.[6]

In a nutshell: the Axiom of Countability makes perfectly good paraconsistent sense, even within the context of ZF. And it provides a radically new possible perspective on the universe of sets.

Acknowledgements

The author is grateful an anonymous referee for some helpful comments, and to Hamdi Mlika for kind permission to republish this article, which first appeared as *Al-Mukhatabat*, 2012, 1:27–31.

References

[1] J. Myhill. The hypothesis that all Classes are Nameable. *Proceedings of the National Academy of Sciences of the United States of America*, 38(11):979–981, 1952.

[2] G. Priest. *In Contradiction: A Study of the Transconsistent*. Oxford University Press, 2nd edition, 2006.

[3] Z. Weber. Transfinite Cardinals in Paraconsistent Set Theory. *Review of Symbolic Logic*, 5(2):269–293, 2012.

Received: 7 April 2016

---

[6]It was Zach Weber who first suggested to me that one might see the transfinite in this way. See [3, sec. 8].

Topologies and Sheaves Appeared as Syntax and Semantics of Natural Language

Oleg Prosorov

St. Petersburg Department of V. A. Steklov Institute of Mathematics RAS
27, nab. r. Fontanki, St. Petersburg 191023, Russia

prosorov@pdmi.ras.ru

Abstract: In a sheaf-theoretic framework, we describe the process of interpretation of a text written in some unspecified natural language, say in English. We consider only texts written for human understanding, those we call *admissible*. A *meaning* of a part of a text is accepted as the communicative content grasped in a reading process following the reader's interpretive initiative formalized by the term *sense*. For the meaningfulness correlative with an idealized reader's linguistic competence, the set of all meaningful parts of an admissible text is stable under arbitrary unions and finite intersections, and hence it defines a topology that we call *phonocentric*. We interpret syntactic notions in terms of topology and order; it is a kind of *topological formal syntax*. The connectedness and the T_0-separability of such a phonocentric topology are *linguistic universals*. According to a particular sense of reading, we assign to each meaningful fragment of a given text the set of all its meanings those may be grasped in all possible readings in this sense. This way, to any sense of reading, we assign a sheaf of fragmentary meanings. All such sheaves constitute a category, in terms of which we develop a *sheaf-theoretic formal semantics*. It allows us to generalize Frege's compositionality and contextuality principles related with the *Frege duality* between the category of all sheaves of fragmentary meanings and the category of all bundles of contextual meanings. The acceptance of one of these principles implies the acceptance of the other. This Frege duality gives rise to a representation of fragmentary meanings by continuous functions. Finally, we develop a kind of *dynamic semantics* that describes how the interpretation proceeds as a stepwise extension of a meaning representation function from the initial meaningful fragment to the whole interpreted text.

Keywords: sense, meaning, phonocentric topology, linguistic universals, sheaf of fragmentary meanings, compositionality principle, contextuality principle, bundle of contextual meanings, Frege duality, dynamic semantics.

1. Introduction and informal outline

In this work, we apply rigorous mathematical methods in studying the process in which the understanding of a written text or an uttered discourse is reached. Our aim is to present a formal model for the understanding of a text or a discourse in a natural language communication process.

Any natural language serves as a means of communication between members of a community that shares this language. The life of a human society, primitive or developed, ancient or contemporary would be impossible without linguistic communication. When we communicate with each other, we are involved in the activity of exchange with two complementary sides, that is, the production and the understanding of language messages in oral or in written form. Any linguistic communication presupposes the emitting activity that produces a message and the receiving activity that produces an understanding. The message is an externalization of thoughts either by utterance or by writing. As a linguistic message unit, a single stand-alone sentence (or phrase) does not suffice to express the variety of thoughts and ideas that people need to communicate. The minimal exchange units that serve as messages in linguistic communications are written texts and uttered discourses. Linguistics is a discipline that studies the use of a language; for empirical objects, it has, therefore, texts and discourses as the units of human interaction, and not stand-alone words or phrases favoured by traditional grammars and the logic in the wake of Aristotelian tradition primarily concerned with questions of reference and truth.

The main parts of traditional grammars are syntax and semantics. A traditional syntax is a study of sentence structures in a given language, specifically in terms of word order. A semantics, of whatever kind, is the study of relationships between the linguistic expressions and their meanings. Traditional approaches are very restrictive or even inadequate to extend grammatical concepts and theories to the level of text or discourse in order to describe linguistic communication in all its forms.

The present work proposes a mathematical framework that generalizes syntax and semantics of a natural language from the traditional level of a stand-alone sentence or phrase to the level of written or spoken discourse. We propose a kind of a discourse analysis that describes the process of a natural language message interpretation in a uniform manner at all semantic levels.

The paper is organized as follows:

- In the next Sect. 2, we discuss in details our acceptance of basic semantic notions *meaning*, *sense*, and *reference*. We study the interpretation of a text in a certain unspecified natural language, say in English, considered as a means of linguistic communication (mostly in written form). We consider the class of minimal communicative units of a language as made up of texts, and thus it is broader than the class of all stand-alone sentences studied in traditional logical and grammatical theories.

From the set-theoretic point of view, any text is a sequence of its constituent sentences.[1] But from the theoretic point of view on linguistic communication, do we need to define somehow what is a genuine text? It seems useless to set some formal criteria of *textuality* those, likewise to formal criteria of *grammaticality*, would decide that a given sequence of sentences is a well-formed text. Although some particular sequence of words or sentences does not appear to be well-formed, nobody can guarantee the contrary for the future, because a natural language is always open for changes. However, the ethics of linguistic communication presupposes that a genuine text is written by its author(s) as a message intended to be understood by a reader. That is why, instead of adopting any criterion of textuality, we restrict the domain of our study to texts that we assume to be written 'with good grace' as messages intended for human understanding; those we call *admissible*. All sequences of words written in order to imitate some human writings are cast aside as irrelevant to the linguistic communication.

A *meaning* of a part of text is accepted as the communicative content grasped in a particular reading of this part following the idealized reader's attitude, presupposition and intention put together in the term *sense*. We adopt this acceptance of terms 'sense' and 'meaning' because it is close to the ordinary usage of these words in everyday English. The advantage of such a choice of terminology is that we can use words 'sense' and 'meaning' sometimes as linguistic terms, sometimes as ordinary words without specifying each time their mode of use. Otherwise, we were to accept in the use their definitions that we reject in the theory. Thus, we may ask, e.g., "What does this word

---

[1] It is clear that any such a sequence is made up of so-called 'sentence-tokens', not of so-called 'sentence-types'. Likewise, a sentence is a sequence of word-tokens, and a word is a sequence of morpheme-tokens. Nevertheless, in speaking further about a sequence of certain language units, we shall sometimes omit the word 'token', in order to not overload the terminology.

(or expression, sentence, text) mean in the literal (or metaphorical, allegorical, moral, Platonic, Fregean, narrow, wide, common, etc.) sense?" So, our acceptance of terms *sense* and *meaning* differs from *Sinn* and *Bedeutung* of Frege's famous paper of 1892. We discuss the difference further.

• In Sect. 3, we discuss topology and order structures underlying an admissible text considered as a means of communication. The linguistic communication may be adequately modelled by a formalism that takes as its object of study texts and discourses in their production and interpretation.

Whatever the human language is, the speaker produces an utterance when putting words one after another in an acoustic string. The listener is forced to interpret such a chain of sounds without the possibility of suspending its course with the purpose to return or to make a leap forward. Everyone knows this property empirically, owing to personal experience of speaker and listener; it should undoubtedly be taken into account by everyone who writes a text intended for a human understanding. We argue that such a fundamental feature of linguistic behaviour enables us to endow an admissible text X with the structure of a finite T_0 topological space where the set of opens $\mathfrak{O}(X)$ is the set of all meaningful parts of a given text X. We call *phonocentric* such a topology defined on the text X.

It is well known that the category **FinTOP**$_0$ of finite T_0 topological spaces with continuous maps is isomorphic to the category **FinORD** of finite partial ordered sets (*posets*, for short) with order preserving maps.[2] We consider two functors L and Q establishing such an isomorphism between these categories. It allows us to define on an admissible text topological and order structures, both of *deep* and *surface* kinds. The writing process consists in endowing the text with the surface structure of so-called linear 'word order' (and corresponding topology). The process of interpretation consists in a backward recovering of the deep structure of the specialization order (and corresponding phonocentric topology) on the text.

Thereafter, we define a phonocentric topology in a similar manner at each semantic level of an admissible text. The mathematical interpretation of different linguistic notions in terms of topology and order is a kind of *topological formal syntax*.

• In Sect. 4, we elaborate in mathematical details the aforesaid topological formal syntax. We argue that the T_0-*separability* and the *connectedness* of a

---

[2]See, for instance, [8, 23].

phonocentric topology are two *linguistic universals* of a topological nature.

- In Sect. 5, we study the process of understanding of an admissible text considered as a means of communication. To understand a text or a compound expression is to grasp what it means, i.e., what communicative content it conveys. Thus, the understanding of a text during its reading is a dynamic process that develops gradually as the reading progresses over the time.

On the other hand, a speaker (a writer) uses words as a preexisting means to express thoughts, and one combines them to convey thoughts one wants to communicate. So the meaning of a compound expression is determined by the meanings of its (meaningful) constituents, as well as the meaning of the whole text is determined by the meanings of its (meaningful) parts.

In the traditional hermeneutics, the relationship between the understanding of (meaningful) parts and the understanding of the whole text was conceived as a fundamental principle of text interpretation called the *hermeneutic circle*. As its counterpart in linguistic theories, there is a need for some principles those describe how the passage from the meanings of parts to the meaning of the whole and the passage in the reverse direction are proceeding. In logic, linguistics and philosophy of language, there exist such two complementary principles both traditionally ascribed to Frege, namely the *compositionality principle* and the *contextuality principle*, those manifest itself in different terms following a particular theoretical framework.

According to J. F. Pelletier [26, p. 89], R. Carnap was the first to attribute the compositionality principle explicitly to Frege in *Meaning and Necessity* [3], where he stated this principle in terms of a functional dependence. The majority of researchers followed him when formulating their definitions of Frege's compositionality principle in the mathematical paradigm of a function. To illustrate this, we cite a few definitions:

> [...] the meaning (semantical interpretation) of a complex expression is a function of the meanings (semantical interpretations) of its constituent expressions. (J. Hintikka [14, p. 31])

> Like Frege, we seek to do this [...] in such a way that [...] the assignment to a compound will be a function of the entities assigned to its components. (R. Montague [24, p. 217])

> [...] The meaning of a whole is a function of the meanings of the parts. (B. H. Partee [25, p. 313])

In many similar definitions, the meaning of a compound expression is set to be a function of the meanings of its parts, whereas what the meanings are differs substantially. Also, these definitions remain reticent about the explicit form of a function concerned. In sharpening her definition, B. H. Partee notices that "the Principle of Compositionality requires a notion of partwhole structure that is based on syntactic structure", and then she modifies the latter definition to the following one:

> The meaning of a whole is a function of the meanings of the parts and of the way they are syntactically combined. (B. H. Partee [25, p. 313])

Nevertheless, the modified definition of the compositionality principle remains implicit with regard to the function it refers to. In fact, the pages subsequent to definitions of compositionality principle in [25, p. 313] are devoted to the discussion of how one may explicitly define the input values (arguments) of such a function, and describe how this function acts on its arguments, and what it returns as output values. On this way, B. H. Partee leads the reader to the formal definitions given in the Montague's seminal paper [24].

To sum up our discussion, we have to note that in agreement with the tradition going back to Carnap, almost all generally accepted definitions of the compositionality principle convey the mathematical concept of a function in a set-theoretic paradigm.

In the contemporary mathematics, there are different formalizations of the concept of a function and functional dependence. In a prevailing set-theoretic paradigm, a function (map, mapping) is identified with its *graph*. Formally, a function $f\colon X \to Y$ is a set of ordered pairs $f \subseteq X \times Y$ (a *graph*) that satisfies the following two Claims:

1° For *every* argument's value $x \in X$, *there exists* a function's value $y \in Y$ such that $\langle x, y \rangle \in f$;

2° This function's value y is *unique* as such, that is, whenever $\langle x, y \rangle$ and $\langle x, z \rangle$ are members of f, then $y = z$. Thus, all functions are single-valued.

Intuitively, for an ordered pair $\langle x, y \rangle \in f$, a function f is a 'rule' that assigns the element y to the element x. This y is the value of f for the argument x, that is denoted usually as $y = f(x)$.

What is a function in the set-theoretic paradigm is understood in an unambiguous manner by all the scientific community, and the rigorous definition of a function is therefore imposed on any attempt to clarify a vague notion that bears in germ the idea of functional dependence. This is also true for the notion of compositionality in natural language semantics. Any attempt to define explicitly the principle of compositionality as a function $f\colon X \to Y$ in the set-theoretic paradigm meets with serious technical problems to explain what are these sets X, Y, and how is defined the functional graph $f \subseteq X \times Y$. This is a difficult task and even a trap for any attempt to translate literally the set-theoretic notion of a function into the linguistic notion of a compositionality.

The aim of an adequate semantic theory is to conceptualize how the understandings of parts are integrated during the process of reading to produce the understanding of the whole. However, any semantic theory that combines the compositionality defined as the functionality (meant in the 'function as graph' paradigm) with the non-postponed understanding (meant as a dynamic process that develops step by step while the reading progresses over the time) should be obviously inconsistent.

There are two main directions in which the solution of this apparent conflict might be sought:

○ either one conserves the compositionality meant as a set-theoretic functionality but refuses to take into account the process of text understanding over the time, and then establishes a kind of *static semantics*;

○ or otherwise, one renounces of compositionality meant as a set-theoretic functionality, or somehow redefines it, and then studies the process of text understanding over the time, in order to establish a kind of *dynamic semantics*.

If the semantic compositionality is taken to be the functionality in a set-theoretic paradigm, then it imposes the almost indubitable conclusion that Frege had never explicitly stated (in this way) the principle of semantic compositionality generally ascribed to him, whatever it were, the compositionality of *Sinn* or the compositionality of *Bedeutung*. In several papers, T. M. V. Janssen had carefully analyzed the development of Frege's views on such a semantic compositionality during his long scientific career, and then concluded, as a result, that Frege "would always be against compositionality" [15, p. 19]. Another point of view is expressed by F. J. Pelletier who writes in a solid historical research that "Frege may have believed the principle of semantic compositionality, although there is no straightforward evidence for it and in any case it does not play any central role in any writing of his [...]." [26, p. 111].

However, another theoretical view on the part-whole text structure without prejudice to define the compositionality as a kind of the set-theoretic functionality allows us to interpret Frege's views on the subject in a different way. We notice that in the unpublished work *Logic in Mathematics* of 1914, Frege writes:

> As a sentence is generally a complex sign, so the thought expressed by it is complex too: in fact it is put together in such a way that parts of the thought correspond to parts of the sentence. So as a general rule when a group of signs occurs in a sentence it will have a sense which is part of the thought expressed. (G. Frege [10, pp. 207–208])

In this translation, the expression 'will have a sense' concerning a group of signs should really mean 'will be understandable'. In fact, it is an implicit expression of the hermeneutic circle principle in the particular case of a stand-alone sentence. In a general case, this principle prescribes **'to understand a part in accordance with the understanding of the whole'**. It means that Frege believed the hermeneutic circle principle at the semantic level of a stand-alone sentence. As a logician, Frege was interested primarily in a particular case of sentences, that is, in judgements. It does not really matter whether Frege was familiar with the philological discipline of hermeneutics or not. The principle of hermeneutic circle reveals one of key cognitive operations involved in a natural language text (or discourse) understanding process, and so it is implicitly known by any competent language user. We argue that the hermeneutic circle principle carries in germ the mathematical concept of a *sheaf*, which expresses a passage from a local data to the global one, and which is very close to the idea of a functional dependence. From the sheaf-theoretic point of view, one can revise the aforesaid Frege's quotation like this: 'a family of compatible understandings of parts of the sentence are composable into the understanding of the whole sentence'. However, Frege considered words as being elementary units of a sentence, and he believed in the contextuality principle, bearing today his name, in accordance with which words have no meanings in isolation, "but only in the context of a sentence" [9]. We hypothesize that the reluctance to be got involved into the confusion between elements and parts of a whole (between "words [...] in isolation" and "parts of the sentence" in his formulations) prevented Frege from stating explicitly what would be called the compositionality principle. Surely, a meaningful sentence has some meaningful parts, the meanings of which are constitutive to the meaning of this sentence as a whole; but not every of word-tokens may be found among such meaningful parts. This is a kind of the type difference between an element and a subset of a given set.

For an adopted sense $\mathscr{F}$ of reading of a given text X, to each non-empty open (that is to say, meaningful) part $U \subseteq X$ we assign the set $\mathscr{F}(U)$ of all its meanings that may be grasped in all its possible readings in this sense. In fact, it assigns naturally a presheaf $\mathscr{F}$ of fragmentary meanings to the adopted sense of reading. In the beginning of Sect. 5, we argue that such a presheaf $\mathscr{F}$ should satisfy to both Claims **S** and **C** needed for a presheaf to be a sheaf. Thus, the presheaf $\mathscr{F}(U)$ of fragmentary meanings attached to a sense (mode of reading) of an admissible text is really a sheaf. This statement is our generalization of Frege's compositionality principle in the sheaf-theoretic framework. The issuing *sheaf-theoretic formal semantics* takes its departure from another formalization of a functional dependence that is based on the mathematical concept of a sheaf. We use this revised concept of functional dependence in order to define explicitly what is, or rather what should be the compositionality of fragmentary meanings. In this generalized concept of functionality, the arguments and their numbers are not given in advance (one takes for arguments any family of locally compatible *sheaf sections*); but due to the Claim **C**, *for every* such a family of arguments, *there exists* the global sheaf section that becomes their composition; and due to the Claim **S**, this composition is *unique* as such. In the Subsect. 5.1 we show that these Claims **C** and **S** are analogous to those Claims **1°** and **2°** in the aforesaid formal definition of a function in a set-theoretic paradigm.

- So far, we have considered only the meanings of open sets in the phonocentric topology that we have defined in Sect. 3 at any semantic level. Then, in Sect. 6, we describe how we have to define the meanings of points in the phonocentric topology at any semantic level. For this goal, we recast a famous Frege's contextuality principle in order to define the set of contextual meanings of any point x that belongs to the phonocentric topological space X of some semantic level, whatever this point x may be, a word, a sentence, a paragraph, etc., when considered as an element of a syntactic entity of the higher type. For any semantic level, it is the distinction between the notion of a contextual meaning of a primitive element (a point) at this level and the notion of a fragmentary meaning of a part (a subset) of the whole at this level, that is, of the whole space endowed with a phonocentric topology. The contextual meaning of a point x is defined to be the *inductive limit* of fragmentary meanings s of different open neighbourhoods $U \ni x$ those are got identified on some smaller common open neighbourhood of x. Finally, we generalize Frege's contextuality principle in the categorical terms of *bundles* of contextual meanings.

- In Sect. 7, we show that these generalized Frege's compositionality and contextuality principles are related by a *duality* that we formulate in terms of category theory, and that we name after Frege. This sheaf-theoretic duality sheds new light on the delicate relation between Frege's compositionality and contextuality principles, in revealing that the acceptance of one of them implies the acceptance of the other. It resolves Frege's embarrassing situation with the reconciliation of two principles those bear now his name. As two sides of the same coin, Frege's compositionality and contextuality principles express indeed two complementary parts of the hermeneutic circle principle. That is why they always come together in philosophy, linguistics, and logic. *Grosso modo*, the compositionality principle prescribes to understand a meaningful whole by means of understanding of its meaningful parts, whereas the contextuality principle prescribes to understand the meaning of an entity in accordance with the understanding of its meaningful neighbourhoods.

- Once explicitly stated, Frege duality gives rise to a functional representation of fragmentary meanings. In Sect. 8, this functional representation enables us to develop a kind of *compositional dynamic semantics* that describes how the interpretation proceeds over the time as the step-by-step extension of a meaning representation function, from the initial meaningful fragment to the whole interpreted text. Defined in the proposed sheaf-theoretic framework, such a dynamic semantics conceptualizes the compositionality in a uniform manner at each semantic level: word, clause, sentence, paragraph, section, chapter, text as a whole. Moreover, it treats the polysemy in a realistic manner as one of the essential features of a natural language. This sheaf-theoretic dynamic semantics provides the mathematical model of a text interpretation process, while rejecting attempts to codify interpretative practice as a kind of calculus. We call such a mathematical model of a natural language text interpretation process as *formal hermeneutics* (see, e.g., [29, 31, 32]).

- Then, in Sect. 9, we compare the compositional dynamic semantics proposed in our sheaf-theoretic framework with several algebraic compositional semantics. We notice that an algebraic semantic, of whatever kind, is always static because the meaning of the whole sentence is calculated just after the calculation of meanings of all its syntactic components was done. Algebraic semantic theories are appropriate to study the synonymy, but their irremovable drawback is the inability to describe the polysemy. Any kind of formal grammar that formalizes the compositionality as the functionality in a set-theoretic paradigm shares this fallacy with an algebraic semantics described by T. M. V. Janssen in [15] as "a homomorphism from syntax to semantics".

By contrast, the proposed mathematical framework formalizes the compositionality of fragmentary meanings in a sheaf-theoretic paradigm of functional dependence. In this formal framework, the dynamic semantics describes how the interpretation is incrementally built up as a meaning representation function stepwise extension from the initial meaningful fragment to the whole text. Moreover, in this approach the process of a natural language text interpretation is modelled in a similar manner at all semantic levels.

- The present article culminates in the final Sect. 10 devoted to the statement of a *sheaf-theoretic formal hermeneutics* that describes a natural language in the category of *textual spaces* **Logos**. Appeared as syntax and semantics of a natural language, phonocentric topologies and sheaves of fragmentary meanings constitute together an adequate mathematical framework to formalize different linguistic phenomena in our works, such as linguistic universals of geometric nature in [29], as dynamic semantics in [34], as interpretations of one text by the others, as text summarization and abstracting, as well as many other aspects of intertextuality in [31].

2. Basic semantic concepts

Concerning the linguistic terminology to be used in this work, we have certain difficulties because the sciences of language do not have a unified terminology. According to F. Rastier [37], two traditions seem dominant in the sciences of language: (1) the grammatical tradition centered on the issue of the sign, that confines itself to the word and the sentence; (2) the rhetoric and hermeneutic tradition centered on the communication, that privileges the text and the discourse. Based on different conceptions, these two traditions differ in problematic and in terminology. When using the definition of a technical term proper to one doctrine, we have to privilege this doctrine compared with others, that would not be our goal. The aim of our work is to discern the mathematical structures underlying the process of reading, with the purpose to design a semantic theory that formalizes a natural language understanding process in a uniform manner at all semantic levels (word, sentence, text). We are therefore obliged to accept a terminology based on distinctions that are valid at all semantic levels of an admissible text. In this perspective, we have to study only those spoken or written language segments that are admissible as units of linguistic communication. Therefore, we keep to the hermeneutic tradition in the analysis of a text understanding process. We recognize that there are different scientific trends in discourse analysis; that is why we have to clarify basic semantic terms we use in the present paper. The technical acceptance of terms *meaning*, *sense*, and *reference* as these are used in the present paper may be explained as follows:

Meaning. The term *fragmentary meaning* of some fragment of a given text X is accepted as the communicative content grasped in some particular situation of reading. In this terminological acceptance, a *fragmentary meaning* is immanent not in a given fragment of a text, but in the interpretative process of its reading based on the linguistic competence, which is rooted in the social practice of communication with others through the medium of a language. Any reading is really an interpretative process where the historicity of the reader and the historicity of the text are involved. The understanding of meaning is based not only on the shared language but also on the shared experience as a common life-world, and it deals so with the reality. According to Gadamer, this being-with-each-other is a general building principle both in life and in language. The understanding of a natural language text results from being together in a common world. This understanding as a presumed agreement on 'what this fragment $U \subseteq X$ wants to say' becomes for the reader its fragmentary meaning s. In this acceptance, the meaning of an expression is the communicative content that a competent reader grasps when s/he understands it; and such an understanding can be reached regardless of the ontological status of its *reference*.

The process of coming to some fragmentary meaning s of a fragment $U \subseteq X$ demonstrates a human communicative ability in action. When we qualify some fragment as being meaningful, we state that an idealized competent reader can understand a communicative content that this fragment conveys; the understanding manifests itself as the ability of the reader to express at once this content in other words or in another language (e.g., if the reader is bilingual).

The fact of having such an understanding may be labelled with a certain abstract entity s called *fragmentary meaning* of U. When someone acknowledges the fact that a meaning of U has been understood, this situation may be described by saying that 'this fragment U has the fragmentary meaning s'; it presumes implicitly that the understanding of the meaning s of the fragment U is arrived at through some linguistic communication, direct or mediated. This meaning may be shared in a dialogue with another native speaker, and such a possibility describes the ontological status of the meaning s as being some abstract entity subtracted from the linguistic communication. This situation may be summed up by an external observer as 'the understanding of the fragmentary meaning s of a fragment U', where the 'meaning' may be perceived as a linguistic term in our technical acceptance, and also as an ordinary word of English language. So, our use of the term *fragmentary meaning* corresponds well to the common English usage.

We have noticed above that for any admissible text X, one should distinguish a fragmentary meaning of a meaningful part $U \subseteq X$ and a contextual meaning of an element (point) $x \in X$. It expresses the fact that clauses are parts of a sentence, but idioms and words are its indivisible elements. A fragmentary meaning s is assigned to the part $U \subseteq X$, and this s conveys some part of the communicative content of the whole X in a concrete situation of linguistic communication. This part U is a sequence of primitive elements (tokens) x those have contextual meanings in the context of U.

In the situation of linguistic communication, a unit that is proper to convey a communicative content may be some text or its fragment, some sentence or its clause, some elliptic expression, and yet a word or an exclamation in certain cases of communication. Thus, a meaning is related to the communicative content, regardless of its possible truth value, whatever it may be: true, false or indefinite.

However, the linguistic communication, either spoken or written, consists of the use of words in a conventional way. It is quite difficult to trace the history of how a single word enters the lexicon (vocabulary) of a language. Taken beyond the situation of linguistic communication, a single word is not a discourse nor a part of it, and this word says nothing to nobody. But this word had entered the lexicon in the process of repeated participation in a variety of situations of linguistic communication, with the result that native speakers of the language have a clear idea of the situations in which the use of a particular word is appropriate, and what it then means. These so-called literal meanings of words are recorded in the dictionaries and thesauruses. Generally, by means of examples, these dictionaries allow us to understand what meaning is associated with the use of each word in several standard situations of its use. In this way, dictionaries define the abstract objects those are called the literal meanings of words. Such definitions carry the entire history of the language and the experience of the numerous uses of the words in the specific situations of communication. The dictionaries thereby demonstrate that the relationship of each word with the set of its possible meanings in specific contexts had gained a normative value. This usage is normative for native speakers of a particular linguistic community, in a particular historic period. These descriptions are aimed to help for a competent reader to adjust better the orientation of his/her efforts to grasp a meaning. In this terminological acceptance, a word, a fragment, a text has a specific meaning only in the situation of linguistic communication, direct or mediated.

However, when using a particular expression in a particular situation of linguistic communication, each interlocutor establishes his/her own connection between this expression and its meaning, which is a mental concept (signified), grasped by means of this expression used in this particular situation of communication. This meaning is the mental concept concerning either some physical objects of the world, or some ideas, or some fictional entity, but this meaning is not itself a referred object in the world (in contrast to Frege's *Bedeutung*). As the mental concept, this meaning is apprehended as a being of intersubjective nature because it may be shared with native speakers of the same linguistic community. We equate the 'meaning' with the 'communicative content' because a message (in spoken or written form) is intended by its author as a carrier of a certain communicative content to be grasped by the addressee, that is, as a carrier of a certain meaning to be understood.

Let us take for example the word 'wolf'. A hunter, a scientist zoologist, an adult urban dweller who have never seen of living wolves, or a child who is familiar with them only by fairy tales, they all have different concepts conceived in connection with the word 'wolf'. The ostensive definition of the meaning of this word by pointing out wolves in a zoo, and its definition by dictionaries as a 'wild, flesh-eating animal of the dog family' are conveying different concepts. It implies certainly that an adequate semantic theory should take into account that a lexicon of a competent reader counts not only one but several literal meanings of the word 'wolf'. Every competent native speaker knows also about the use of this word in one of figurative senses, for example, in the moral sense of the proverb: "Who lives with the wolves should howl like a wolf".

It is, therefore, the intention of the reader that controls the choice of meanings during the reading. Which of possible meanings of a particular expression is grasped by the reader depends on the specific situation of reading guided by the reader's intention in the interpretative process, presuppositions and preferences, that we denominate by the term *sense* (or *mode of reading*).

Sense. In our acceptance, the term *sense* (or *mode of reading*) denotes a kind of semantic orientation in the interpretative process that relates to the whole text or its meaningful fragment, to some sentence or its syntagma, and involves the reader's subjective premises that what is to be understood constitutes a meaningful whole. Concerning a word-token of a phrase, one may ask a question "What does this word mean here in a literal sense?", and as we have argued above, an answer consists of the choice of only one meaning from the set of many possible ones. Likewise for a question, "What could it mean in a metaphoric sense?", as for many similar questions in a reading process. In such an acceptance, the term 'sense' is correlative to the intentionality of our interpretative efforts; that is, a sense is not immanent to the text we read, but in some way, it may even precede the reading process. For example, one may intend to read a fable in the moral sense yet in advance of its reading. But when the reading unfolds in time, one still controls own intentions following the current reading situation. These examples illustrate the acceptance of the term 'sense' as the reader's interpretative intention, and the acceptance of the term 'meaning' as the content actualized during the process of communication.

To some extent, our acceptance of the term 'sense' is close to the exegetic conception of four senses of the Holy Scripture. The traditional presentation of this conception of biblical hermeneutics is summarized by the famous distich of Augustine of Dacia: "Littera gesta docet, quid credas allegoria, moralis quid agas, quo tendas anagogia."[3]

According to the biblical hermeneutics, the readings of the Scripture in literal, allegorical, moral, and anagogical senses are coherent in each of its parts. Suppose we read the whole text of the Scripture by fragments, where each fragment was read in one of four senses: literal, allegorical, moral, or anagogical, but the choice of sense was not the same for all fragments. The composition of these four senses is a method of interpretation that gives rise

---

[3] Augustine of Dacia, *Rotulus pugillaris*, I: ed. A. Walz: Angelicum 6 (1929) p. 256. The distich is translated in English as: "The letter tells us what went down, the allegory what faith is sound, the moral how to act well – the anagogy where our course is bound."

to a large number of senses of the whole text. Indeed, the overall sense $\mathscr{F}$, as the integral intention in the reading process, is the result of all local intentions taken during these partial readings.

But what guides the subsequent choice of local intentions of an empirical reader? Following Fathers of Church, it is the presence of the Holy Spirit that guides the soul of the individual believer who reads the text of the Scripture. But for a secular text, how can we characterize in linguistic terms the possibility to join these partial senses? It is the presumed sincerity and a goodwill on the part of the author, whom we suppose to be of sound mind and perfect memory, while writing this text intended to communicate something to an alleged reader.

However, the local intentions those were taken in the writing process were got integrated into an overall intention of an empirical author; so, these partial writings are consistent to satisfy a certain *gluing condition* of the type that we discuss further in Sect. 5.4. Since the empirical author is almost always inaccessible for a dialogue, how can we understand what does the text mean by virtue of its textual coherence denoted by U. Eco as the *intentio operis*? According to U. Eco [7, p. 65], "it is possible to speak of the text's intention only as the result of a conjecture on the part of the reader. The initiative of the reader basically consists in making a conjecture about the text's intention." He asks further, "How to prove a conjecture about the *intentio operis*?", and he responds: "The only way is to check it upon the text as a coherent whole." He continues then that this idea comes from *De doctrina Christiana* of St. Augustine:

> [...] any interpretation given of a certain portion of a text can be accepted if it is confirmed by, and must be rejected if it is challenged by, another portion of the same text. [7, p. 65]

According to St. Augustine, the presumed textual coherence controls the partial interpretations that are made by an empirical reader. Therefore, in the process of reading, all these local intentions to understand a text have also to verify the *gluing condition* of the type that we discuss further in the Sect. 5.4.

In the process of actual communication, a mere consistency of the local interpretations would be insufficient. The inference on the speaker's intention is essential here for the understanding; the contact of interlocutors allows them to get into the coordination between the intention of the sender and the intention of the recipient.

With regard to a text produced not for a single recipient, but for a community of readers, the strategy of a *model author* is to lead his *model reader* to speculate about the text. Among these leading indexes, the central place is held by the *semantic isotopy* that A. J. Greimas defines as "a complex of plural semantic categories which makes possible the uniform reading of a story." [12, p. 188]. Concerning the notion of isotopy, U. Eco notices in [6, pp. 189–190] that "The category would then have the function of textual or transsentential disambiguation, but on various occasions Greimas furnishes examples dealing with sentences and outright noun phrases."

Following B. Pottier, the seme does not exist in isolation but as a part of a sememe, or as the set of coexisting semes.

> Le sémème, l'être de langue (en compétence), s'actualise dans le discours [...]. Le sémème donne le sens (l'orientation sémantique), et la mise en discours le transforme en signification.[4] [27, pp. 66, 67]

From this definition, we retain the acceptance of the term *sense* as the semantic orientation of the reader's intentions provoked by a sememe, and the fact that a meaning is actualized in the discourse. The reader's conjecture on the subject discussed in a text determines the first interpretive intention that will be clarified in the course of the reading when the recognition of a semantic isotopy becomes possible owing to the context that is more and more revealed. Following U. Eco,

> The first movement toward the recognition of a semantic isotopy is a conjecture about the topic of a given discourse: once this conjecture has been attempted, the recognition of a possible constant semantic isotopy is the textual proof of the 'aboutness' of the discourse in question. [7, p. 63]

In *Two Problems in Textual Interpretation* published in 1980, U. Eco describes the interpretative process as based on the reader's interpretive cooperation:

> Between the theory that the interpretation is wholly determined by the author's intention and the theory that it is wholly determined by the will of the interpreter there is undoubtedly a third way. Interpretive cooperation is an act in the course of which the reader of a text, through successive abductive inferences, proposes topics, ways of reading, and hypotheses of coherence, on the basis of suitable encyclopedic competence; but this interpretive initiative of his is, in a way, determined by the nature of the text. [2, pp. 43–44]

But later in 1992, in the analysis of so-called *superinterpretation*, U. Eco raises again the problem of a reader's conjectures about the empirical author's

---

[4]Our translation of this quotation is: "The sememe, the entity of language (in competence), is actualized in the discourse [...]. The sememe gives the sense (the semantic orientation), and the putting into discourse transforms it into meaning."

intention during the reading. His updated conception of the interpretation of texts "makes the notion of the intention of an empirical author radically unnecessary" [7, p. 60]. He defends this thesis with the support of his own experience as a writer who has discussed with his readers a few different interpretations of his novels.

To summarize now our acceptance of the term *sense* (or *mode of reading*), we have to say that it is close to the latter acceptance described by U. Eco. The term *sense* concerns the reader's initiative in the interpretation of the text; it is wholly determined by the reader's intention to understand possible meanings of the text. In Sect. 5, we identify a particular sense $\mathscr{F}$ (in our acceptance) with the assignment to each meaningful fragment U of a given text X the set of all its meanings $\mathscr{F}(U)$ that may be grasped in all possible readings of U in this sense $\mathscr{F}$. This way, to any *sense* (or *mode of reading*), we assign a *sheaf* of fragmentary meanings.

Remark. It should be noticed that our terminological acceptance of basic semantic notions of *sense* and *meaning* differs from their acceptance in the theories developed within the tradition that goes back to Carnap's semantic theory, sometimes called the theory of "intension and extension". In such theories, expressions of different syntactic kinds refer to entities of different kinds as their extensions, and also refer to entities of different kinds as their intensions. The terms *intension, intensional* are not to be confused with the terms *intention, intentional* we have discussed above. The notions 'intension', 'intensional' primarily concern the domain of logic, whereas 'intention', 'intentional' concern the philosophy of mind. According to A. R. Lacey, "Intuitively extensions can be thought of as the extents which certain kinds of terms range over and intensions as that in virtue of which they do so." [18, p. 164], whereas the intentionality is "that feature of certain mental states by which they are directed at or about objects and states of affairs in the world" [18, p. 50].

Reference. Certainly, the referential function of a language is important in the linguistic communication, which concerns the world where the interlocutors live. A natural language has a huge arsenal of denoting expressions to designate real and imaginary objects during communication. The linguistic competence is characterized by the know-how in production and comprehension of natural language expressions realizing the referential relationship called *reference* or *denotation*. In the analytic philosophy of language, the study of denoting expressions plays a considerable role, because the reference to objects with an uncertain ontological status is responsible for some logical paradoxes.

In the present work, we assume a total referential competence of an idealized reader who knows the lexicon of a language and follows the rules of common usage. In short, we assume that the reader has a total language skill, combined with a general knowledge. Such a reader meets no problems to understand the meaning of denotative expressions and the ontological status of objects so defined.

3. Topologies appeared as syntax

The author of an admissible text doesn't suppose that the reader's understanding will be suspended until the end of reading because everybody knows that the words already read trigger intellectual mechanisms of interpretation based on the indissoluble links between the signifier and the signified. To be understood in linguistic communication, one must take it into account and organize one's writing in such a way that the reader's understanding at every moment may be arrived at on the basis of what has been already read. It seems that the primacy of speech over writing is a cause that implies in writing the subordination of graphic expressions to acoustic ones. A spoken utterance is a temporal series of sounds produced by a speaker using a human articulatory apparatus. When written, an acoustic signal is converted into a series of signs whose positions are linearly ordered following an adopted convention; in English, it is from left to right within the lines, and from top to bottom between them. Once a particular sign is taken as the initial, it allows us to specify the position of the following signs by enumeration. From the mathematical point of view, the whole segment may be considered as a finite sequence when the last sign is specified. Thus, we ought to consider a text X as a finite sequence $(x_1, x_2, x_3, \ldots, x_n)$ of its constituent sentences x_i, and so it is formally identified with the graph of a function $i \mapsto x_i$ defined on some interval of natural numbers. When reading a particular fragment of the text X, we delete mentally the other sentences but follow the induced order of remaining ones. Important is the induced order of their reading and not the concrete index numbers of their occupied places. Thus, any part of the text is a subsequence whose graph is a subset of the whole sequence graph. Likewise for a sentence considered as a finite sequence of its words.

While reading a text, the understanding is not postponed until the final sentence. So the text should have the meaningful parts, and the meanings of these parts determine the meaning of the whole as it is postulated by the hermeneutic circle principle. For the meaningfulness conveying an idealized reader's linguistic competence, a meaning of a meaningful part is the communicative content grasped in a particular reading of this part guided by the reader's presuppositions and preferences in the interpretative process, that is, guided by the sense (or mode) of reading.

Certainly, there are many meaningful fragments in the text. A simple example of a meaningful fragment is supplied by the interval including all sentences, from the first x_1 until the last x_n. Anybody reads the text as if it would be a written transcription of the story uttered by the author. When telling or writing a story, an author should take into account that the understanding can't be postponed, for "the texts never know the suspense of interpretation. It is compulsive and uncontrollable", as it is noticed by F. Rastier in [36]. If the author don't want to be misunderstood, s/he has to organize the text in such a way that any sentence x is preceded by certain sentences those provide a necessary context for the understanding of x. Thus, any meaningful part contains each sentence together with some its context, and this is characteristic of any part to be meaningful. It is clear that this property fails for a part including, e.g., all sentences x_i whose placehold number i is divisible by 100, and that is why this part is meaningless, and nobody try to read the text in such a manner. In [28, 31–33], we argue that in agreement with our linguistic intuition, the set of all meaningful parts of any admissible text should satisfy two properties:

(t_1) *The union of any set of meaningful parts is a meaningful part.*

(t_2) *The non-empty intersection of two meaningful parts is a meaningful part.*

The first property (t_1) is taken for granted, because it expresses the precept of generally accepted hermeneutic circle principle, which ensures us to understand the union of a given set of meaningful parts through the understanding of all its constitutive members. In the union of any set of meaningful parts, each part contains every its sentence together with some its context, whence the union itself is a part that has such a property. To be more accurate, we have to take into account that the meaning s of a meaningful part U isn't immanent to this part itself, but this meaning is grasped in the reading process following a sense (mode of reading) $\mathscr{F}$ guided by the reader's interpretative intentions. Thus, in the statement (t_1), some sense (or mode of reading) $\mathscr{F}$ is implicitly presumed to be the same for all members of the union. In the following Sect. 5–8, we discuss in details how the resulting meaning of the whole is obtained via the meanings of its constitutive parts.

The second property (t_2) expresses the contextuality of understanding. To understand a meaningful part U of the text X is to understand contextually all sentences $x \in U$, where the context of a particular sentence x is some meaningful part W such that $x \in W \subseteq U$. In the standard process of reading (i.e., from the beginning up to x), this part W should contain a subsequence of sentences those precede x and provide a necessary context for the understanding of x in the sense $\mathscr{F}$. For a particular sense $\mathscr{F}$, there should exist a smaller subsequence $(x_{i_1}, \ldots, x_{i_m}) \subseteq W$ whose sentences have been understood during the reading, and then have been taken into account at the moment when the reader understands a meaning of x grasped in the sense $\mathscr{F}$. Let us denote $U_x = (x_{i_1}, \ldots, x_{i_m})$. The tokens x_{i_k} of U_x may be consecutive or dispersed among other tokens of W, it does not matter, but they should be read before the reading of x.

Consider first the case of one session process of reading of X in some sense $\mathscr{F}$. When the part U_x belongs to any meaningful part $W \subseteq X$ such that $x \in W$. Let U, V be two meaningful parts such that $x \in U \cap V$. According to our premises, $x \in U_x \subseteq U$ and $x \in U_x \subseteq V$; hence $x \in U_x \subseteq U \cap V$.

Consider now the case when $x \in U \cap V$, and parts U, V were read in two different sessions of reading, but in the same sense $\mathscr{F}$. This means that the reader is self-identical, and the reading is guided by the same intentionality. It implies that $U_x \subseteq U$ and $U_x \subseteq V$. Hence $x \in U_x \subseteq U \cap V$.

Thus in both cases, $U \cap V$ is meaningful because $U \cap V = \bigcup_{x \in U \cap V} U_x$ is the union of meaningful parts, due to (t_1).

Since an admissible text X is supposed to be meaningful as a whole by the very definition, it remains only to define formally the meaning of its empty part (for example, as a singleton) in order to satisfy the third property:

(t_3) *The whole admissible text and the empty part are meaningful.*

This enables us to endow an admissible text X with some *topology* in a strict mathematical sense, where the set $\mathfrak{O}(X)$ of open sets is defined to be the set of all meaningful parts. We call the topology so defined *phonocentric topology* to indicate in its name the subordination of graphic expressions to phonetic ones.

An admissible text X gives rise to a finite space; hence an arbitrary intersection of its open sets is open and so it is an *Alexandrov space*.

In general, a topology on a set X is defined by specifying the set $\mathfrak{O}(X)$ of open subsets of X satisfying axioms similar to ours (t_1), (t_2), and (t_3). But almost always it is impossible to enumerate all the open subsets. Instead, a topology is usually defined by specifying a smaller set of open subsets, called a *basis*, and then generating all the open subsets from this basis.

Likewise, when studying the process of interpretation of an admissible text X, many of linguistic concepts may be well expressed in terms of the phonocentric topology on X that is defined by specifying the set of open subsets $\mathfrak{O}(X)$ to be the set of all meaningful parts satisfying properties (t_1), (t_2), and (t_3). However, it will be more convenient and useful to develop the theory in more concrete, say even constructive, terms of empirically given meaningful parts those constitute a basis for a phonocentric topology.

Fortunately, the set of all meaningful parts $\mathfrak{O}(X)$ of a given text X may be described by specifying a class of fairly simple meaningful parts given as an empirical data related to a reading process. In the reading of a particular text X, the reader is practically concerned with a smaller class of meaningful parts $(U_x)_{x \in X}$, where each part U_x contains a sentence x and provides the smallest context that is necessary for a reader to grasp a particular meaning of x. Because the phonocentric topology $\mathfrak{O}(X)$ is finite, for each x, there exists such a smallest open neighbourhood U_x that is defined as the intersection of all open neighbourhoods of x.

For a given sentence x, the understanding of a whole U_x requires the grasping of meanings of all constitutive sentences of U_x; hence, for any sentence $y \in U_x$, its smallest context U_y should be a part of U_x. Suppose now that we are given two smallest meaningful parts U_x and U_y such that $U_x \cap U_y \neq \emptyset$. Then for each $z \in U_x \cap U_y$, we have $U_z \subseteq U_x$ and $U_z \subseteq U_y$; hence $U_z \subseteq U_x \cap U_y$. Therefore, the set $\mathfrak{B}(X) = \{U_x : x \in X\}$ is the set of meaningful parts of X satisfying two properties:

(b_1) For each $x \in X$, there exists $U_x \in \mathfrak{B}(X)$ such that $x \in U_x$.

(b_2) For every two $U_x, U_y \in \mathfrak{B}(X)$ such that $U_x \cap U_y \neq \emptyset$, and for each sentence $z \in U_x \cap U_y$, there exists $U_z \in \mathfrak{B}(X)$ such that $z \in U_z$ and $U_z \subseteq U_x \cap U_y$.

So, the set $\mathfrak{B}(X)$ is a basis for a phonocentric topology on X, because any meaningful part (i.e., open) $V \subseteq X$ is the union $V = \bigcup_{x \in V} U_x$ of the members of some subset of $\mathfrak{B}(X)$. Recall that a set $\mathfrak{B}$ of open sets of a topological space X is called a *basis* for its topology if and only if every open set U of X is the union of the members of a subset of $\mathfrak{B}$. Thus, the class of open sets $\mathfrak{O}(X)$ in a phonocentric topology on X is defined by the subclass $\mathfrak{B}(X)$ of all open sets of the type U_x, that is, a phonocentric topology on X is defined by the empirical data $\mathfrak{B}(X)$.

Any explicitly stated *concept of meaning* or a *criterion of meaningfulness* satisfying conditions (t_1), (t_2), and (t_3) allows us to define some type of *discursive topology* on texts, and then to interpret several problems of discourse analysis in topological terms [31]. In what follows, we consider only admissible texts endowed with a phonocentric topology that is a particular type of discursive topology corresponding to the criterion of meaningfulness conveying the linguistic competence of an idealized reader, meant as the ability to grasp a communicative content.

3.1. Phonocentric topology and partial order

In the ordinary process of reading, any sentence x of a text X should be understood on the basis of the part already read because the interpretation of a natural language text cannot be postponed, although it may be made more precise and corrected in further reading and rereading. In [36], F. Rastier describes this fundamental feature of a competent reader's linguistic behaviour as the following:

> Alors que le régime herméneutique des langages formels est celui du suspens, car leur interprétation peut se déployer après le calcul, les textes ne connaissent jamais le suspens de l'interprétation. Elle est compulsive et incoercible. Par exemple, les mots inconnus, les noms propres, voire les non-mots sont interprétés, validement ou non, peu importe.[5] [36, pp. 165, 166]

Thus, for every pair of distinct sentences x, y of X, there exists an open part U containing one of them (to be read first in the natural order $\leq$ of sentences reading) but not the other. This means explicitly that the phonocentric topology satisfies the *separation axiom T_0* of Kolmogorov.

For a sentence $x \in X$, we have defined the open neighbourhood U_x to be the intersection of all the meaningful parts those contain x, that is the smallest open neighbourhood of x. The *specialization relation* $x \preceq y$ (read as 'x is

---

[5]Our translation of this quotation is: "While the hermeneutic regime of formal languages is that of suspense, because their interpretation can be deployed after the calculation, the texts never know the suspense of interpretation. It is compulsive and uncontrollable. For example, unknown words, proper names, even non-words are interpreted, valid or not, whatever."

more special than y') on a topological space X is defined by setting $x \preceq y$ if and only if $x \in U_y$ or, equivalently, $U_x \subseteq U_y$. It is clear that $x \in U_y$ if and only if $y \in \mathrm{cl}(\{x\})$, where $\mathrm{cl}(\{x\})$ denotes the topological closure of a one-point set $\{x\}$.

Key properties of these notions are summarized in the Propositions 1, 2 those are linguistic versions of general mathematical results concerning the interplay of topological and order structures defined on a finite set. The proofs may be found in many sources, as for example, in [23].

Proposition 1. *For an admissible text X, the set of all smallest opens $\{U_x : x \in X\}$ is a basis for a phonocentric topology on X. Since the phonocentric topology on X satisfies the separation axiom T_0, it defines a partial order $\preceq$ on X by means of the specialization relation. The initial phonocentric topology can be recovered from this partial order $\preceq$ in a unique way as the topology with the basis made up of all sets of the kind $U_x = \{z : z \preceq x\}$.*

Proposition 2. *Let X, Y be admissible texts endowed with phonocentric topologies. Then the following statements are equivalent:*
1. *The function $f \colon X \to Y$ is continuous.*
2. *For each $x \in X$, the function f maps a basis set into a basis set, that is $f(U_x) \subseteq U_{f(x)}$.*
3. *The function f preserves the specialization order, that is $x \preceq y$ implies $f(x) \preceq f(y)$.*

Example. A continuous function $f_1 \colon X_2 \to X_1$ arises in writing process when an author goes from a first plan X_1 of some future text to its more detailed plan X_2, where a sentence x_d of X_1 is substituted by some passage $(x_{d_1}, \ldots, x_{d_m})$. And so on, in going to more and more detailed texts $X_3, \ldots, X_n$, one gets a sequence of continuous functions

$$X_n \xrightarrow{f_{n-1}} X_{n-1} \xrightarrow{f_{n-2}} \ldots \xrightarrow{f_3} X_3 \xrightarrow{f_2} X_2 \xrightarrow{f_1} X_1.$$

3.2. Deep structures and surface structures

Let **FinTOP**$_0$ be the category of finite T_0-topological spaces and continuous maps, and let **FinORD** be the category of finite partially ordered sets (posets) and their monotone maps.

Given a finite partially ordered set $(X, \leq)$, one defines a T_0-topology τ on X by means of the basis for τ made up of all *low sets* $\{z : z \leq x\}$.

Thus, one obtains a functor $L\colon \mathbf{FinORD} \to \mathbf{FinTOP}_0$ acting identically on the maps of underlying set. Conversely, one defines the specialization functor $Q\colon \mathbf{FinTOP}_0 \to \mathbf{FinORD}$, assigning to each finite T_0-topological space (X, τ) a poset $(X, \preceq)$ with the specialization order $\preceq$, and acting identically on the maps of underlying set. Thus, the functors L and Q establish the isomorphism between the category $\mathbf{FinTOP}_0$ and the category $\mathbf{FinORD}$. From the mathematical point of view, the study of one of these two categories is equivalent to the study of the other.

Now we generalize and summarize the considerations of the mathematical structures of topology and order underlying an admissible text:

The considerations in the beginning of Sect. 3 may be slightly modified in order to define a *phonocentric* topology at the semantic level of sentence and even word [31]. Thus, at each semantic level, there exist two topological structures:

(i) *the natural phonocentric topology at a considered semantic level;*

(ii) *the topology defined by applying the functor L to the linear order $x \leq y$ of reading.*

At an arbitrary semantic level (where the whole is a sequence of primitive elements), the difference between topologies can be summed up so that in the phonocentric topology the least neighbourhood U_x of a primitive element x contains only such primitive elements that precede x in the linear order of writing and provide the context necessary to understand the meaning of x in the adopted sense $\mathscr{F}$; whereas in the topology defined by the functor L applied to $(X, \leq)$, the least neighbourhood U_x of a primitive element x contains all primitive elements that precede x in the linear order of writing.

Note that the explicit definition of the phonocentric topology at the semantic level of sentence requires more delicate work in treatment of different grammatical types of sentences due to the lack of space, so to speak. Here there is a certain analogy with the topological classification of varieties that turns out to be more difficult in dimensions 3 and 4 than in lower and in higher dimensions.

On the other hand, at each semantic level, there exist two order structures:

(i′) *the specialization order $x \preceq y$ defined by applying the specialization functor Q to the natural phonocentric topology of a considered semantic level;*

(ii′) *the linear order $x \leq y$ of ordinary text reading.*

Similar to a generative grammar, we will qualify the equivalent structures of (i) and (i′) as *deep structures* compared to the equivalent structures of (ii) and (ii′) qualified as *surface structures*. We notice that this denomination has nothing to do with the acceptance of these terms in a generative grammar.

Remark. The relation $x \preceq y$ implies obviously the relation $x \leq y$, for all the primitive units x, y of the same semantic level. In particular, at the level of text, where the sentences are primitive units, the map $\mathrm{id} \colon L(X, \preceq) \to L(X, \leq)$, which acts as identity $x \mapsto x$ of the underlying set, is a continuous map of topological spaces. Thus, the necessary linearization during the writing process, that is the passage from $(X, \preceq)$ to $(X, \leq)$, results in weakening of the phonocentric topology by transition from $L(X, \preceq)$ to $L(X, \leq)$. The process of interpretation consists in a backward recovering of the phonocentric topology (or equally, of the specialization order) on the text.

3.3. Phonocentric topology at the level of text

There is a simple intuitive tool for graphical representation of a finite poset, called Hasse diagram. For a poset $(X, \preceq)$, the *cover relation* $x \prec y$ (read as 'x is covered by y') is defined by setting $x \prec y$ if and only if $x \preceq y$ and there is no other z such that $x \preceq z \preceq y$. For a given poset $(X, \preceq)$, its Hasse diagram is defined as the graph whose vertices are the elements of X and whose edges are those pairs $\langle x, y \rangle$ for which $x \prec y$. In the picture, the vertices of Hasse diagram are labeled by the elements of X and the edge $\langle x, y \rangle$ is drawn by an arrow going from x to y (or sometimes by an indirected line connecting x and y, but in this case the vertex y is displayed lower than the vertex x); moreover, the vertices are displayed in such a way that each line meets only two vertices.

The usage of some kind of Hasse diagram named *Leitfaden* is widely spread in the mathematical textbooks to facilitate the understanding of logical dependence of its chapters or paragraphs. Mostly, the poset is constituted of all chapters of the book. So, in *Local Fields* by J.-P. Serre [39] and in *A Mathematical Logic* by Yu. I. Manin [21], there are such diagrams.

These diagrams may surely be 'split' in order to draw the corresponding ones whose vertices are all the paragraphs, like it is done directly in *Differential Forms in Algebraic Topology* by R. Bott and L. W. Tu [1], where authors suppose indeed the linear reading of paragraphs 1-6, 8-11, 13-16 and 20-22, but it may be drawn explicitly. These three Hasse diagrams are shown in the Fig. 1.

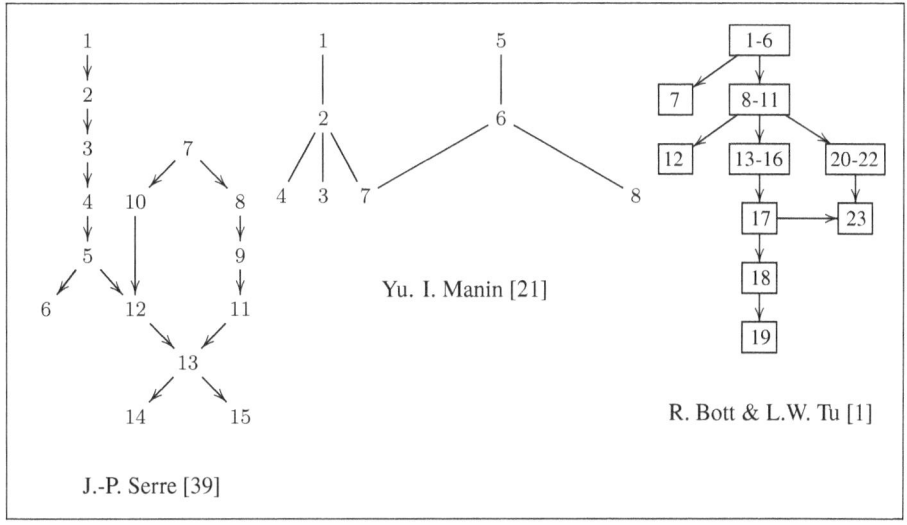

Figure 1: *Leitfiden* of J.-P. Serre [39], Yu. I. Manin [21], R. Bott & L. W. Tu [1].

This way, one may go further and do the next step. For every sentence x of a given admissible text X, one can find a basis open set of the kind U_x in order to define the phonocentric topology at the semantic level of text (where points are sentences), and then to draw the Hasse diagram of the corresponding poset.

In [31], we describe how one may interpret this way the most of diagrams from the *Rhetorical Structure Theory* (RST) conceived in the 1980s the by W. C. Mann and S. A. Thompson [22]. Since then, RST has seen a great development, especially in the computational linguistics, where it is often used for the automatic generation of coherent texts, as well as for the automatic analysis of the structure of texts. The RST aims to describe an arbitrary *coherent text*, which is not the random sequence of sentences. The textual coherence demands that for every part of a coherent text there exists a reason for its presence, which is obvious to a competent reader. It seems that RST notion of a coherent text is similar to our notion of an admissible text. In [31], we show that the RST analysis of contextual dependencies between sentences of certain small textual fragments represented as RST diagram may be redrawn as the Hasse diagram for the partial order structure of the corresponding specialization relationship. But the RST diagram may be drawn only for certain small textual fragments such that their sentences are *nucleus* and *satellite* in the sense of the RST. On the other hand, it is not the case when such a fragment is a part of a larger text. Then, according to the RST, there will be no link between a sentence x belonging to such a fragment and any other sentence y that is far enough in the text, because rhetorical relations can only bind adjacent segments. While in our approach, such a link is possible in the specialization relation (of deep order). This link is seen on the corresponding Hasse diagram as a direct edge $\langle x, y \rangle$ or as a sequence of edges that link these two sentences x and y. Thus, our approach is more general than this one of the RST.

3.4. Phonocentric topology at the level of sentence

In order to define a phonocentric topology at the semantic level of sentence, we must distinguish there the meaningful fragments that are similar to meaningful fragments at the level of text. Let x, y be any two word-tokens such that $x \preceq y$ in the specialization order at the level of sentence that is similar to the specialization order coming from the 'logical relations among the different chapters' in a text. This relation $x \preceq y$ means that the word-token x should necessary be an element of the set of word-tokens U_y required to understand the meaning of the word-token y in the interpreted sentence. So we have $x \leq y$ in the order of writing and there should be some syntactic dependence between them. It means that a grammar in which the notion of dependence between pairs of words plays an essential role will be closer to our topological theory than a grammar of Chomsky's type.

There are many formal grammars focused on links between words. The history of this stream of ideas is described by S. Kahane in a detailed review [16]. We think that the theoretical approach of the *special link grammar* of D. Sleator and D. Temperley is most appropriate to define a phonocentric topology at the level of sentence, because in whose formalism "[t]he grammar is distributed among the words" [40, p. 3], and "the links are not allowed to form cycles" [40, p. 13] comparing with *dependency grammars* that draw syntactic structure of sentence as a planar tree with one distinguished root word.

For a given sentence s, the link grammar assigns to it a syntactic structure (called *linkage diagram*) that consists of a set of labeled links connecting pairs of words. We use these diagrams to define all phonocentric topologies on this sentence s.

Example. To explain how to define phonocentric topologies on a particular sentence, let us borrow from [42] the following example of an ambiguous sentence:

(1) John saw the girl with a telescope.

We had yet considered this sentence in [29] by using Chomsky's generative grammar, and also in [31] by using link grammar. The analysis of this sentence by means of the *Link Parser 4.0* of D. Temperley, D. Sleator, and J. Lafferty [41] gives two linkage diagrams shown in the Fig. 2.

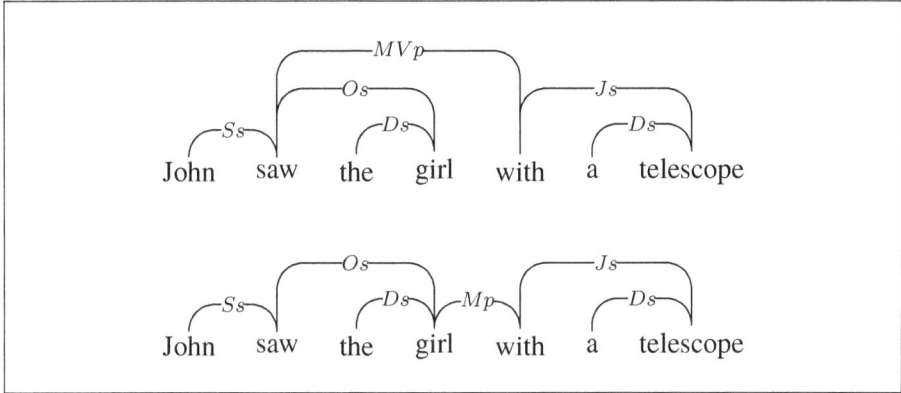

Figure 2: Two linkage diagrams with connector names.

These two diagrams rewritten with arrows that indicate the direction in which the connectors match (instead of connector name) have the appearance shown in the Fig. 3.

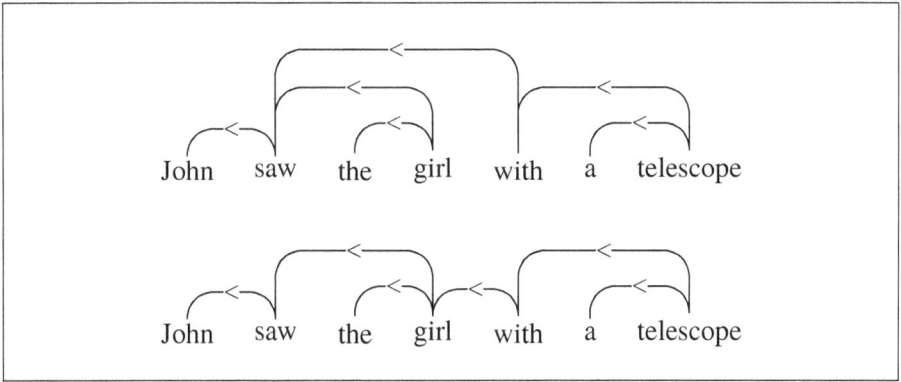

Figure 3: Two linkage diagrams with arrows instead of connector names.

It is clear that the transitive closure $x \preceq y$ of this relation $<$ between pairs of words defines two partial order structures on the sentence (1). By applying the functor L defined in Sect. 3.2, we can endow the sentence (1) with a phonocentric topology in two different ways. The Hasse diagrams of corresponding posets are shown in the Fig. 4.

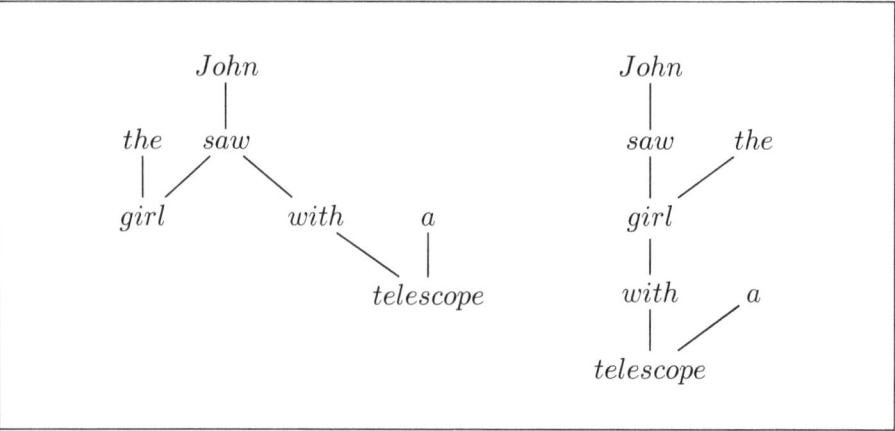

Figure 4: Two Hasse diagrams of the sentence (1) as displayed in [29, 31].

To understand the sentence (1), the reader has to do the *ambiguity resolution* when arriving to the word-token $x =$ "with" by choosing only one of two possible basis sets:

$U_x = \{\langle 1, \text{John}\rangle, \langle 2, \text{saw}\rangle, \langle 5, \text{with}\rangle\}$;
$U_x = \{\langle 1, \text{John}\rangle, \langle 2, \text{saw}\rangle, \langle 3, \text{the}\rangle, \langle 4, \text{girl}\rangle, \langle 5, \text{with}\rangle\}$.

In the general case, the step by step choice of an appropriate context $U_x = \{z \colon z \preceq x\}$ for each word x results in endowing the interpreted sentence with a particular phonocentric topology among many possible.

In [31], we have shown how to define a phonocentric topology at the level of word considered as a sequence of morphemes.

We summarize the results of our analysis presented in Sect. 3 as the following:

Slogan (Phonocentric Topologies as Syntax). Once the phonocentric topology and the corresponding specialization order are determined at a given semantic level, the systematic interpretation of linguistic phenomena in terms of topology and specialization order, and their mathematical study is a formal syntax at this level.

4. Linguistic universals of a topological nature

Throughout the history of scientific study of human languages, researchers are interested in discovering linguistic universals, that is, particular traits common to all languages. Because it is impossible to recognize everything about all languages, it is necessary to first decide where and how to look for linguistic universals. It appears that our sheaf-theoretic approach makes here a small contribution.

By its very origin, a human language is used for linguistic communication; for that reason, written texts and uttered discourses should be considered as communicative units. We must therefore look for linguistic universals, not only in terms of word as it is done by J. H. Greenberg [11] and his successors, but especially in terms of text. A true linguistic universals at the level of text (or discourse) must have a corresponding counterpart at the level of sentence.

By linguistic universals, we understand the characteristic properties of texts those are admissible as messages having communicative purposes, regardless of the language in which they are written. The question is, therefore, reduced to this: What criteria should we accept to be sure that a particular characteristic is truly shared by all admissible texts in any natural language? One can adopt a statistical criterion ensuring, to a certain extent, that if some property is shared by hundreds of natural languages, it is likely that it is shared by all. Such an approach is taken up in the classical works of J. H. Greenberg. But there are no guarantees that a particular trait of the languages already studied is also shared by the language of a lost Indian tribe that escaped the statistical body of research.

To our deep conviction, the way to avoid counter-examples is to adopt a criterion based not only on statistical considerations, but mainly on the analysis of the communicative function of languages. In our talk [30] at the 39th Annual Meeting of *SLE*, we argued that the properties of a phonocentric topology to satisfy the separation axiom T_0 of Kolmogorov and to be connected are linguistic universals. These properties should be required of the underlying phonocentric topology on any text written for the purpose to be understood in the linguistic communication.

A correct translation of an admissible text from one language into another is done by successive translation of each sentence in a manner to conserve their contextual relations. It results in a bijection between the original text and its translation, and also in a homeomorphism between corresponding topological spaces.

It is clear that a phonocentric topology on an admissible text written in one language (as well as the corresponding Hasse diagram) is invariant under translation into another language. Hence, a phonocentric topology on a text X and its properties and geometric invariants (say T_0-separability, connectedness, homology groups, etc.) are stable under translation from one language into another (i.e., under homeomorphism), and so they are formal invariants of the text X.

The properties those are shared by all texts in all natural languages are absolute linguistic universals. In [30–32], we argue that the T_0-*separability*, the *connectedness* of a phonocentric topology, and the *acyclicity* of corresponding Hasse diagram are features shared by the majority of languages.

4.1. Kolmogorov's axiom T_0 as a linguistic universal

One important example of a topological linguistic universal seems to be the *separation axiom T_0* of Kolmogorov. In the Sect. 3, we argued for the relevance of the separation axiom T_0 to all semantic levels of an admissible text on the base of a lucid formulation by F. Rastier [36]. Anyway, there is an essential difference between the hermeneutic regime of formal languages and that one of natural languages; it is important for us that texts written in a natural language "never know the suspense of interpretation" [36, p. 166]. It's still the same idea that Origen expresses in the biblical hermeneutics regarding the non-understanding. According to Origen, yet for an *imbulatum*, there is a meaning as a sign of divine presence in the text.

Such an empirical truth that everyone knows from his/her own experience of reader still deserves a more nuanced discussion. Firstly, this property of understanding of texts in natural language is obviously taken into account by everyone who writes a text intended for human understanding, whether he/she is a professional writer or not; the rule is accepted as that one of a 'writing game', so to speak.

If we do not want to be misunderstood, we do not propose the reader to suspend understanding until the end of writing because we know that the words already read trigger intellectual interpretation mechanisms based on indissoluble links between signifier and signified. This is well expressed by the colourful Russian saying: *A word is not a sparrow; you can't catch it when it flies away!* In order to be understood, we must organize our writing in such a way that the reader's understanding would always be based on the part of text already read, in total ignorance of its future development.

The second reading (as all subsequent readings) is governed by the same rule, despite the fact that we already know the whole text. The repetitive reading respects the unpredictability of the future; while reading at the time being, we are being in the 'here and now', that leads us to identify the physical real time with the time of the narrative. What lies in the pages that follow makes no context for the understanding of what has been read. In particular, this rule is just applicable to scientific texts.

A question arises: What is the reason for this indisputable empirical phenomenon? It seems to us that it is the primacy of speech over writing, which causes the subordination of graphic expressions to phonetic ones.

Preliterate civilizations existed thousands of years before the advent of writing and even still exist somewhere else. Even today there are thousands of people who cannot read. The cultural history of the human species is repeated in the personal history of each individual because we learn to speak before we learn to read and write. But as a physical phenomenon, a phonetic expression exists in the dimension of time, and here the physiological properties of our speech organs are just involved.

In a conversation, the interlocutors have access only to whatever is already said, because the future remains unpredictable. Once said, the spoken word is flying away and the only chance to get by in such a situation is to understand on the spot all that is said by the others.

For anybody speaking, this attitude quickly becomes a habit and even a conditioned reflex on the situation of linguistic communication. As functional and even physiological in origin, this property of the oral communication is inherited by the written communication. So it becomes a linguistic universal because it is specific to understanding in linguistic communication, regardless of the natural language concerned. In our formalism, this linguistic universal is expressed by the statement that the topological space underlying any semantic level of an admissible text satisfies the separation axiom T_0 of Kolmogorov.

4.2. Topological connectedness as a linguistic universal

In Sect. 3, we have considered some examples of phonocentric topologies at various levels of semantic description of an admissible text. In all these examples, we see that their underlying topological spaces are connected. This shows empirically an important topological property of all genuine natural language texts, namely the connectedness, in the mathematical sense, of their phonocentric topology. The reasons for it aren't accidental, but it reveals a very important topological property of genuine natural language texts. At the conference [30], we presented arguments that the topological connectedness is one of the linguistic universals.

Any literary work has a property to be the communicative unity of meaning. So, for any two novels X and Y yet of the same kind, say historical, detective or biographical, their concatenation Z under one and the same cover doesn't constitute a new one. What does it mean, topologically speaking? We see that for any $x \in X$ there exists an open neighbourhood U of x that doesn't meet Y, and for any $y \in Y$ there exists an open neighbourhood V of y that doesn't meet X. Therefore, $Z = X \bigsqcup Y$ (i.e., Z is a disjoint union of two non-empty open subsets X and Y); hence, Z isn't connected. Thus, a property of a literary work to be the communicative unity of meaning may be expressed as a connectedness of a topological space related to text.

Recall that a space X is said to be *connected* if it is not the disjoint union of two non-empty open subsets. It is the same to say that X and $\varnothing$ are the only subsets opened and closed at a time. Such a property is called the connectedness of the space X. In any topological space X, a connected set is a subset U of X that is a connected space for the induced topology. It is clear that the union of connected parts having one point in common is also a connected part.

Define on a topological space X the relation $\sim$ by setting $x \sim y$ if and only if x and y belong to a connected subset of X. It is immediate that this relation is an equivalence; the equivalence class containing a point x is a connected part that is called *connected component* of x. It is clear that a topological space X is the disjoint union of its connected components, and any connected part is contained in exactly one component. If $f \colon X \to Y$ is a continuous mapping of topological spaces where the space X is connected, then $f(X)$ is a connected subset of Y.

Let X be an Alexandrov topological space. It is clear that for all $x \in X$, the smallest open U_x is connected. So, each open set U_x of the basis $\mathfrak{B}(X)$ of a phonocentric topology is connected.

For all $x, y \in X$ such that $x \neq y$, the subspace $\{x, y\}$ is connected if and only if $x \in U_y$ or $y \in U_x$; in terms of the specialization order, this amounts to saying that $x \preceq y$ or $y \preceq x$. The following well-known proposition (see, e.g., [23, p. 8]) characterizes connected Alexandrov topological spaces:

Proposition 3. *Let X be a connected Alexandrov topological space. Then for every pair of points x, y of X, there exists a finite sequence $(z_1, \ldots, z_s)$ of points in X such that $z_1 = x$, $z_s = y$ and each $\{z_i, z_{i+1}\}$ is connected (i.e., $z_i \preceq z_{i+1}$ or $z_i \succeq z_{i+1}$) for all $i = 1, \ldots, s-1$.*

Indeed, let Z be a set of points accessible by a finite sequence $(z_1, \ldots, z_s)$ of points in X starting from $x = z_1$, such that each set $\{z_i, z_{i+1}\}$ is connected for $i = 1, \ldots, s-1$. For each $z \in Z$, we have $U_z \subseteq Z$ because any element $y \in U_z$ is itself also accessible by a chain $(z_1, \ldots, z, y)$. We have $Z \subseteq \bigcup_{z \in Z} U_z \subseteq Z$; hence Z is open. For each $z \in Z$, we have also $\mathrm{cl}(\{z\}) \subseteq Z$ because, for all $y \in \mathrm{cl}(\{z\})$, any neighbourhood of y, including U_y, contains z. This implies $z \preceq y$ and $y \in Z$. We have $Z \subseteq \bigcup_{z \in Z} \mathrm{cl}(\{z\}) \subseteq Z$; therefore Z is closed because X is an Alexandrov space. Now, the set Z is non-empty because $x \in Z$, opened and closed subset of the connected space X. Hence, $Z = X$.

It should be noticed that the formulation and the proof of the Proposition 3 are valid regardless of the (finite or infinite) number of points in the space X.

Since the relation $x \preceq y$ is transitive, we can, in the assertion of Proposition 3, exclude unnecessary elements of the finite sequence $(z_1, \ldots, z_s)$. Namely, after excluding repetitive elements, we can reduce each subsequence $z_i \prec z_{i+1} \prec z_{i+2}$ to $z_i \prec z_{i+2}$ if any exists, and we can reduce each subsequence $z_j \succ z_{j+1} \succ z_{j+2}$ to subsequence $z_j \succ z_{j+2}$ if any exists.

After a finite number of such steps of reduction, we have a sequence $(z_1, \ldots, z_r)$, such that in this sequence, the relations $\prec$ and $\succ$ follow one after the other, namely:

if $z_i \prec z_{i+1}$, then $z_{i-1} \succ z_i \prec z_{i+1}$ for all i such that $1 < i < s$;
if $z_i \succ z_{i+1}$, then $z_{i-1} \prec z_i \succ z_{i+1}$ for all i such that $1 < i < s$.

Example. In the Hasse diagram of the book [21], one immediately sees such a sequence $(4 \succ 2 \prec 7 \succ 6 \prec 8)$, which connects the Chapter 4 with the Chapter 8, that is shown in the Fig. 5.

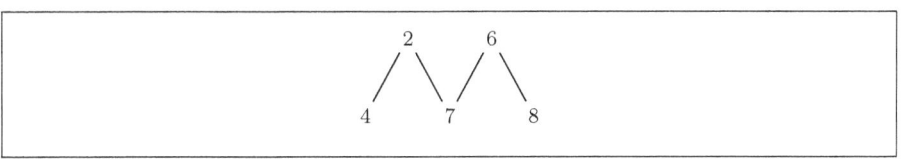

Figure 5: A Khalimsky arc traced in the Leitfaden of [21] shown in the Fig. 1.

The Hasse diagram of the type shown in the Fig. 5 is called *Khalimsky arc*.

We define now the *Khalimsky topology* by means of a structure that differs slightly from the original definition of [17]. Let us first define the partition $\mathbb{R} = \bigcup_{m \in \mathbb{Z}} P_m$ of Euclidean line of real numbers $\mathbb{R}$ by setting:

$P_m = [m - \frac{1}{2}, m + \frac{1}{2}]$, closed interval of real numbers $\{t \colon m - \frac{1}{2} \leqslant x \leqslant m + \frac{1}{2}\}$, for each even integer $m \in \mathbb{Z}$;

$P_m =]m - \frac{1}{2}, m + \frac{1}{2}[$, open interval of real numbers $\{t \colon m - \frac{1}{2} < x < m + \frac{1}{2}\}$, for each odd integer $m \in \mathbb{Z}$.

Recall the notion of a *quotient topology*. Let X be a topological space, and let P be an equivalence relation on X. The quotient topology on the quotient set X/P is the finest topology making continuous the canonical projection $X \to X/\mathsf{P}$ that associates to each element of X its equivalence class. That is, the set of equivalence classes of X/P is open in the quotient topology if and only if its inverse image is open in X.

Let P be an equivalence relation on $\mathbb{R}$ associated with the partition $\mathbb{R} = \bigcup_{m \in \mathbb{Z}} P_m$. We then define a quotient topology on X/P. By identifying $P_m \in X/\mathsf{P}$ with $m \in \mathbb{Z}$, we define the *Khalimsky topology* on $\mathbb{Z}$. The set of integers $\mathbb{Z}$ endowed with the Khalimsky topology is called the *Khalimsky line*. Since $\mathbb{R}$ is connected, the Khalimsky line is connected as well.

It is immediate that an even point is closed, and that an odd point is open. Concerning the smallest neighbourhoods, we have $U_m = \{m\}$ if m is odd, and we have $U_m = \{m-1, m, m+1\}$ if m is even. For integers $m \leqslant n$, we define a *Khalimsky interval* to be the interval $[m, n] \cap \mathbb{Z}$ with the topology induced from Khalimsky line, and we denote it by $[m, n]_\mathbb{Z}$. We call a *Khalimsky arc* any topological space that is homeomorphic to a Khalimsky interval $[m, n]_\mathbb{Z}$. We say that the points that are images of m and n are connected by a Khalimsky arc. Now it is clear that the Proposition 3 is equivalent to the following:

Proposition 4. *An Alexandrov topological space X is connected if and only if for every pair of points x, y of X, there exists a Khalimsky arc that connects them.*

In other words, for an Alexandrov space, the connectedness and the connectedness by a Khalimsky arc are equal.

It is obvious that all topological spaces whose Hasse diagrams are shown in the Fig. 1 are connected. It is difficult to imagine a book in which there is a single chapter that has no contextual links to other chapters. The same holds not only at the semantic level where primitive elements are chapters, but also at the semantic level where primitive elements are sentences of the text (such a level is called the semantic level of text). If at the end of the reading, we realize that a sentence x has nothing to do with the reminder of the text, we have a feeling that 'a noise crept into the message' because the reading of the text is finished, but the sentence x remains to be its completely strange ingredient.

On the contrary, if during the reading we meet a sentence that does not have direct contextual links with the sentences already read (like the item 7 in the Hasse diagram of the textbook [39] as shown in the Fig. 1), we have a feeling to be on a turning point in the narrative, and that the author prepares the reader for the future development, where the suspended sentence will be necessary for the understanding. For an admissible text, these considerations confirm that the connectedness of the underlying topological space expresses mathematically the necessary requirement of a textuality in the sense one understands this concept in the semiotics of text.

This explains why a basic unit that is pertinent as a message in the situation of linguistic communication should be an admissible text (or discourse) whose underlying topological space is connected! It is a connected unit because, after having communicated such a message, the transmitter (author, sender) may become silent to give the floor to its receptor (reader, receiver).

At the level of text, the connectedness of message is also a requirement specific to the kind of linguistic communication qualified as a dialogue, that is, to a bi-directional communication with others. If somebody produces, as a message, a series of phrases that disintegrates into pieces that have no links between, it reveals the disregard for the interlocutor, or the absence of the desire to communicate, or the use of a language for purely expressive purposes without a desire to communicate.

It is the same at the semantic level of sentence with regard to connectedness, although the formal definition of a phonocentric topology at the level of sentence needs more delicate work.

Remark. It should be noticed that for an admissible text, the corresponding Hasse diagram with directed edges is acyclic at any semantic level. It is clear that this property of a phonocentric topology is stable under homeomorphism. This means that the *acyclicity* of the Hasse diagram corresponding to the phonocentric topology is yet another linguistic universal of a topological nature.

5. Sheaves of meanings appeared as semantics

Let X be an admissible text endowed with a phonocentric topology, and let $\mathscr{F}$ be an adopted sense of reading. In a Platonic manner, for each non-empty open (that is meaningful) part $U \subseteq X$, we collect in the set $\mathscr{F}(U)$ all fragmentary meanings of this part U read in the sense $\mathscr{F}$; also we define $\mathscr{F}(\varnothing)$ to be a singleton pt. Thus, we are given a map

$$U \mapsto \mathscr{F}(U) \tag{1}$$

defined on the set $\mathfrak{O}(X)$ of all open sets in a phonocentric topology on X.

Following the precept of hermeneutic circle '**to understand a part in accordance with the understanding of the whole**', for each inclusion $U \subseteq V$ of non-empty opens, the adopted sense of reading $\mathscr{F}$ gives rise to restriction map $\mathrm{res}_{V,U} \colon \mathscr{F}(V) \to \mathscr{F}(U)$. We will consider the inclusion of sets $U \subseteq V$ as being the canonical injection map $U \xhookrightarrow{\mathrm{inj}} V$. Thus, we are also given a map

$$\{\, U \xhookrightarrow{\mathrm{inj}} V \,\} \mapsto \{\, \mathscr{F}(V) \xrightarrow{\mathrm{res}_{V,U}} \mathscr{F}(U) \,\} \tag{2}$$

with the properties:

(i) $\mathrm{id}_V \mapsto \mathrm{id}_{\mathscr{F}(V)}$ for all opens V of X;

(ii) $\mathrm{res}_{V,U} \circ \mathrm{res}_{W,V} = \mathrm{res}_{W,U}$ for all nested opens $U \subseteq V \subseteq W$ of X.

The first property means that the restriction $\mathrm{res}_{V,U}$ respects identity inclusions. The second property means that two consecutive restrictions may be done by one step.

As for the empty part $\varnothing$ of X, the restriction maps $\mathrm{res}_{\varnothing,\varnothing}$ and $\mathrm{res}_{V,\varnothing}$ with the same properties are obviously defined.

Let $(X, \mathfrak{O}(X))$ be a topological space. We can consider its topology $\mathfrak{O}(X)$ as the category $\mathbf{Open}_X$ whose objects are open sets of X, and where for two open sets $U, V \in \mathfrak{O}(X)$, the class of morphisms $\mathrm{Mor}(U, V)$ is empty if $U \not\subseteq V$, and $\mathrm{Mor}(U, V)$ is the set reduced to the canonical injection $U \xhookrightarrow{\mathrm{inj}} V$ if $U \subseteq V$. The composition of morphisms is defined as the composition of canonical injections.

From the mathematical point of view, the assignments (1) and (2) give rise to a presheaf $\mathscr{F}$ defined as a contravariant functor from the category $\mathbf{Open}_X$

to the category **Set** of sets and maps

$$\mathscr{F}: \mathbf{Open}_X \to \mathbf{Set}, \tag{3}$$

acting on objects as defined by (1), and acting on morphisms as defined by (2).

In sheaf theory, an element $s \in \mathscr{F}(V)$ is called *section* (*over V*); sections over the whole space X are said to be *global*.

We consider the reading process of an open fragment U as its covering by some family of open subfragments $(U_j)_{j \in J}$ already read, that is $U = \bigcup_{j \in J} U_j$.

Following Quine, **"There is no entity without identity"** [35]. We argue that two fragmentary meanings should be equal globally if and only if they are equal locally. It motivates the following *identity criterion*:

Claim S (Separability). *Let X be an admissible text, and let U be an open fragment of X. Suppose that $s, t \in \mathscr{F}(U)$ are two fragmentary meanings of U and there is an open covering $U = \bigcup_{j \in J} U_j$ such that $\mathrm{res}_{U, U_j}(s) = \mathrm{res}_{U, U_j}(t)$ for all fragments U_j. Then $s = t$.*

According to the precept of hermeneutic circle, **'to understand the whole by means of understandings of its parts'**, a presheaf $\mathscr{F}$ of fragmentary meanings satisfies the following:

Claim C (Compositionality). *Let X be an admissible text, and let U be an open fragment of X. Suppose that $U = \bigcup_{j \in J} U_j$ is an open covering of U; suppose we are given a family $(s_j)_{j \in J}$ of fragmentary meanings, $s_j \in \mathscr{F}(U_j)$ for all fragments U_j, such that $\mathrm{res}_{U_i, U_i \cap U_j}(s_i) = \mathrm{res}_{U_j, U_i \cap U_j}(s_j)$. Then there exists some meaning s of the whole fragment U such that $\mathrm{res}_{U, U_j}(s) = s_j$ for all fragments U_j.*

Thus, any presheaf of fragmentary meanings defined as above should satisfy both Claims **S** and **C**, and so it is a *sheaf* by the very definition. This motivates the following definition:

Frege's Generalized Compositionality Principle. *A presheaf of fragmentary meanings naturally attached to any sense (mode of reading) of an admissible text is really a sheaf; its sections over a meaningful fragment of the text are its fragmentary meanings; its global sections are the meanings of the text as a whole.*

Traditionally attributed to Frege, the compositionality principle arises in logic, linguistics and philosophy of language in many different formulations, which all however convey the concept of functionality.

We note that the Claim **S** guarantees the meaning s (whose existence is stated by the Claim **C**) to be unique as such. It is not so hard to see that these two conditions **C** and **S** needed for a presheaf to be a sheaf are analogous to those two conditions **1°** and **2°** needed for a binary relation to be a function.

5.1. Sheaf-theoretic conception of a functional dependence

Formally, for a function f of n variables, it is set that: **1°** *for any* family of variables' values $(s_1, \ldots, s_n)$, *there exists* a function's value $f(s_1, \ldots, s_n)$ being dependent on them, and **2°** this function's value is *unique*. Likewise, for a sheaf $\mathscr{F}$, it is set that: (due to **C**) *for any* family of sections $(s_i)_{i \in I}$ those are locally compatible on an open U, *there exists* a section s being their composition dependent on them, and (due to **S**) this composition s is *unique* as such. In this generalized (sheaf-theoretic) conception of a functional dependence, the variables and their number are not fixed in advance (we consider an arbitrary family of pairwise compatible sections as variables), but for any such a family of variables, *there exists* the glued section considered as their composition (analogous to the function's value in a given family of variables) and such a section is *unique*. So the true formulation of Frege's compositionality principle does not demand a set-theoretic functionality, but demands its sheaf-theoretic generalization stating that any presheaf of fragmentary meanings naturally attached to an admissible text ought *de facto* to be a sheaf. The sheaves arise whenever some consistent local data glues into a global one.

5.2. Schleiermacher category of sheaves of fragmentary meanings

The reader should become at home with the senses treated as functors although we call them sometimes as 'modes of readings' instead of 'senses' not only to emphasize the character of intentionality of each actual process of reading but rather to avoid a possible confusion that may be caused by another technical acceptance of the term 'sense'. So one can think, for example, about the historical sense $\mathscr{F}$ and the moral sense $\mathscr{G}$ of some biographical text.

Let us consider now any two senses (modes of reading) $\mathscr{F}, \mathscr{G}$ of a given text X, and let $U \subseteq V$ be two arbitrary meaningful fragments of the text X. It seems to be very natural to consider that any meaning s of fragment V understood in the historical sense $\mathscr{F}$ gives a certain well-defined meaning $\phi(V)(s)$ of the same fragment V understood in the moral sense $\mathscr{G}$. Hence, for each $V \subseteq X$, we are given a map $\phi(V)\colon \mathscr{F}(V) \to \mathscr{G}(V)$. To transfer from the meaning s of V in the historical sense to its meaning $\phi(V)(s)$ in the moral sense and then to restrict the latter to a subfragment $U \subseteq V$ is the same operation as to make first the restriction from V to U of the meaning s in the historical sense, and to make then a change of the historical sense to the moral one. This kind of transfer from the understanding in one sense $\mathscr{F}$ to the understanding in another sense $\mathscr{G}$ is a usual matter of linguistic communication. In the Christian theology, the possibility of such a transfer from one of four senses of any biblical verse to some another its sense is considered as the cornerstone method of exegesis.

Formally, this idea is well expressed by the notion of *morphism* of the corresponding sheaves $\phi\colon \mathscr{F} \mapsto \mathscr{F}'$ defined as a family of maps $\phi(V)\colon \mathscr{F}(V) \to \mathscr{F}'(V)$ those commute with restrictions for all opens $U \subseteq V$, that is, $\operatorname{res}'_{V,U} \circ \phi(V) = \phi(U) \circ \operatorname{res}_{V,U}$. This can be expressed in a simple way by saying that the following diagram

$$\begin{array}{ccc} \mathscr{F}(V) & \xrightarrow{\phi(V)} & \mathscr{G}(V) \\ {\scriptstyle \operatorname{res}_{V,U}}\downarrow & & \downarrow{\scriptstyle \operatorname{res}'_{V,U}} \\ \mathscr{F}(U) & \xrightarrow[\phi(U)]{} & \mathscr{G}(U) \end{array}$$

commutes for all opens $U \subseteq V$ of X.

This notion of morphism is very near to that of *incorporeal transformation* of G. Deleuze and F. Guattari illustrated by several examples, one of which we quote:

> In an airplane hijacking, the threat of a hijacker brandishing a revolver is obviously an action; so is the execution of the hostages, if it occurs. But the transformation of the passengers into hostages, and of the plane-body into a prison-body, is an instantaneous incorporeal transformation, a "mass media act" in the sense in which the English speak of "speech acts." [5, p. 102]

To adapt this example, we need only to transform it into some written story about a hijacking. Hence, the family of maps $(\phi(V))_{V \in \mathfrak{D}(X)}$ defines a change of mode of reading of a given text X, or simply a morphism $\phi\colon \mathscr{F} \mapsto \mathscr{G}$. It is obvious that a family of identical maps $\operatorname{id}_{\mathscr{F}(V)}\colon \mathscr{F}(V) \to \mathscr{F}(V)$ given for each open $V \subseteq X$ defines the identical morphism of the sheaf $\mathscr{F}$ that will be denoted as $\operatorname{id}_{\mathscr{F}}$. The composition of morphisms is defined in an obvious manner: For two arbitrary morphisms $\phi\colon \mathscr{F} \mapsto \mathscr{G}$, $\psi\colon \mathscr{G} \mapsto \mathscr{H}$, we define

$(\psi \circ \phi)(V) = \psi(V) \circ \phi(V)$. It is clear that this composition is associative every time it may be defined.

Thus, given an admissible text X, the data of all sheaves $\mathscr{F}$ of fragmentary meanings together with all its morphisms constitutes some *category* in a strict mathematical sense of the term. We name this category of particular sheaves describing the exegesis of the text X as *category of Schleiermacher* and denote it as $\mathbf{Schl}(X)$ because he is generally considered to be the author of the cornerstone principle of a natural language text understanding, called later by Dilthey as the *hermeneutic circle*. The parts are understood in terms of the whole, and the whole is understood in terms of the parts. This part-whole structure in the understanding, he claimed, is principal in the matter of interpretation of any text in natural language.

The theoretical principle of *hermeneutic circle* is a precursor to Frege's principles of *compositionality* and *contextuality* formulated later. The succeeded development of hermeneutics has confirmed the importance of Schleiermacher's concept of circularity in text understanding. From our point of view, the concept of part-whole structure expressed by Schleiermacher in 1829 as the hermeneutic circle principle reveals, in the linguistic form, the fundamental mathematical concept of a sheaf formulated by Leray in 1945, more than a hundred years later. This justifies us to name the particular category of sheaves $\mathbf{Schl}(X)$ after Schleiermacher.

5.3. Building a sheaf of fragmentary meanings from local data

An admissible text X is endowed with a phonocentric topology in such a way that the set $\mathfrak{O}(X)$ of all open sets of this topology is made up of all meaningful parts of X. The Hasse diagram presents a perfect visualization of this topological structure but its construction requires a lot of analytical work. It seems that the author has such a representation about his/her proper text, as well as the structure of text may be rebuild after philological considerations. But for a reader, how this topological structure is obtained during the reading? Obviously, the understanding is manifested in the reader's conscience as an empirical fact of having grasped the meaning of a sentence read in the present moment. Thus, the meaningful parts that are most clearly manifested during the reading process are the opens U_x of the phonocentric topology basis $\mathfrak{B}(X)$ that is defined in the Sect. 3. These meaningful parts U_x provide the set of contexts for the understanding of the whole text.

The Proposition 1 states that the set of all these fragments U_x constitutes the minimal basis for a phonocentric topology. Formally, this means that any arbitrary open set is the union of a family of these basis sets U_x. The liberty in choice of basis sets whose union gives an open set $U \subseteq X$ makes us doubtful whether it would be too strong to impose the satisfaction of Claims **S** and **C** for all opens of the topology $\mathfrak{O}(X)$. Would it be more convenient and more useful to develop the very theory in more concrete terms, say even in constructive terms of opens U_x of the minimal basis $\mathfrak{B}(X)$ of a phonocentric topology on X? The answer is plain and simple: From the psychological point of view, yes, perhaps; but from the mathematical point of view, this approach will be formally equivalent but less technically convenient! Moreover, a general truth is sometimes more understandable that a mass of concrete data. In what follows, we will present formal arguments to justify this point of view.

A topological space $(X, \mathfrak{O}(X))$ may be considered as the category $\mathbf{Open}_X$ with open sets $U \in \mathfrak{O}(X)$ as objects, and injection maps $U \xhookrightarrow{\text{inj}} V$ as morphisms.

Let $\mathfrak{B}(X)$ be a basis for the topology $\mathfrak{O}(X)$ of X. It is obvious that the basis $\mathfrak{B}(X)$ gives rise to a category defined in the same way that we consider the topology $\mathfrak{O}(X)$ as being the category $\mathbf{Open}_X$. By a slight abuse of notations, we will also denote such a category as $\mathfrak{B}(X)$. In the same manner as above, we define a presheaf $\mathscr{F}$ of sets on the topology basis $\mathfrak{B}$ as a *contravariant functor* on the category $\mathfrak{B}$ with values in the category of sets **Set**.

Namely, for every basis open $U \in \mathfrak{B}(X)$, the presheaf $\mathscr{F}$ attaches a set $\mathscr{F}(U)$, and so we are given a map

$$U \mapsto \mathscr{F}(U) \qquad (4)$$

defined on the basis $\mathfrak{B}(X)$ for a topology on X. Also, for every pair of opens $U, V \in \mathfrak{B}(X)$ such that $U \subseteq V$, the presheaf $\mathscr{F}$ attaches a map $\text{res}_{V,U} \colon \mathscr{F}(V) \to \mathscr{F}(U)$, and so we are given a map

$$\{U \subseteq V\} \mapsto \{\text{res}_{V,U} \colon \mathscr{F}(V) \to \mathscr{F}(U)\} \qquad (5)$$

with the properties of identity preserving and transitivity:

(i) $\text{id}_V \mapsto \text{id}_{\mathscr{F}(V)}$ for all opens $V \in \mathfrak{B}(X)$;

(ii) $\text{res}_{V,U} \circ \text{res}_{W,V} = \text{res}_{W,U}$ for all nested basis opens $U \subseteq V \subseteq W$ of $\mathfrak{B}(X)$.

Given a basis $\mathfrak{B}(X)$ for a topology on X, the data of $(\mathscr{F}(V), \operatorname{res}_{V,U})_{V,U \in \mathfrak{B}(X)}$ satisfying these properties is called presheaf of sets over the basis $\mathfrak{B}(X)$ for the topology on X. In the case of an admissible text X, the topological basis $\mathfrak{B}(X)$ consists of all fragments of the kind U_x, that may be considered as empirical data.

Let $\mathscr{F}$ be a presheaf of sets over a basis $\mathfrak{B}(X)$ for the topology $\mathfrak{O}(X)$ on X. This presheaf $\mathscr{F}$ is said to be a *sheaf* over the topological basis $\mathfrak{B}(X)$ if the following Claims **Sb** and **Cb** are satisfied:

Claim Sb. *Let U be any open of the basis $\mathfrak{B}(X)$ for the topology on X, and let $s, t \in \mathscr{F}(U)$ be two elements of U. If there exists an open covering $U = \bigcup_{j \in J} U_j$ by basis open sets $U_j \in \mathfrak{B}(X)$ such that for each U_j of this covering, we have $\operatorname{res}_{U, U_j}(s) = \operatorname{res}_{U, U_j}(t)$. Then $s = t$.*

Claim Cb. *Let U be any open of the basis $\mathfrak{B}(X)$ for the topology on X, and let $U = \bigcup_{j \in J} U_j$ be a covering of U by basis open sets $U_j \in \mathfrak{B}(X)$. Suppose we are given a family $(s_j)_{j \in J}$ of elements $s_j \in \mathscr{F}(U_j)$ such that $\operatorname{res}_{U_i, U_i \cap U_j}(s_i) = \operatorname{res}_{U_j, U_i \cap U_j}(s_j)$. Then there exists an element $s \in \mathscr{F}(U)$ such that $\operatorname{res}_{U, U_j}(s) = s_j$ for each open U_j.*

It is obvious that the Claims **Sb** and **Cb** are similar to the Claims **S** and **C** in the definition of a sheaf over a topological space.

Let $\mathscr{F}$ be a presheaf of sets over the basis $\mathfrak{B}(X)$ for a topological space X. For any open $U \in \mathfrak{O}(X)$, the sets $(\mathscr{F}(V))_{\mathfrak{B} \ni V \subseteq U}$ together with maps $\operatorname{res}_{W,V}$ (where $W \in \mathfrak{B}(X)$, $V \in \mathfrak{B}(X)$ such that $V \subseteq W \subseteq U$) form a projective system. We can associate with $\mathscr{F}$ a presheaf of sets $\mathscr{F}'$ over X in the ordinary sense by assigning to any open $U \in \mathfrak{O}(X)$, the projective limit

$$\mathscr{F}'(U) = \varprojlim_{\mathfrak{B} \ni V \subseteq U} \mathscr{F}(V), \tag{6}$$

where V are running the set (ordered by $\subseteq$) of all opens $V \in \mathfrak{B}(X)$ such that $V \subseteq U$.

In the *EGA* of A. Grothendieck and J. A. Dieudonné [13, p. 75], there is a general proposition that, for a presheaf with values in the category of sets, is interpreted as the following result.

Proposition 5. *For the presheaf $\mathscr{F}'$ defined over topology $\mathfrak{O}(X)$ by (6) to be a sheaf, that is, to verify the Claims **S** and **C**, it is necessary and sufficient that the presheaf $\mathscr{F}$ defined over the basis $\mathfrak{B}(X)$ for $\mathfrak{O}(X)$ verifies the Claims **Sb** and **Cb**.*

Let $\mathscr{F}, \mathscr{G}$ be two presheaves of fragmentary meanings defined over the topological basis $\mathfrak{B}(X)$. We define a morphism $\theta\colon \mathscr{F} \to \mathscr{G}$ as a family $(\theta(V))_{V \in \mathfrak{B}}$ of maps $\theta(V)\colon \mathscr{F}(V) \to \mathscr{G}(V)$ satisfying the conditions of compatibility with the corresponding restriction morphisms. With the notation of Proposition 5, we deduce a morphism $\theta'\colon \mathscr{F}' \to \mathscr{G}'$ of presheaves of fragmentary meanings defined on all opens $U \in \mathfrak{O}(X)$ by taking $\theta'(U)$ to be the projective limit of $\theta(V)$ for $V \in \mathfrak{B}(X)$ and $V \subseteq U \in \mathfrak{O}(X)$.

Let $\mathscr{F}$ be a sheaf of fragmentary meanings over $\mathfrak{O}(X)$, and let $\mathscr{F}_1$ be a sheaf over $\mathfrak{B}(X)$ defined by the restriction of $\mathscr{F}$ to $\mathfrak{B}(X)$. Then, the sheaf $\mathscr{F}_1'$ over $\mathbf{Open}_X$ that we obtained from $\mathscr{F}_1$ according to the Proposition 5 is canonically isomorphic to $\mathscr{F}$, because of the claims **S** and **C**, and by the uniqueness of the projective limit. Usually, we identify $\mathscr{F}$ and $\mathscr{F}_1'$.

For two sheaves $\mathscr{F}, \mathscr{G}$ defined over $\mathfrak{O}(X)$ and a morphism $\theta\colon \mathscr{F} \to \mathscr{G}$, one can show that the data of $\theta(V)\colon \mathscr{F}(V) \to \mathscr{G}(V)$ given only for $V \in \mathfrak{B}(X)$ determines completely the morphism θ. For more details, see *EGA* of A. Grothendieck and J. A. Dieudonné [13, p. 76].

Theoretically speaking, this means that we have a good reason to move the considerations from the level of empirical data, where a phonocentric topology is revealed by the minimal basis $(U_x)_{x \in X}$, to the general level, more abstract but more simple, where a phonocentric topology is defined by the set $\mathfrak{O}(X)$ of all opens according to the classical Hausdorff axioms (t_1), (t_2), and (t_3) of a topological space. In mathematics, the axiomatic view on a topology is particularly useful in all sorts of reasonings where topological structures are concerned. Once we have defined a topological space in terms of its basis, we may continue the reasoning in terms of all open sets of this topology.

5.4. Compositionality of locally defined modes of reading

Note that the class of objects in the category $\mathbf{Schl}(X)$ is not limited to a modest list of sheaves corresponding to *literal, allegoric, moral, psychoanalytical* and other senses mentioned above. In the process of text interpretation, the reader's semantic intentionality changes from time to time, with the result that there is some compositionality (or gluing) of these locally defined sheaves of fragmentary meanings, which we consider in details in [31]. There is a standard way to name the result of such a gluing as, for example, this is the case of *Freudo-Marxist* sense.

As the reader's intentionality to interpret an arbitrary text in a certain sense

$\mathscr{F}$, this particular sense $\mathscr{F}$ yet precedes a reading; for example, one may intend to read a story in a moral sense. But for a given text X, the *intentional object* 'sense $\mathscr{F}$' is represented by the sheaf of sets $(\mathscr{F}(V), \text{res}_{V,U})_{V,U \in \mathfrak{D}(X)}$ of fragmentary meanings.

To analyze the compositionality of senses (or modes of reading) in our sheaf-theoretic formalism, we recall firstly the notion of *induced sheaf*. Let X be a topological space, let U be an open set of X, and let $i \colon U \hookrightarrow X$ be the canonical injection of the open U in X. Then, for any sheaf $\mathscr{F}$ of sets over X, one can define a sheaf of sets over U, which is called *sheaf induced by $\mathscr{F}$ on U*, and which is denoted as $\mathscr{F}|_U$, by setting:

$(\mathscr{F}|_U)(V) = \mathscr{F}(i(V))$ for any open $V \subseteq U$;
$(\text{res}|_U)_{W,V} = \text{res}_{i(W), i(V)}$ for all opens $V, W \subseteq U$ such that $V \subseteq W$.

For any morphism $\theta \colon \mathscr{F} \to \mathscr{G}$ of sheaves of sets over X, we note by $\theta|_U$ the morphism $\mathscr{F}|_U \to \mathscr{G}|_U$ consisting of maps $\theta(i(V))$ for opens $V \subseteq U$.

We have a reason to assume that the reading of the whole text X in a sense $\mathscr{F}$ is represented by an open covering $(U_\lambda)_{\lambda \in L}$ of the text X, where each fragment U_λ is read in a sense $\mathscr{F}_\lambda$ that is defined as $\mathscr{F}_\lambda = \mathscr{F}|_{U_\lambda}$.

The obvious concordance of these senses $\mathscr{F}_\lambda$ means that for all pairs of open fragments $U_\lambda, U_\mu \subseteq X$, we have an isomorphism

$$\theta_{\lambda\mu} \colon \mathscr{F}_\mu|_{(U_\lambda \cap U_\mu)} \xrightarrow{\sim} \mathscr{F}_\lambda|_{(U_\lambda \cap U_\mu)}. \tag{7}$$

In other words, in the interpretation of the common part $U_\lambda \cap U_\mu$, we can change the sense $\mathscr{F}_\lambda$ to the sense $\mathscr{F}_\mu$ and vice versa.

It is useful to denote $U_{\lambda\mu} = U_\lambda \cap U_\mu$ and $U_{\lambda\mu\nu} = U_\lambda \cap U_\mu \cap U_\nu$. Then, in this notation, the family of isomorphisms

$$\theta_{\lambda\mu} \colon \mathscr{F}_\mu|_{U_{\lambda\mu}} \xrightarrow{\sim} \mathscr{F}_\lambda|_{U_{\lambda\mu}} \tag{8}$$

satisfies the condition:

$$(\text{for all } U_\lambda, U_\mu, U_\nu) \quad \theta_{\lambda\mu} \circ \theta_{\mu\nu} = \theta_{\lambda\nu} \text{ on } U_{\lambda\mu\nu}. \tag{9}$$

In the theory of sheaves, there is a theorem stating that a family of isomorphisms satisfying the condition (9) allows us to rebuild the sheaf $\mathscr{F}$ uniquely. The following proposition is a linguistic version of this general mathematical result:

Proposition 6. *Let $(U_\lambda)_{\lambda \in L}$ be an open covering of the text X, where each fragment U_λ is read in a sense $\mathscr{F}_\lambda$. Let for each pair of fragments U_λ, U_μ of $(U_\lambda)_{\lambda \in L}$ be given an isomorphism $\theta_{\lambda\mu} \colon \mathscr{F}_\mu|_{U_{\lambda\mu}} \xrightarrow{\sim} \mathscr{F}_\lambda|_{U_{\lambda\mu}}$ of sheaves over $U_{\lambda\mu}$. Assume these isomorphisms are satisfying the condition that for all U_λ, U_μ, U_ν of the covering:*

$$\theta_{\lambda\mu} \circ \theta_{\mu\nu} = \theta_{\lambda\nu} \text{ on } U_{\lambda\mu\nu}. \tag{10}$$

Then, there exists a sheaf $\mathscr{F}$ over X, and for each U_λ of the covering $(U_\lambda)_{\lambda \in L}$ there exists an isomorphism $\theta_\lambda \colon \mathscr{F}|_{U_\lambda} \xrightarrow{\sim} \mathscr{F}_\lambda$ such that $\theta_\mu = \theta_{\mu\lambda} \circ \theta_\lambda$ for U_λ, U_μ of the covering $(U_\lambda)_{\lambda \in L}$. Moreover, $(\mathscr{F}, (\theta_\lambda)_{\lambda \in L})$ is unique up to unique isomorphism.

For the proof, see *EGA* of A. Grothendieck and J. A. Dieudonné [13, p. 77].

The family of isomorphisms $(\theta_{\lambda\mu})$ satisfying the *gluing condition* (10) is called a 1-*cocycle*. One says that the sheaf $\mathscr{F}$ is obtained by *gluing of* sheaves $(\mathscr{F}_\lambda)_{\lambda \in L}$ by means of $\theta_{\lambda\mu}$, and usually one identifies $\mathscr{F}_\lambda$ and $\mathscr{F}|_{U_\lambda}$ by means of θ_λ.

For a finite family of sheaves $(\mathscr{F}_\lambda)_{\lambda \in L}$ and their isomorphisms $\theta_{\lambda\mu}$ satisfying the condition of gluing (10), the sheaf $\mathscr{F}$ is called to be their composition obtained by the gluing of sheaves $(\mathscr{F}_\lambda)_{\lambda \in L}$ by means of the $\theta_{\lambda\mu}$; this describes how we define the compositionality of locally defined modes of reading (senses) understood as sheaves of fragmentary meanings.

The gluing of sheaves is a compositionality method that enables us to obtain a large number of globally defined sheaves from a small number of locally defined ones. In fact, the sense $\mathscr{F}$ as a global mode of reading (or an integral intention during the interpretation of the whole text) is composed of all local modes of reading taken during interpretations of parts.

Example. According to the biblical hermeneutics, the readings of the Scripture in the *literal, allegorical, moral,* and *anagogical* senses are consistent over each fragment of the type U_x. Suppose that we have read the whole text of the Scripture by fragments, where each fragment was read in one of these four senses (literal, allegorical, moral, anagogical). These partial readings satisfy the gluing condition (10) above. There exists therefore a sense $\mathscr{F}$ of reading of the whole text of the Scripture such that for each of its sentence, there are a neighbourhood and one of these four senses (literal, allegorical, moral, anagogical) that is consistent with the reading of this neighbourhood in the sense $\mathscr{F}$. The sense $\mathscr{F}$ is a composition of these four senses (literal, allegorical, moral, anagogical), but globally it differs from each of these four senses being applied to the whole text. Hence, for the text E of the Scripture, the class of objects of the category of Schleiermacher $\mathbf{Schl}(E)$ contains not only these four senses (literal, allegorical, moral, and anagogical) but much more their compound senses, where each compound sense $\mathscr{F}$ is defined by gluing a family of these four senses according to a particular covering of text by fragments, such that each fragment is read in only one sense of these four.

We summarize the results of our analysis presented in Sect. 5 as the following:

Slogan (Sheaves of Fragmentary Meanings as Semantics). The mathematical study of a natural language texts interpretation in terms of the category of sheaves of fragmentary meanings and their morphisms is a sheaf-theoretic formal semantics.

6. Bundles of contextual meanings

So far, we have considered only the meanings of open sets in the phonocentric topology at any semantic level. In this section, we describe how we have to define the meanings of points in the phonocentric topology at a given semantic level. It should be noticed that in general, not every singleton x is open in T_0-topology, and if this is the case, the meaning of such a point x has not yet been defined.

In 1884, Frege wrote in the *Die Grundlagen der Arithmetik* [9, p. X]: "nach der Bedeutung der Wörter muss im Satzzusammenhange, nicht in ihrer Vereinzelung gefragt werden;" This declaration is traditionally named as Frege's principle of contextuality. Frege stated it eight years before he pointed out his theoretic distinction between *Sinn* and *Bedeutung*; that is why the word 'Bedeutung' here is usually translated in English as 'meaning': "Never ask for the meaning of a word in isolation, but only in the context of a sentence". As we have yet seen in the Sect. 3.4, the context of a whole sentence is the greatest possible at the semantic level of sentence. We may also ask for the meaning of a word x in the context of a clause to which it belongs, or in the context of some lesser part of this clause as, e.g., of the smallest part U_x. This restatement makes Frege's definition more precise. If we try to recast such a contextuality principle to the level of text, then we would have to say: **"Never ask for the meaning of a sentence in isolation, but only in the context of some meaningful fragment of a text"**. Such a fragment may be chosen in many ways to induce the same contextual meaning of the sentence.

To formalize this definition, let us consider the phonocentric topology at the level of text. Let a sentence x belongs to meaningful fragments (opens) U and V. Then fragmentary meanings $s \in \mathscr{F}(U)$, $t \in \mathscr{F}(V)$ are said to *induce the same contextual meaning of a sentence* $x \in U \cap V$ if there exists some open neighbourhood W of x, such that $W \subseteq U \cap V$ and $\text{res}_{U,W}(s) = \text{res}_{V,W}(t) \in \mathscr{F}(W)$. The identity of fragmentary meanings is understood here accordingly to the criterion claimed by **S**.

This relation '*fragmentary meanings s, t induce the same contextual meaning of the sentence x*' is clearly an equivalence relation. The equivalence class so defined by a fragmentary meaning s is called a *germ* at x of this s, and is denoted by $\text{germ}_x(s)$. The equivalence class of fragmentary meanings agreeing in some open neighbourhood of a sentence x is natural to define as a *contextual meaning* of x. Let $\mathscr{F}_x$ be the set of all contextual meanings of x. Following S. Mac Lane and I. Moerdijk [20, pp. 83,84], this $\mathscr{F}_x$ is nothing else but the inductive limit $\mathscr{F}_x = \varinjlim(\mathscr{F}(V), \text{res}_{V,U})_{V,U \in \mathfrak{O}(x)}$, where $\mathfrak{O}(x)$ is the set of all open neighbourhoods of x.

In the bundle-theoretic terms, we summarize the aforesaid as the following:

Frege's Generalized Contextuality Principle. *Let $\mathscr{F}$ be an adopted sense of reading of a fragment U of an admissible text X. For a sentence $x \in U \subseteq X$, its contextual meaning is defined as a* $\text{germ}_x(s)$ *at x of some fragmentary meaning* $s \in \mathscr{F}(U)$. *The set $\mathscr{F}_x$ of all contextual meanings of a sentence $x \in X$ is defined as the inductive limit* $\mathscr{F}_x = \varinjlim(\mathscr{F}(V), \text{res}_{V,U})_{V,U \in \mathfrak{O}(x)}$, *where $\mathfrak{O}(x)$ is the set of all open neighbourhoods of x, that is the set of all meaningful fragments containing x.*

Remark. Note that for an open singleton $\{x\}$, we may canonically identify $\mathscr{F}_x = \mathscr{F}(\{x\})$.

For the coproduct $F = \bigsqcup_{x \in X} \mathscr{F}_x$, we define now a *projection* map $p\colon F \to X$ by setting $p(\text{germ}_x s) = x$. Every fragmentary meaning $s \in \mathscr{F}(U)$ determines a genuine function $\dot{s}\colon x \mapsto \text{germ}_x s$ to be well-defined on U.

We define the topology on F by taking as a basis for this topology all the image sets $\dot{s}(U) \subseteq F$. For an open $U \subseteq X$, a continuous function $t\colon U \to F$ such that $t(x) \in p^{-1}(x)$ for all $x \in U$ is called a *cross-section*. The topology defined on F makes p and every cross-section of the kind of $\dot{s}$ to be continuous.

For a given topological space X, we have so defined a topological spaces F and a continuous surjection $p\colon F \to X$. In topology, this data (F, p) is called a *bundle over the base space X*. A *morphism* of bundles from $p\colon F \to X$ to

$q\colon G \to X$ is a continuous map $h\colon F \to G$ such that the diagram

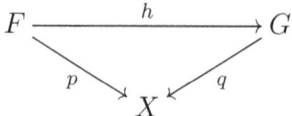

commutes, that is, $q \circ h = p$.

Thus, we have defined a category of bundles over X. A bundle (F, p) over X is called *étale* if $p\colon F \to X$ is a local homeomorphism. It is immediately seen that a bundle of contextual meanings $(\bigsqcup_{x \in X} \mathscr{F}_x, p)$ constructed as above from a given sheaf $\mathscr{F}$ of fragmentary meanings is étale. Thus, for an admissible text X, we have defined the category **Context**(X) of étale bundles (of contextual meanings) over X as a framework for the generalized contextuality principle at the level of text.

The similar definition may be formulated at each semantic level. The definition formulated at the level of sentence returns Frege's classic contextuality principle. Once a semantic level is given, the definition of a contextual meaning for a point x of the corresponding topological space X is stated as germ$_x s$, where s is some fragmentary meaning defined on some neighbourhood U of x.

7. Frege duality

For a given admissible text X, we have defined two categories formalizing the interpretation process, that is, the Schleiermacher category **Schl**(X) of *sheaves of fragmentary meanings* and the category **Context**(X) of *étale bundles of contextual meanings*. Our intention now is to relate them to each other.

We will firstly define a so-called *germ-functor*

$$\Lambda\colon \mathbf{Schl}(X) \to \mathbf{Context}(X).$$

For each sheaf $\mathscr{F}$, it assigns an étale bundle $\Lambda(\mathscr{F}) = (\bigsqcup_{x \in X} \mathscr{F}_x, p)$, where the projection p is defined as above. For a morphism of sheaves $\phi\colon \mathscr{F} \to \mathscr{F}'$, the induced map of fibers $\phi_x\colon \mathscr{F}_x \to \mathscr{F}'_x$ gives rise to a continuous map $\Lambda(\phi)\colon \bigsqcup_{x \in X} \mathscr{F}_x \to \bigsqcup_{x \in X} \mathscr{F}'_x$ such that $p' \circ \Lambda(\phi) = p$; hence $\Lambda(\phi)$ defines a morphism of bundles. Given another morphism of sheaves ψ, one sees easily that $\Lambda(\psi \circ \phi) = \Lambda(\psi) \circ \Lambda(\phi)$ and $\Lambda(\mathrm{id}_{\mathscr{F}}) = \mathrm{id}_F$. Thus, we have constructed a desired germ-functor $\Lambda\colon \mathbf{Schl}(X) \to \mathbf{Context}(X)$.

We will now define a so-called *section-functor*

$$\Gamma \colon \mathbf{Context}(X) \to \mathbf{Schl}(X).$$

We denote a bundle (F, p) over X simply by F. For a bundle F, we denote the set of all its cross-sections over U by $\Gamma(U, F)$. If $U \subseteq V$ are open sets, one has a restriction map $\mathrm{res}_{V,U} \colon \Gamma(V, F) \to \Gamma(U, F)$ that operates as $s \mapsto s|_U$, where $s|_U(x) = s(x)$ for all $x \in U$. It is clear that $\mathrm{res}_{U,U} = \mathrm{id}_{\Gamma(U,F)}$ for any open U, and that the transitivity $\mathrm{res}_{V,U} \circ \mathrm{res}_{W,V} = \mathrm{res}_{W,U}$ holds for all nested opens $U \subseteq V \subseteq W$. So we have constructed obviously a sheaf $(\Gamma(V, F), \mathrm{res}_{V,U})$.

Then for any given morphism of bundles $h \colon E \longrightarrow F$, we have a map $\Gamma(h)(U) \colon \Gamma(U, E) \to \Gamma(U, F)$ defined as $\Gamma(h)(U) \colon s \mapsto h \circ s$, which is obviously a morphism of sheaves. Thus, we have constructed a desired section-functor $\Gamma \colon \mathbf{Context}(X) \to \mathbf{Schl}(X)$.

The fundamental theorem of topology states that the section-functor Γ and the germ-functor Λ establish a *dual adjunction* between the category of presheaves and the category of bundles (over the same topological space); this dual adjunction restricts to a *dual equivalence* of categories (or *duality*) between corresponding full subcategories of sheaves and of étale bundles (see, e.g., [19, p. 179] or [20, p. 89]). Transferred to linguistics in our [28], it yields the following result:

Theorem (Frege Duality). *The generalized compositionality and contextuality principles are formulated in terms of categories those are in natural duality*

$$\mathbf{Schl}(X) \xrightleftharpoons[\Gamma]{\Lambda} \mathbf{Context}(X)$$

established by the section-functor Γ and the germ-functor Λ, the pair of adjoint functors.

Each fragmentary meaning $s \in \mathscr{F}(U)$ determines a function $\dot{s} \colon x \mapsto \mathrm{germ}_x s$ to be well-defined on U; for each $x \in U$, its value $\dot{s}(x)$ is taken in the stalk $\mathscr{F}_x$. This gives rise to a functional representation

$$\eta(U) \colon s \mapsto \dot{s} \qquad (11)$$

defined for all fragmentary meanings $s \in \mathscr{F}(U)$. This representation of a fragmentary meaning s as a genuine function $\dot{s}$ provides an insight into the nature of fragmentary meanings. Each fragmentary meaning $s \in \mathscr{F}(U)$, which has been described in Sect. 5 as an abstract entity, may now be thought of as a genuine function $\dot{s}$ defined on the fragment U of a given text. At the argument (sentence) $x \in U$, this function $\dot{s}$ (representing s) takes its value $\dot{s}(x)$ to be the contextual meaning $\mathrm{germ}_x s$ of this sentence x

$$x \mapsto \dot{s}(x) = \mathrm{germ}_x s \tag{12}$$

Remark. Due to the functional representation (11), the Frege duality is of a great theoretical importance because it allows us to consider any fragmentary meaning s as a genuine function $\dot{s}\colon x_i \mapsto \mathrm{germ}_{x_i} s$ that assigns to each sentence $x_i \in U$ its contextual meaning $\mathrm{germ}_{x_i} s$, and which is continuous on U. It allows us to develop a kind of *dynamic theory of meaning* [28,31,34] describing how, during the reading of the text $X = (x_1, x_2, x_3, \ldots, x_n)$, the understanding proceeds through the discrete time $i = 1, 2, 3, \ldots, n$ as a sequence of grasped contextual meanings $(\dot{s}(x_1), \dot{s}(x_2), \dot{s}(x_3), \ldots, \dot{s}(x_n))$. That gives rise to a genuine function $\dot{s}$ on X representing some $s \in \mathscr{F}(X)$; this s is one of possible meanings of the whole text X interpreted in the sense $\mathscr{F}$.

Moreover, this duality gives a solution to an old problem concerning delicate relations between Frege's compositionality and contextuality principles, in revealing that the acceptance of one of them implies the acceptance of the other (see, e.g., [31]).

8. Sheaf-theoretic dynamic semantics

We sketch now a formal model of a natural language text understanding, which is a kind of dynamic semantics we proposed in [29,31,34]. Our approach describes the dynamics of interpretation process that results in the understanding of a certain meaning of the whole text in its integrity. With the notations used above, for a given text $X = (x_1, \ldots, x_n)$ interpreted in a sense $\mathscr{F}$, we have to describe how a reader finally grasps some global section $s \in \mathscr{F}(X)$ of a sheaf $\mathscr{F}$ of fragmentary meanings.

We consider first a particular case of reading from the very beginning of an admissible text $X = (x_1, x_2, x_3, \ldots, x_n)$ whose size is short enough to allow a reading at one sitting. The general case will be reduced to this particular case by means of the generalized Frege's compositionality principle.

The first sentence x_1 in the order $\leq$ of writing must obviously be understood in the context that consists of its own data. This means that a first sentence

x_1 constitutes an open one-point set $\{x_1\}$. Thus $U_{x_1} = \{x_1\}$, and hence the sentence x_1 should be a minimal element in the specialization order; therefore $\mathscr{F}_{x_1} = \mathscr{F}(\{x_1\})$.

This means that the grasping of a contextual meaning of x_1 is equivalent to the grasping of a fragmentary meaning of the fragment $\{x_1\}$ reduced to this sentence x_1. It is obviously equivalent to the grasping of a global meaning of this sentence x_1 at the semantic level of a sentence considered as a sequence of words. We understand first the theme (topic) of this sentence x_1, and then we understand the rheme (comment) as what is being said in the sense $\mathscr{F}$ concerning this theme. Thus, we have done a descent from the level of text to the level of sentence. In our reasoning, it is the *basis of induction*.

Let us now do the *induction step*. Let us suppose that we have read and understood the text X in the sense $\mathscr{F}$ from the beginning x_1 up to the sentence x_k, $1 < k < n$. That is, we suppose that we have already endowed $X = (x_1, \ldots, x_k)$ with a phonocentric topology and we have built a suite $(\dot{s}_{x_1}, \ldots, \dot{s}_{x_k})$ of contextual meanings of sentences of the open set $U = (x_1, \ldots, x_k)$ of a given text $X = (x_1, \ldots, x_k, \ldots, x_n)$. The suite $(\dot{s}_{x_1}, \ldots, \dot{s}_{x_k})$ of contextual meanings is a continuous function that represents some fragmentary meaning $s \in \mathscr{F}(U)$.

We consider the interpretation process at its $(k+1)$-th step as the choice of an appropriate context $U_{x_{k+1}}$ for x_{k+1} that endows the initial segment $(x_1, \ldots x_{k+1})$ with a particular phonocentric topology among many possible, and allows us to extend the function s defined on the open $(x_1, \ldots, x_k)$ to a function defined on the open $(x_1, \ldots, x_{k+1})$.

The phrase x_{k+1} is read in the context of the fragment $(x_1, \ldots, x_{k+1})$ of the text X. This neighbourhood is the most large context among possible ones we dispose to understand the contextual meaning of x_{k+1}. To grasp the same contextual meaning of x_{k+1}, it suffices to understand only its minimal neighbourhood $U_{x_{k+1}}$. It may be two cases:

Case 1°. It may happen that the understanding of the sentence x_{k+1} is independent of the understanding of $U = (x_1, \ldots, x_k)$, for it constitutes alone its own context $\{x_{k+1}\} = U_{x_{k+1}}$ because there is here a turning point in the narrative, what may be confirmed by various morphologic markers such as the beginning of a new chapter, etc. The contextual meaning $\dot{s}_{x_{k+1}}$ is defined at a point x_{k+1}, and as such it is a continuous function because $\{x_{k+1}\}$ constitutes an open set.

The process of understanding of x_{k+1} is therefore conducted in the same way as that one of the first sentence x_1 whose case we have considered above as the basis of induction.

Note that the interval $U = (x_1, \ldots, x_k)$ is open. We can therefore extend the suite $(\dot{s}_{x_1}, \ldots, \dot{s}_{x_k})$ we supposed to be a continuous function on $U = (x_1, \ldots, x_k)$ to the suite $(\dot{s}_{x_1}, \ldots, \dot{s}_{x_{k+1}})$ that is a continuous function on $(x_1, \ldots, x_{k+1})$.

Case 2°. The understanding of x_{k+1} is reached with the support of the understanding of the preceding sentences of the interval $U = (x_1, \ldots, x_k)$. Not all the sentences in $U = (x_1, \ldots, x_k)$ are required to determine the understanding of x_{k+1}, but only some subsequence of U. Let V be a subsequence of U, such that V contains only sentences those are required for the understanding of x_{k+1}. We define a phonocentric topology on $(x_1, \ldots, x_{k+1})$ by defining $U_{x_{k+1}} = V \cup \{x_{k+1}\}$.

Now we transform the subsequence V into one sentence in such a way that each sentence of V, except the first in the order $\leq$ of writing, begins with "and then" that assembles it to the preceding sentence in order to get a compound sentence. This single lengthy sentence x is made up of all sentences of V in order to get the thematic context that allows the sentence x_{k+1} to express its communicative content. Finally, we join x_{k+1} to x by means of "and then" inserted at the beginning of the sentence x_{k+1}, that transforms x_{k+1} into another sentence x'_{k+1}.

In the text $(x_1, \ldots, x_k, x'_{k+1})$ so defined, the sentence x'_{k+1} constitutes an open one-point set $\{x'_{k+1}\}$ that is understandable in the context of its own data. A contextual meaning of x'_{k+1} is grasped when we understand the rheme of x_{k+1} as being what is said in the sense $\mathscr{F}$ concerning the theme of x_{k+1} in the context defined by the sentences of V. But obviously the contextual meaning of x'_{k+1} is the same as the contextual meaning of x_{k+1}. So we have extended the sequence of contextual meanings $(\dot{s}_{x_1}, \ldots, \dot{s}_{x_k})$ to the sequence $(\dot{s}_{x_1}, \ldots, \dot{s}_{x_{k+1}})$.

Thus, we have done a descent from the level of text to the level of sentence. This trick is inspired by Russell's work *How I write* [38], where he discuss advises he received at the beginning of his career of a writer.

We consider now a general case of reading of an admissible text X whose size does not allow us to finish reading at one sitting. In this case, we consider the reading process of a text X as its covering by some family of meaningful fragments $(U_j)_{j \in J}$ already read, that is $X = \bigcup_{j \in J} U_j$ is an open covering.

Let us suppose given a family $(s_j)_{j \in J}$, where $s_j \in \mathscr{F}(U_j)$ such that all genuine functions $\dot{s}_j \colon x \mapsto \mathrm{germ}_x s_j$ of the corresponding family $(\dot{s}_j)_{j \in J}$ are pairwise compatible, that is $\dot{s}_i \big|_{U_i \cap U_j}(x) = \dot{s}_j \big|_{U_i \cap U_j}(x)$ for all $x \in U_i \cap U_j$.

Let us define the function t on $X = \bigcup_{j \in J} U_j$ as $t(x) = \dot{s}_j(x)$ if $x \in U_j$ for some j. The Frege duality theorem states that $t = \dot{s}$ where $s \in \mathscr{F}(X)$ is a composition of the family $(s_j)_{j \in J}$, whose existence is ensured by the generalized Frege's compositionality principle.

The formalization of the interpretation process as an extension of a function introduces a *dynamic* view of semantics, and its theory deserves the term *inductive* because the domain of a considered function is naturally endowed with two order structures, that is, the linear order of writing $\leq$ and the specialization order $\preceq$ of context-dependence. We have outlined so a sheaf-theoretic framework for the dynamic semantics of a natural language, where the understanding of a text X in some sense $\mathscr{F}$ is described as a process of step-by-step grasping for each sentence x_i of only one contextual meaning $\dot{s}(x_i)$ from the fiber $\mathscr{F}_{x_i}$ lying over x_i in the étale bundle $\mathbf{Context}(X)$ of contextual meanings.

9. Algebraic semantics versus sheaf-theoretic semantics

According to T. M. V. Janssen, the compositionality principle is a basis for *Montague grammar, Generalized phrase structure grammar, Categorial grammar* and *Lexicalized tree adjoining grammar*. These theories propose the different notions of meaning, but follow the compositionality principle in its *standard interpretation*:

> A technical description of the standard interpretation is that syntax and semantics are algebras, and meaning assignment is a homomorphism from syntax to semantics. (T. M. V. Janssen [15, p. 116])

Let us consider this conception of standard interpretation as an algebraic homomorphism $f \colon A \to B$, where the algebra A is representing Syntax, and the algebra B is representing Semantics.

Whatever the algebras A and B would be, the homomorphism f is a function in a set-theoretic paradigm. Given the function f, we define the relation q on A so that $\langle x, y \rangle \in$ q, if and only if $f(x) = f(y)$. Clearly, this q is an equivalence relation on A. Any given element $a \in A$ lies in precisely one equivalence class; if $f(a) = b \in B$, then the equivalence class of a is $f^{-1}(b)$. The set of equivalence classes is denoted by $A/\mathfrak{q}$ and called the *quotient set* of A by $\mathfrak{q}$. Let the equivalence classes of a be denoted by $a^{\mathfrak{q}}$. If with each $x \in A$ we associate $x^{\mathfrak{q}}$, we obtain a function $\varepsilon \colon A \to A/\mathfrak{q}$, called the identification associated with $\mathfrak{q}$. Clearly the function ε is surjective, by definition. Following the Theorem 3.1 of [4, p. 15], there is a decomposition of f:

$$\begin{array}{ccc} A & \xrightarrow{f} & B \\ \varepsilon \downarrow & & \uparrow \mu \\ A/\mathfrak{q} & \xrightarrow{f'} & f(A), \end{array}$$

where $\varepsilon \colon A \to A/\mathfrak{q}$ is a surjection, $f' \colon A/\mathfrak{q} \to f(A)$ is a bijection, and $\mu \colon f(A) \to B$ is an injection.

In the category of algebras, an injective homomorphism is called a monomorphism; a surjective homomorphism is called an epimorphism; every bijective homomorphism should be an isomorphism (usually defined as an invertible homomorphism). The above decomposition theorem remains valid in the category of algebras; moreover, $A/\mathfrak{q}$ and $f(A)$ may be endowed with the structures of algebras in such a way that ε, f', μ become homomorphisms.

Linguistically speaking, the Syntax and the Semantics should not be one and the same theory. Thus, the meaning assignment homomorphism $f \colon A \to B$ should not be an isomorphism. Nor should this homomorphism f be a monomorphism; otherwise the Syntax A would be isomorph with a proper part of the Semantics B. Hence, f should be an epimorphism with a non-trivial kernel that is defined to be the congruence relation $\mathfrak{q}$ described above. Two different elements of an algebra A representing Syntax are congruent if and only if they are mapped to the same element of an algebra B representing Semantics. Thus, the different syntactical objects will have one and the same meaning as their value under such a homomorphism $f \colon A \to B$. Thus, an algebraic approach is pertinent in the study of synonymy, but the problems of polysemy do resist to algebraic semantic theories. Moreover, an algebraic semantic, of whatever kind, is always static because the meaning $f(x) \in B$ of a syntactic element $x \in A$ under the homomorphism f is calculated in the algebra B just after the calculation of meanings of all syntactic components of x was done.

However, when studying the process of interpretation of a natural language text, we are confronted with a quite another situation. Any admissible text is really a great universe of meanings to be disclosed or reconstructed in the process of reading and interpretation. But these multiple meanings are offered to a reader as got identified in a single text. Thus, in the process of interpretation of a natural language text, the reader is confronted with a surjection Semantics → Syntax. Note that we have turned the arrow round, and this is a **paradigmatic turn**.

From a sheaf-theoretic point of view, a discourse interpretation activity proceeds as the following: The text X under interpretation is a given sequence of its sentences $x_1, x_2, x_3, \ldots, x_n$; this is a finite combinatorial object from the universe of Syntax. Over these sentences, there is another sequence of stalks of their contextual meanings $\mathscr{F}_{x_1}, \mathscr{F}_{x_2}, \mathscr{F}_{x_3}, \ldots, \mathscr{F}_{x_n}$; this is a potentially infinite and, in some degree, a virtual object from the universe of Semantics. The total disjoint union of all these stalks, that is, the coproduct $F = \bigsqcup_{x \in X} \mathscr{F}_x$ is projected by a local homeomorphism p on the text X. Thus, we have the surjective projection $p \colon F \to X$ from Semantics to Syntax. The challenge of text interpretation is to create a global cross-section s of the projection p; this s is constructed as a sequence of grasped step-by-step contextual sentences' meanings $(\dot{s}(x_1), \dot{s}(x_2), \dot{s}(x_3), \ldots, \dot{s}(x_n))$; it gives rise to a genuine function $\dot{s}$ on X representing some global cross-section $s \in \mathscr{F}(X)$; this s is one of all possible meanings of the whole text X interpreted in the sense $\mathscr{F}$.

The proposed sheaf-theoretic semantics answers to crucial questions about *what* the fragmentary meanings are and *how* they are formally composed. That is, we consider the reading process of a fragment U in a sense $\mathscr{F}$ as its covering by some family of subfragments $(U_j)_{j \in J}$, each read in a unique session. Any family $(s_j)_{j \in J}$ of pairwise compatible fragmentary meanings $s_j \in \mathscr{F}(U_j)$ under a functional representation (11) gives rise to a family $(\dot{s}_j)_{j \in J}$ of genuine functions (where each $\dot{s}_j$ is defined on U_j by (12)), those are pairwise compatible in the sense that $\dot{s}_i \big|_{U_i \cap U_j}(x) = \dot{s}_j \big|_{U_i \cap U_j}(x)$ for all $x \in U_i \cap U_j$. Let a cross-section s be defined on $U = \bigcup_{j \in J} U_j$ as $s(x) = \dot{s}_j(x)$ if $x \in U_j$ for some j. Then this cross-section s over U is clearly a composition of the family $(\dot{s}_j)_{j \in J}$ as it is claimed by the generalized Frege's compositionality principle.

The sheaf-theoretic conception of compositionality serves as the basis for the dynamic semantics we discussed in the Sect. 8. This approach has an advantage because **1°** it extends the area of semantics from the level of sentence or phrase to the level of text or discourse, and it gives a uniform treatment of discourse interpretation at each semantic level (word, sentence, paragraph, text); **2°** it takes into theoretical consideration the polysemy of words, sentences and texts.

10. Sheaf-theoretic formal hermeneutics

Our approach provides a mathematical model of a text interpretation process while rejecting attempts to codify interpretative practice as a kind of calculus. In a series of previous papers [28, 29, 31–33], we named this text interpretation theory as *formal hermeneutics*. It presents a formal framework for syntax and semantics of texts written in some unspecified natural language, say for us English, French, German, Russian considered as a means of communication. The object of study in this formal hermeneutics are couples $(X, \mathscr{F})$ made up of an admissible text X and a sheaf $\mathscr{F}$ of its fragmentary meanings; we call any such a couple *textual space*. But this representation is possible only in the realm of a language following the famous slogan of Wittgenstein, "to understand a text is to understand a language". Rigorously, this claim may be formulated in the frame of category theory. Likewise, the present sheaf-theoretic formal semantics describes a natural language in the *category of textual spaces* **Logos**. The objects of this category are couples $(X, \mathscr{F})$, where X is a topological space naturally attached to an admissible text and $\mathscr{F}$ is a sheaf of fragmentary meanings defined on X; the morphisms are couples $(f, \theta) \colon (X, \mathscr{F}) \to (Y, \mathscr{G})$ made up of a continuous map $f \colon X \to Y$ and a f-morphism of sheaves θ that respects the concerned sheaves; such an f-morphism is formally defined as $\theta \colon \mathscr{G} \to f_*\mathscr{F}$, where f_* is a well-known *direct image* functor (see, e.g., [31]).

Given any admissible text E considered to be fixed forever as, for instance, the Scripture, it yields a full subcategory **Schl**(E) in the category **Logos** of all textual spaces. Named after Schleiermacher, the category **Schl**(E) describes the exegesis of this particular text.

The *topological syntax* and the *dynamic sheaf-theoretic semantics* based on Frege duality, as well as different categories and functors related to discourse and text interpretation process are the principal objects of study in the *sheaf-theoretic formal hermeneutics* as we understand it.

References

[1] R. Bott and L. W. Tu. *Differential Forms in Algebraic Topology*. Springer-Verlag, Berlin-Heidelberg-New York, 1982.

[2] R. Capozzi. *Reading Eco: An Anthology*. Advances in Semiotics. Indiana University Press, Bloomington and Indianapolis, 1997.

[3] R. Carnap. *Meaning and Necessity*. The University of Chicago Press, Chicago, 1947.

[4] P. M. Cohn. *Universal Algebra*. D. Reidel Publishing Company, Dordrecht, 1981.

[5] G. Deleuze and F. Guattari. *A Thousand Plateaus: Capitalism and Schizophrenia, Vol. 2*. University of Minnesota Press, Minneapolis, MN, 1987.

[6] U. Eco. *Semiotics and the Philosophy of Language*. Advances in Semiotics. Indiana University Press, Bloomington, 1986.

[7] U. Eco, R. Rorty, J. Culler, and Ch. Brook-Rose. Overinterpreting texts. In S. Collini, editor, *Interpretation and overinterpretation*. Cambridge University Press, UK, 1992.

[8] M. Erné. The ABC of Order and Topology. In H. Herrlich and H.-E. Porst, editors, *Research and Exposition in Mathematics Vol. 18. Category Theory at Work*, pages 57–83. Heldermann, Berlin, 1991.

[9] G. Frege. *Die Grundlagen der Arithmetik. Eine logisch mathematische Untersuchung über den Begriff der Zahl*. Verlag von W. Koebner, Breslau, 1884.

[10] G. Frege. Logic in mathematics. In H. Hermes, F. Kambartel, and F. Kaulbach, editors, *Posthumous Writings*, pages 203–252. University of Chicago Press, Chicago, IL, 1979.

[11] J. H. Greenberg. Some Universals of Grammar with Particular Reference to the Order of Meaningful Elements. In J. H. Greenberg, editor, *Universals of Language*, pages 73–113. MIT Press, Cambridge, MA, 1963.

[12] A. J. Greimas. *Du sens*. Seuil, Paris, 1970.

[13] A. Grothendieck and J. A. Dieudonné. *Eléments de Géométrie Algébrique,* chap. I (2e éd.). Springer-Verlag, Berlin-Heidelberg-New York, 1971.

[14] J. Hintikka. A hundred years later: The rise and fall of Frege's influence in language theory. *Synthese*, 59(1):27–49, April 1984.

[15] T. M. V. Janssen. Frege, contextuality and compositionality. *Journal of Logic, Language, and Information*, 10:115–136, 2001.

[16] S. Kahane. Grammaires de dependance formelles et theorie Sens-Texte, Tutoriel. In *Actes TALN'2001*, volume 2, Tours, France, 2001. Université François-Rabelais.

[17] E. Khalimsky, R. Kopperman, and P. Meyer. Computer graphics and connected topologies on finite ordered sets. *Topology and its Applications*, 36:1–17, 1990.

[18] A. R. Lacey. *A Dictionary of Philosophy*. Routledge, London, 3 edition, 1996.

[19] J. Lambek and P. S. Scott. *Introduction to Higher Order Categorical Logic*. Cambridge University Press, Cambridge, 1986.

[20] S. Mac Lane and I. Moerdijk. *Sheaves in Geometry and Logic. A First Introduction to Topos Theory*. Springer-Verlag, Berlin-Heidelberg-New York, 1992.

[21] Yu. I. Manin. *A Course in Mathematical Logic*. Springer-Verlag, New York Heidelberg Berlin, 1977.

[22] W. C. Mann and S. A. Thompson. Rhetorical Structure Theory: A Theory of Text Organization. URL = <http://www.sfu.ca/rst/pdfs/Mann_Thompson_1987.pdf>, University of Southern California, Information Sciences Institute (ISI), Jun. 1987. ISI/RS-87-190.

[23] J. P. May. Finite Topological Spaces. URL = <http://www.math.uchicago.edu/~may/MISC/FiniteSpaces.pdf>, 2003.

[24] R. Montague. English as a formal language. In B. Visentini, et al., editor, *Linguaggi nella Società e nella Tecnica*, page 217. Edizioni di Comunità, Milan, 1970.

[25] B. H. Partee. *An Invitation to Cognitive Science, Vol. 1: Language*, chapter 11, Lexical Semantics and Compositionality, pages 311–360. The MIT Press, Cambridge, Massachusetts, 2nd edition, 1995.

[26] F. J. Pelletier. Did Frege Believe Frege's Principle? *Journal of Logic, Language, and Information*, 10:87–114, 2001.

[27] B. Pottier. *Théorie et analyse en linguistique*. Hachette, Paris, 1992.

[28] O. Prosorov. Compositionality and Contextuality as Adjoint Principles. In M. Werning, E. Machery, and G. Schurz, editors, *The Compositionality of Meaning and Content, (Vol. II : Applications to Linguistics, Psychology and Neuroscience)*, pages 149–174. Ontos-Verlag, Frankfurt, 2005.

[29] O. Prosorov. Sheaf-theoretic formal semantics. *TRAMES Journal of the Humanities and Social Sciences*, 10(1):57–80, 2006a. ISSN 1406-0922.

[30] O. Prosorov. Semantic Topologies as Linguistic Universals. In T. Stolz, editor, *The 39th Annual Meeting of the Societas Linguistica Europaea (SLE). Relativism and Universalism in Linguistcs*, pages 109–110, Bremen, Germany, 2006b. IAAS.

[31] O. Prosorov. Topologies et faisceaux en sémantique des textes. Pour une herméneutique formelle. URL = <http://www.theses.fr/2008PA100145>, Université Paris X, Nanterre, France, Dec. 2008. PhD thesis.

[32] O. Prosorov. Formal hermeneutics as semantics of texts. In Yu. N. Solonin, editor, *Philosophy in the dialogue of cultures*, pages 351–359. St. Petersburg State University, Russia, 2010. In Russian.

[33] O. Prosorov. Topologies and Sheaves in Linguistics. In R. Bhatia, editor, *The International Congress of Mathematicians (ICM 2010). Abstracts of Short Communications and Posters*, pages 623–624, Hyderabad, India, 2010. Hindustan Book Agency.

[34] O. Prosorov. A Sheaf-Theoretic Framework for Dynamic Semantics. In A. Butler, editor, *Proceedings of the Eight International Workshop of Logic and Engineering of Natural Language Semantics (LENLS 8)*, pages 52–67, Takamatsu, Japan, 2011. JSAI.

[35] W. V. Quine. *Theories and Things*. Harvard University Press, Cambridge, 1981.
[36] F. Rastier. Communication ou transmission. *Césure*, 8:151–195, 1995.
[37] F. Rastier. Problématiques du signe et du texte. *Intellectica*, 23:11–52, 1996.
[38] B. Russell. How I Write. In A. G. Bone, editor, *The Collected Papers of Bertrand Russell (Volume 28): Man's Peril, 1954-55*, pages 102–104. Routledge, London, 1983.
[39] J.-P. Serre. *Local Fields*. Springer-Verlag, New York, 1979.
[40] D. Sleator and D. Temperley. Parsing English with a Link Grammar. Computer Science technical report CMU-CS-91-196, Carnegie Mellon University, Pittsburgh, Oct. 1991.
[41] D. Temperley, D. Sleator, and J. Lafferty. Link Grammar. URL = <http://www.link.cs.cmu.edu/link/>, 2008.
[42] M. Werning. The Reasons for Semantic Compositionality. First Düsseldorf Summer Workshop "Philosophy and Cognitive Science", 2003.

Received: 19 April 2017

Venus Homotopically

Andrei Rodin

Institute of Philosophy RAS - St. Petersburg State University, Russia
andrei@philomatica.org

Abstract: The identity concept developed in the Homotopy Type theory (HoTT) supports an analysis of Frege's famous *Venus* example, which explains how empirical evidences justify judgements about identities. In the context of this analysis we consider the traditional distinction between the extension and the intension of concepts as it appears in HoTT, discuss an ontological significance of this distinction and, finally, provide a homotopical reconstruction of a basic kinematic scheme, which is used in the Classical Mechanics, and discuss its relevance in the Quantum Mechanics.

Keywords: identity, homotopy type theory, intension, kinematics.

1. Introduction

According to Frege

> Identity is a relation given to us in such a specific form that it is inconceivable that various kinds of it should occur [7, p. 254].[1]

In the second half of the 20th century this view was challenged by Peter Geach [11] who developed a theory of what he called the *relative identity*. Contrary to Frege, Geach holds that the identity concept allows for specifications, which depend on certain associated sortals.[2]

Geach's unorthodox view on identity has been never developed into an independent formal logical system and remain today rather marginal [2]. However

---

I thank Danielle Macbeth for very useful comments and discussion.

[1] "Die Identitaet ist eine so bestimmt gegebene Beziehung, dass nicht abzusehen ist, wie bei ihr verschiedene Arten vorkommen können."

[2] Let a, b be parallel lines on Euclidean plane, in symbols $a//b$. Given that $//$ is an equivalence relation, Frege suggests to "take this relation as identity" (in symbols $a = b$) and thus obtain a new abstract object called *direction* [8, p. 74e]; (for a more detailed reconstruction of Frege's abstraction see [25]). Geach's analysis of the same example is different: according to Geach $a = b$ reads "a and b are the same *as direction*" even if a and b are different *as lines*.

© The Author(s) and College Publications 2017

the idea that, contrary to Frege's view, the identity concept can and should be diversified more recently reappeared in a different form in Martin-Löf's Constructive Type theory (MLTT) [15] and in the yet more recent geometrical interpretation of MLTT called Homotopy Type theory (HoTT) [17]. Unlike Geach's original proposal, which has hardly had any influence outside the philosophical logic, HoTT is a piece of new interesting mathematics and mathematical logic closely relevant to Computer Sciences.

The aim of this paper is to analyze some of Frege's ideas about identity in terms of the identity concept as it appears in MLTT and HoTT. In this way I hope to make the technical MLTT-HoTT identity concept more philosophically meaningful and apt to possible applications in science.

The rest of the paper is organized as follows. In the next Section I present Frege's *Venus* example and overview its analysis by the author. In the following three Sections I introduce a basic fragment of MLTT and HoTT and discuss the difference between extensional and intensional versions of these theories. Then I present a reconstruction of Frege's *Venus* with HoTT and discuss in this context an ontological impact of the distinction between extensions and intensions. Finally, I extend my reconstruction of *Venus* to what I call the Basic Kinematic Scheme used in the Classical Mechanics and briefly discuss its relevance in the Quantum Mechanics.

2. How identity statements are known?

Some identity statements are trivial and non-informative while some other are highly informative and in some cases very hard to prove. For example "$2 = 2$" (in words "two is two") is trivial, "2 is the only even prime number" is somewhat more informative but easy (since it follows immediately from the definitions of "even" and "prime"), while "2 is the biggest power n such that the equation $x^n + y^n = z^n$ has a solution in natural numbers" is both informative and highly non-trivial (it is a famous theorem conjectured by Pièrre Fermat in 1637 and proved by Andrew Wiles in 1994).

A non-mathematical example of the same kind is given by Frege in his classical *On Sense and Reference* [5] (English translation [6]). Frege considers three different names - *Venus, Morning Star* and *Evening Star* - which all refer to the same planet. Frege wonders how it is possible that while the identity statement

$$\textit{Morning Star} \text{ is } \textit{Morning Star} \qquad (1)$$

and the identity statement,

$$\textit{Morning Star} \text{ is } \textit{Venus} \qquad (2)$$

(which expresses a mere linguistic convention according to which "Venus" is an alternative name of *Morning Star*) are <u>trivial</u> the statement

$$\textit{Evening Star} \text{ is } \textit{Morning Star} \qquad (3)$$

is a non-obvious astronomical fact that needs an accurate justification, which involves both a solid theoretical background and appropriate observational data.[3]

Where does the difference between informative and non-informative identity statements come from? Frege does not provide a full answer to this question but does provide a theoretical framework for answering it. For this end he distinguishes between the *sense* and the *reference* of any given linguistic expression.[4]

Whether an identity statement is informative or not depends on its sense (and hence on the sense of its constituents[5]) but not on its reference. Thus there is no mystery in the fact that statements of the form $a = a$ are always trivial (assuming that both the sense and the reference of "a" is fixed), while statements of the form $a = b$ can be either trivial (when terms a, b have the same sense) or non-trivial (when terms a, b have different senses). In expressions (1) and (2) both terms have the same meaning (even if in (2) these terms differ linguistically) but in (3) the senses of two terms are different. This is why (1) and (2) are trivial but (3) is not.

---

[3]Instead of talking about trivial and non-trivial statements Frege uses here a Kantian distinction between synthetic and analytic judgements and talk about the "cognitive value" of the corresponding "thoughts". I shall not use Frege's original way of expressing these ideas in my presentation.

[4]Some writers who want to stress the originality of Frege's logical ideas leave Frege's German terms for *sense* and *reference* (Sinn und Bedeutung) without translation even if they write in English. I use standard English translations instead.

[5]This follows from a general principle known as the *compositionality* of meaning. Frege is sometimes credited for the alleged invention of this principle but the true history is more complicated [16].

Obviously this is not a complete explanation. Frege's system of symbolic logic aka *Begriffsschrift* [3] does not do full justice to his own distinction between the sense and the reference of a linguistic expression [14]. It provides rules for operating with references of propositions (i.e., with their truth-values) but does not provide rules for operating with their senses. So Frege points to a problem but leaves it largely open. More recently a number of so-called *intensional* logical systems have been developed, some of which have been explicitly motivated by the idea of formalizing certain aspects of Frege's *sense*. The distinction between extensions and intensions of linguistic expressions and logical terms is closely related to Frege's distinction between sense and reference [1]. It has a long history in logic and its philosophy and turns out to be instrumental in MLTT-HoTT, as we shall now see. In the next section I explain the technical meaning of this distinction in MLTT and then discuss its philosophical underpinning.

3. Extension and intension

MLTT [15] comprises two different forms of identity concept.[6] These two forms of identity should look familiar to anyone who has at least a rudimentary experience in programming. It's one thing to assign to a certain symbol or symbolic expression its semantic value (which can be a number, a character, a string of characters and many other things) and it is quite a different thing to state that certain things are equal. (Hereafter I use words "equal" and "identical" interchangeably.) Only in the latter case one forms a *proposition*, which typically has precisely one of the two *Boolean* values: True and False. Outside the context of programming a similar distinction can be made between *naming* or making some more elaborated linguistic convention, on the one hand, and making a *judgement* to the effect that certain things are equal, on the other hand. It is one thing to adopt and use the convention according to which the goddess' name *Venus* is an alias for what is also known as the *Morning Star*, and it is, of course, quite a different thing to judge and state that two apparently different celestial objects known as the *Morning Star* and the *Evening Star* are, in fact, one and the same. In the latter case it is appropriate to ask for a proof.

---

[6]The original version of this theory involves *four* different kinds of identity [15, p. 59]. I simplify the original account by deliberately confusing some syntactic and semantical aspects. Then we are left with the two forms of identity described below in the main text.

Such a demand is obviously pointless in the former case.

The first kind of identity (one related to conventions) Martin-Löf calls *definitional* or *judgmental*; the second kind of identity he calls *propositional*. Following [17] I shall use sign "≡" for the definitional identity and the usual sign "=" for the propositional identity. Further, we should take *typing* into account. In MLTT both kinds of identity apply only to terms of the same type.[7] Typing is expressed in the notation as follows:

$$s, t : A \qquad (4)$$

is a judgment that states that terms s, t are of type A. Formula

$$s \equiv_A t \qquad (5)$$

stands for a judgement, which is tantamount to a convention (aka *definition*) according to which terms s, t of the same type A have the same meaning. Given (5) one says that s, t are *definitionally equal*. The expression

$$s =_A t \qquad (6)$$

in its turn, stands for a *proposition* saying that terms s, t of type A are equal. Unlike (5) formula (6) by itself does not express a judgment but only represents a *type*. Under the intended proof-theoretic semantic of MLTT any term p of this type is thought of as a *proof* of the corresponding proposition; in the proof-theoretic jargon proofs are also called *witnesses* and sometimes *evidences*. So the following judgement

$$p : s =_A t \qquad (7)$$

states that terms s, t are (propositionally) equal as this is evidenced by proof p.

Let us now see what kind of thing such a proof p can possibly be. In MLTT definitionally equal terms are interchangeable *salva veritate* as usual. Under the intended semantic of this theory this means that definitionally equal terms are interchangeable as proofs. This property of ≡ and the reflexivity of = justify the following rule

---

[7]I leave now aside how identity is applied in MLTT to *types* on the formal level. It is sufficient for my present purpose to talk about the "same type" and "different types" in MLTT informally.

$$\frac{s \equiv_A t}{p : s =_A t} \tag{8}$$

where $p \equiv refl_s$ is built canonically [17, p. 46]. In words: the definitional identity (equality) implies the propositional identity (equality).

The converse rule is called the *equality reflection rule* or ER for short:

$$\frac{p : s =_A t}{s \equiv_A t} \tag{ER}$$

In words: the propositional identity implies the definitional identity.

ER does not follow from other principles of MLTT but may be assumed as an independent principle. In this case one obtains a version of MLTT, which is called (definitionally) *extensional*. MLTT without ER is called *intensional*. It can be shown that in the extensional MLTT any (propositional) identity type $s =_A t$ is either empty or has a single term, namely $refl_s$, which is the canonical proof of this identity "by definition".

We see that ER makes the distinction between the definitional and the propositional identity purely formal and epistemologically insignificant. This feature of extensional MLTT can be viewed as a desirable conceptual simplification but it comes with a price. A significant part of this price concerns computational properties in MLTT and is important for applications of this theory in programming: while the intensional MLTT is decidable but the extensional MLTT is not. I shall not discuss this technical feature in this paper. Instead I shall argue that the intensional MLTT has also important epistemic advantages over its extensional cousin.

4. Fixing identities or leaving them evolving?

As we have seen in the extensional MLTT every identity is grounded in a definition. In order to apply this formal theory in reasoning one needs to fix in advance, via appropriate definitions, exact identity conditions for all objects involved in a given reasoning. This logical and epistemic requirement is known in the form of slogan "no entity without identity" due to Quine. It is interesting to notice that Quine himself does not accept this slogan without reservations. In Quine's view the slogan applies only in scientific reasoning and, moreover, only in the contemporary form of scientific reasoning. Bulk terms (aka mass terms) like "water", according to Quine, are remnants of an archaic logical scheme, which does not involve the individuation in its today's form. Quine further speculates that the contemporary "individuative, object-oriented conceptual scheme" can be replaced in a future by a different scheme, that will provide a "yet unimagined pattern beyond individuation" [18, p. 24].[8] In what follows I argue that the *intensional* MLTT along with HoTT provides such a pattern "beyond individuation" or at least a pattern of individuation beyond its usual extensional mode. But beforehand I would like to stress once again that the standard extensional mode of individuation is not sufficient for certain well-recognized and important scientific purposes. Frege's *Venus* example, if one takes it seriously, demonstrates this clearly. Fixing the identity of *Morning Star* and *Evening Star* and *Venus* via a definition is a prerequisite for applying a standard extensional logical scheme in any reasoning about this celestial object. This condition makes it impossible to support with such a scheme a reasoning, in which the identity of the *Morning Star* and the *Evening Star* is established on the basis of certain sufficient evidences.

Frege's example shows that "half-entities inaccessible to identity" [18, p. 23] may look more familiar than Quine's colorful language suggests. In the *Venus* case we deal with a relatively innocent violation of "no entity without identity" requirement. We start with certain well-defined objects such as the *Morning Star* and the *Evening Star* but do not exclude the possibility that these objects can be eventually proved to be the same - even if we know that this fact does not follow from the corresponding definitions. Following Quine one may think of further deviations from the standard extensional individuating scheme and speculate about a possible conceptual scheme, which does not use the definitional form of identity at all. I do not pursue this further project in this paper. Instead I show how the innocent-looking modification of the extensional individuating scheme, which has been just explained, results into a remarkable diversification of the standard identity concept.

---

[8]Here is the full quote:
"[W]e may have in the bulk term a relic, half vestigial and half adapted, of a pre-individuative phase in the evolution of our conceptual scheme. And some day, correspondingly, something of our present individuative talk may in turn end up, half vestigial and half adapted, within a new and as yet unimagined pattern beyond individuation. Transition to some such radically new pattern could occur either through a conscious philosophical enterprise or by slow and unreasoned development along lines of least resistance. A combination of both factors is likeliest [...]." [18, p. 24]

5. Higher identity types

Recall that the intensional version of MLTT has been introduced above via a negative characteristic: it is the core version of MLTT *without* the additional reflexion rule ER.

The absence of ER allows for constructing further identity types as follows. Suppose we have a propositional identity type and a pair of terms of this type:

$$s', t' : s =_A t$$

Terms s', t' witness here the identity of terms s, t. It may now happen that these two witnesses are, in fact, one and the same - as witnessed by two further terms s'', t'':

$$s'', t'' : s' =_{s=_A t} t'$$

Thus we get a tower-like construction, which comprises identity types of two different "levels". It can be further continued indefinitely. In the general case such a construction may have, of course, more than just two elements on each level.

Until the late 1990-ies structural properties of this formal syntactic construction remained opaque. Since the *intentionality* in MLTT is a mere lack of *extensionality*, any model of the extensional MLTT also qualifies as a model of the intensional version of this theory. In 1994–1998 Hofmann and Streicher [12, 13] published the first non-extensional model of MLTT where the first-level identity types were modeled by abstract groupoids. This model allows the first-level identity types (i.e., types of the form $s =_A t$ where A is a type other than identity) to have multiple non-trivial terms (proofs) but does not allow the same for higher identity types. In other words, this model verifies the condition called "extensionality one dimension up". A deeper insight into the structure of higher identity types has been obtained around 2006 when Awodey and Voevodsky independently observed that the abstract groupoids of Hofman and Streicher's model can be thought of as *fundamental groupoids* (i.e., groupoids of all continuous *paths*) of topological spaces and be further extended to homotopy- and higher-homotopy groupoids of the same spaces, which model higher-order identity types of MLTT. Thus the Homotopy theory allows for building models of MLTT, which are "intensional all the way up".

In such models the identity types of all levels are modeled uniformly. This discovery marked the emergence of a new theory known today under the name of Homotopy Type theory and of a closely related foundational project called the Univalent Foundations of mathematics. For a systematic exposition of HoTT I refer the reader to [17].[9]

Unlike Russell's type theories HoTT does not form its hierarchy of types by considering, first, classes of individuals, second, classes of such classes, and so on. The hierarchy of types in HoTT is of a geometric or, more precisely, homotopic nature. Sets are taken to be types of zero level. Terms of 0-types are points having no non-trivial paths between them. Terms of 1-types are points provided with non-trivial paths between them, but not allowing for non-trivial homotopies between these paths. Terms of 2-types allow for paths and non-trivial homotopies but not for non-trivial higher homotopies. And so on.[10]

Notice the cumulative character of the homotopical hierarchy of types described above. Considered in isolation, the identity types $s =_A t$ and $s' =_{s=_A t} t'$ have exactly the same formal properties; correspondingly, there is no intrinsic difference between spaces of points, spaces of paths (aka path spaces) and homotopy spaces of all levels. As usual in the 20-th century geometry one is allowed in HoTT to imagine elements of spaces however one may find it useful - say, as beer mugs after Hilbert's legendary suggestion. However the fact that every path s' is not simply an individual of certain sort but an object with a pair of endpoints s, t, allows for the two-level construction described above. Similarly one obtains n-level constructions by using homotopies and higher homotopies. In order to describe the resulting hierarchy more formally and more precisely we need to complement the bottom-up description used so far but a top-down one. For this end we assume from the outset that every type is a space provided with its infinite-dimensional fundamental groupoid. Then we specify the case of 0-types such that all its paths, homotopies and higher homotopies are trivial; then the case of 1-types such that all its homotopies and higher homotopies (but not paths!) are trivial, and so on.

---

[9]Since this area of research is rapidly developing, the 2013 book [17] does not include certain new results and developments. However it provides an systematic introduction, which is more than sufficient for my present purpose.

[10]Here I follow [17, p. 99–100]. On an alternative count the 0-type is a single point, 1-types are propositional types while sets are 2-types. The count adopted in [17] appears more natural from a logical point view (given the usual understanding of logic) while the latter count used by Voevodsky in his lectures appears more natural from a geometric point of view.

A given n-type can be transformed into its underlying m-type with $m < n$ by forgetting (or, more precisely, by trivializing) its higher-order structure of all levels $> m$. Such an operation is called in HoTT *truncation*. It will play an important role in what follows.

The logical significance and the possible epistemic function of higher identity types in MLTT are not yet well understood. The present work is an attempt of filling a part of this gap. In what follows I consider only 0- and 1-types and leave a study of higher identity types for a future work.

6. Is Frege's *Venus* example linguistic?

Apparently Frege treats his *Venus* example as purely linguistic on equal footing with his other examples, which involve Alexander the Great, Columbus, Napoleon, Kepler dying in misery, Bucephalus and what not. Accordingly, the main result of his classical paper [5,6], namely the distinction between the sense and the reference of a given linguistic expression, belongs primarily to the philosophy of language. Frege scholarship mostly follows Frege in this respect: a linguistic leaning aka *linguistic turn* became a brand mark of the influential *Analytic* branch of the 20th century and today's philosophy. It is quite remarkable, however, that when Frege first introduces and explains the *Venus* problem he does this not only in linguistic terms:

> The discovery that the rising Sun is not new every morning, but always the same, was one of the most fertile astronomical discoveries. Even today the identification of a small planet (i.e., an asteroid - A.R.) or a comet is not always a matter of course. [6, p. 56]

The idea that a logical analysis of ordinary language can be helpful for solving problems of object identification in science in general and in astronomy in particular is based on Frege's strong assumption according to which the identity concept is the same in all these cases, so that "it is inconceivable that various kinds of it should occur" (see the full quote and the reference in the above Introduction). Without trying to challenge this approach on the methodological level I shall provide here an alternative analysis of the same example, which takes its physical content and, even more importantly, its related mathematical form more seriously and applies some basic elements of HoTT introduced in the previous Section. As a matter of course this reconstruction is not intended to be a piece of mathematical physics. Nevertheless it provides a novel formal approach to traditional metaphysical issues concerning the identity through time and motion, which may be possibly helpful for dealing with identity-related problems of modern physics [9, 10].

Frege's remark about the rising Sun quoted above applies both to the *Morning Star* (MS for short) and to the *Evening Star* (ES). These two putative objects are posited as invariants of certain sets of observations made in different places at different times by different people with different astronomical instruments and with the naked eye. However for the sake of the example I leave now this complex underlying structure aside and boldly assume that MS and ES are provided with some appropriate *definitions*, which allow all observers to identify these objects unambiguously. How a *proof* of identity $MS = ES$ may look like in a realistic astronomical context? Classical Celestial Mechanics (CM), or more precisely a very basic fragment of CM that I shall call Basic Kinematic Scheme (BKS) and discuss in more detail in Section 8, provides a definite answer to this question. In order to prove that $MS = ES$ it is necessary and sufficient to present a *continuous path* aka trajectory p, which connects MS and ES and thereby shows that these "two" objects are in fact one and the same. The wanted trajectory p is itself a typical physical object: it is obviously theoretically-laden, it has a canonical mathematical representation, and it is accessible for observations which allow for empirical checks of its theoretically predicted properties. Providing such a proof p amounts to a combination of theoretical work and observation, which is typical in astronomy and any other mature science.[11]

Since proof p has empirical contents it can not be called formal. However it has a mathematical *form*, which is expressed within HoTT straightforwardly. As we shall briefly see, this form qualifies both as logical and geometrical. The fact that in HoTT logical and geometrical forms go together, makes HoTT quite unlike other popular formal systems such as the Classical First-Order Logic

---

[11]The identity conditions of p depend on those of MS, ES, which are left here without a precise specification. If we assume that MS and ES are enduring spatial objects repeatedly appearing on the sky then we should think of p as a fragment of the planet's orbit. Alternatively (and less realistically), if we think of MS and ES as particular spatio-temporal events which occur in a particular morning and a particular evening, then we should think of p as a continuous process that begins with MS and ends with ES. The HoTT-based reconstruction of Frege's *Venus* example given in this Section does not depend on one's specific assumptions about space, time and motion. The idea of identification of spatial objects or spatio-temporal events via continuous paths, which makes part of BKS, is compatible with many different physical theories and many different ontologies.

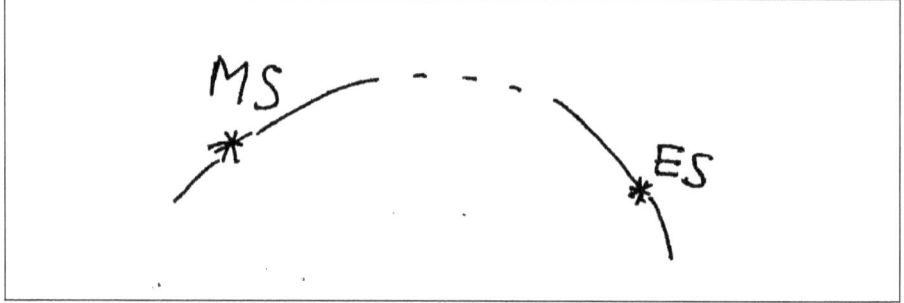

Figure 1: Morning Star and Evening Star are the same

(FOL), see my [19, ch. 7, 10], for a further discussion on this general issue. Remarkably, the geometrical form of p provided by HoTT (namely, a path) and the standard geometrical representation of the same object provided by CM and BKS (namely, a continuous curve) turn out to be alike.[12]

First, we need to specify a type (which under the homotopical interpretation is thought of as a space) where MS and ES belong. Since MS, ES and other celestial bodies are conceived in CM as point-like objects I call the corresponding type/space Pt and think of it as a collection of points:

$$MS, ES : Pt$$

Then we form a new type/space $MS =_{Pt} ES$, which is a space of continuous paths between MS and ES. Finally, we specify a particular path p in this space and form a judgment:

$$p : MS =_{Pt} ES$$

---

[12]In the standard Homotopy theory a *path* is not simply a curve but a parameterized curve. More formally path p with endpoints A, B is a continuous map $[0, 1] \to S$ from the unit interval to space S where points A, B belong, such that $p(0) = A$ and $p(1) = B$. "Paths" about which usually talk HoTT-theorists (as in [17]), cannot be straightforwardly identified with paths of the standard Homotopy theory [22]. But for our purposes the concept of path in the sense of HoTT will suffice: it combines the formalism of HoTT with a mixture of pre-theoretical spatio-temporal intuitions about paths and more elaborated geometrical intuitions (rather than precise concepts) borrowed from the standard Homotopy theory and some other branches of mathematics. By interpreting Frege's *Venus* in terms of HoTT I extend this intuitive part of HoTT with certain additional pre-theoretical intuitions concerning space, time and motion. Conversely, HoTT serves me as a formal tool allowing for putting these pre-theoretical intuitions into an order.

that says that MS and ES are the same as evidenced by p. However little of HoTT's resources we use here, this reconstruction of Frege's example provides some useful lessons as we shall now see.[13]

7. Are intensions real?

Recall Frege's question: What is the difference between the sense of proposition (1) ($MS = MS$) and the sense of proposition (3) ($MS = ES$)? It appears to be in accord with Frege to assume that senses of propositions depend functionally on their corresponding proofs (even if proofs and senses are not exactly the same). Then our reconstruction of *Venus* allows for a precise mathematical answer to Frege's question: while the (unique) proof of (1) is trivial loop $refl_{MS}$, the proof of (3) is a non-trivial path p. In both cases a given proposition has a single proof. However these two proofs essentially differ not only in their intuitive "sense" but also in their geometric representation.

Let us now turn to some ontological issues. Albeit the concept of proof is epistemic *par excellence*, the HoTT-based reconstruction of *Venus* makes it clear that proofs in the standard proof-theoretic semantic of MLTT should not be necessary thought of as purely mental constructions. Thinking about such proofs as *truthmakers* opens a way to various forms of *truthmaker realism* [24]. Whether or not one takes *Venus* and/or its trajectory p to be real entities

---

[13]The proposed HoTT-based reconstruction of Frege's $Venus$ example may not capture some aspects of Frege's volatile notion of sense. This notion may comprise more than HoTT in its existing form is able to detect. For example, arithmetical propositions

$$2 + 2 = 4 \qquad (9)$$

and

$$4 = 4 \qquad (10)$$

arguably have *different* senses. However the standard Peano-style formalization of arithmetic used in HoTT treats both equalities (9) and (10) as definitional and thus doesn't allow for non-trivial proofs of (9), see [17, p. 36 ff]. At the same time, given Frege's specific view on arithmetic as a part of logic developed in his [4], it is not obvious to me that the view that (9) and (10) have one and the *same* sense is indeed untenable in a Fregean conceptual framework. Under this view (9) is a logical truth but $MS = ES$ is a fact of the matter, so the apparent analogy between the two cases should be judged as merely linguistic and superficial. This controversial issue has no bearing on my following argument. I thank an anonymous referee for pointing to this arithmetical example.

depends, of course, on a particular ontology that one may associate with CM or another theory supporting the relevant astronomical observations. In particular, CM allows for a 4-dimensional ontology where atomic entities are points of Classical aka *Neo-Newtonian* space-time [23, p. 202 ff]. In this ontological framework p, seen as a world-line, qualifies as a full-fledged entity while the moving object *Venus* is its momentary slice. I shall not discuss here details of this and rival ontologies but rely on the fact that p of our example allows for natural realistic interpretations.

According to Frege, senses should not be thought of as psychological entities belonging to individual minds [6, p. 38–39]. However he suggests that senses wholly belong to human collective memories stored in existing natural languages. The only way in which a given sense can be possibly related to the non-human parts of our world, according to Frege's account as I understand it, is via the reference (if any) of the corresponding linguistic expression. For example English word "apple" has a sense, which belongs to this language (and arguably is shared by other natural languages) and a reference, which is a real thing that may exist independently of any linguistic and other human activities. English word "unicorn" equally has a sense but has no reference; so this particular sense is detached from any non-human reality.

The above is a rough interpretation of Frege's view but it points to a common idea about linguistic meaning, which is worth being considered here. Since Frege's concept of sense and the logical concept of *intension* are closely related (see the end of Section 2 above), the standard examples of so-called *intensional contexts* apparently provide a further linguistic support to this idea. Such examples always have to do with intentions, beliefs, knowledge and other human-related issues. So these examples square well with Frege's view according to which propositions (1) and (3) have "different cognitive values" because their senses are different - in spite of the fact that their reference (truth-value) is the same.

Our analysis of *Venus* suggests a revision of this view. Since proofs are constituents of senses (of propositions), and since these proofs admit realistic interpretations, such realistic interpretations may extend to senses. What I have in mind is not a justification of some form of Meinongian existence of unicorns but rather the view that the distinction between the sense and the reference of a given linguistic expression must be freed from all ontological commitments altogether. The idea that the reference is the only linguistic anchor that links human languages and the human cognition to non-human realities is hardly justified. Sense and reference and their logical counterparts such as intensions and extensions of concepts all make part of (various versions of) our conceptual apparatus. How this apparatus connects us, humans, to non-human realities is a question, which cannot be answered only by means of logical and conceptual analysis.

I submit that behind the view on meaning, which I purport now to criticize, is the following strong ontological assumption:

All real entities are individuals. (OE)

For further references I shall call this assumption the *ontic extensionality* or OE for short. The reason why I call this assumption *extensionality* becomes clear from a homotopical reconstruction of Frege's distinction between sense and reference, which generalizes upon the above reconstruction of *Venus* as follows. *References* are point-like individuals belonging to classes of alike individuals, which constitute *extensions* of their corresponding concepts. *Senses* are higher-order homotopical structures, which involve spaces of paths and their homotopies (including higher-order homotopies), and constitute *intensions* of the same concepts. As we have already seen, in the *extensional* version of HoTT the higher-order part of the structure is truncated. Hence the name for OE, which allows the truncated higher-order part of the structure to have an epistemic and cognitive value but includes in the ontology only its basic 0-level part.

From this point view it appears reasonable to claim that talks of apples, of unicorns, of Bucephalus and of Alexander the Great have the same logical form, so the words "apple" and "unicorn" both have a sense and a reference. By the reference of "unicorn" I understand here a fictional individual. Propositions about apples and unicorns may well allow for the same forms of truth-evaluation. The difference between merely fictional, legendary and real entities concerns material (contentful) rather than formal features of truth-evaluation. There is no way to distinguish between a fiction, a legend, and a historical fact on purely formal grounds.[14]

---

[14]The Bucephalus example demonstrates this particularly clearly. Bucephalus is a legendary horse belonging to Alexander the Great. According to the legend Bucephalus was born the same day as Alexander and, according to a particular version of the same legend, he also died the same day as Alexander. I don't know about a verdict of today's historical science as to how much of this story (if any) is a historical fact and how much of it is a fiction. I don't believe that any advance in formal logic may help for answering this question.

I can see no a priori reason for assuming that a part of the homotopic structure is more apt to represent reality than any other. For that reason I don't take OE for granted. Moreover that our reconstruction of *Venus* suggests that terms of 1-types (paths) allow for a realistic interpretation as well as terms of 0-types (points). However in the next Section we shall see that the situation is not so simple, and that BKS is compatible with OE after all.

Concluding this Section I would like to remark that OE goes along the view according to which the Classical first-order logic (FOL) should be seen and used as the basic logical tool for scientific reasoning. In this context the suggestion to drop OE and allow for higher-order entities sounds a part of an argument in favor of a higher-order system of logic with a standard class-based semantics. MLTT and HoTT indeed qualify as higher-order systems in a relevant sense but the homotopical semantic used in HoTT is not standard. In HoTT higher types are formed not by the reiteration of the powerset construction (i.e. not by considering classes of classes of ... of individuals) but in the geometric way, which has been briefly explained in Section 5 above. Our homotopical reconstruction of *Venus* given in Section 6 demonstrates how the geometric semantic of HoTT helps one to use this theory as a tool for mathematical modeling *in* science, not only as a tool for a logical analysis *of* science. I believe that this dummy example points to interesting theoretical possibilities in mathematical physics. For serious attempts to use HoTT and its logical structure in physics see [20, 21].

8. Basic kinematic scheme

Here I supplement the homotopical reconstruction of *Venus* from Section 6 with a similar reconstruction of the Basic Kinematic Scheme (BKS), which captures the usual idea of moving particle. The kinematic space K, in which MS and ES live, allows for multiple paths (trajectories) sharing their ending points. I think about K not as a vehicle of moving particles but rather as a collection Pt of such particles provided with appropriate criteria of identity and an additional structure, which represents their relative motions. The motions are represented by paths between the particles as in the *Venus* example. The additional structure is that of groupoid of paths over Pt. I do not include into K homotopies of paths beyond the trivial ones because such things play no role in BKS. Paths in K are assumed to be reversible and composable by concatenation; the composition is

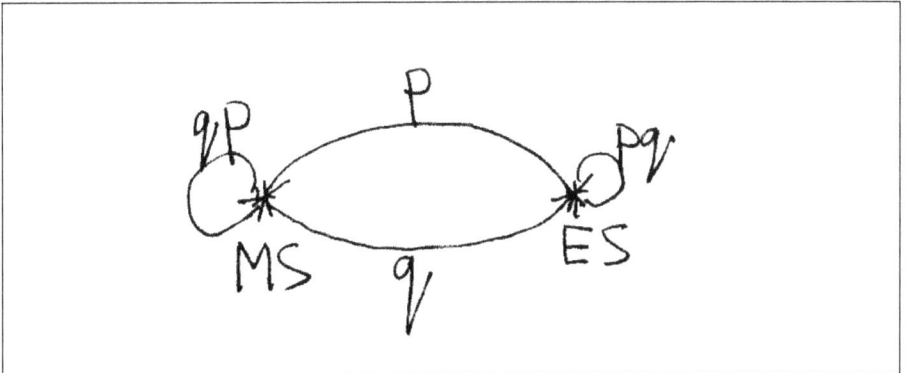

Figure 2: Multiple Paths of Venus

associative.[15] In terms of HoTT K qualifies as a 1-type; Pt is the underlying 0-type of K obtained from K via the (0-)truncation.

Let me now briefly reproduce the above homotopical reconstruction of *Venus* in this slightly extended context. We take two points MS, ES in Pt (and hence in K) and consider the path space $MS =_{Pt} ES$. Then we find in $MS =_{Pt} ES$ a particular path p, which serves us as a proof of identity $MS = ES$. The extended context allows us now to notice an interesting feature of BKS, which so far remained out of the scope of our analysis. Consider the following additional principle, which I'll call the *uniqueness of actual path*:

> There is at most one path between any two given points. (UAP)

Prima facie *Venus* does not verify this principle. Indeed, *Venus*'s orbit, which is a topological circle, admits two different paths p, q between MS and ES and further, via composition, two non-trivial loops qp and pq for MS and ES correspondingly:

The above picture represents MS and ES as apparently different but in fact the same body, which moves along its circular orbit. But neither this picture

---

[15]In the usual Homotopy theory the composition of paths in a given space S is defined only up to homotopy; in order to define such an operation one is obliged to provide an appropriate homotopy aka reparameterization by hand. Since in HoTT homotopy types are primitive objects this issue is treated a bit differently. We stipulate an abstract groupoid K without assuming any ambient space S in advance, and then see how much of BKS can be recovered in this way. This approach allows us to describe the composition of paths in K as concatenation without mentioning homotopies.

nor K construed as above reflects the usual idea that one and the same particle cannot follow two different paths *simultaneously*. This is not particularly surprising since *time* did not feature in our construction of K so far. I am not going now to fill this gap by providing K with an explicit representation of time. Instead, let us consider a model of UAP in the given framework. UAP can be satisfied if we think of MS as "Venus at time t_1" and of ES as "Venus at (later) time t_2". Then during the time period $\Delta = [t_1, t_2]$ Venus follows a unique path p, which can be described as a segment of 'Venus's *worldline* in an appropriate spacetime.[16] This shows that we may use UAP for accounting for a time-related feature of BSK without introducing time explicitly. It is quite remarkable because UAP involves only very basic concepts of HoTT and has a purely formal character; it can be itself easily expressed in HoTT.

If we now add a natural assumption that the propositional identity is an equivalence (which excludes "split" or "branching" identities) then UAP reduces possible forms of K to a trivial spaghetti-like form. In this case each particular connected component or "noodle" of K can be called a *worldline* of its corresponding particle (point). Since every noodle is contractible into a point, in this case K and Pt are homotopically equivalent. They represent the same 0-level homotopy type $K \simeq Pt$ making redundant the very distinction between them. However the distinction between K and Pt becomes useful again when one distinguishes between *actual* and *possible* paths. Indeed, it is plausible to assume that given actual path p with endpoints MS, ES BKS allows for other possible paths with the same endpoints. In other words, BKS allow bodies to follow trajectories, which differ from their actual trajectories. Now we can think of K as groupoid of possible paths where UAP does not hold and distinguish its subgroupoid $A \subset K$ which comprises only actual paths and for which UAP holds. In this case 0-truncation $K \to Pt \simeq A$ becomes non-trivial and represents a *realization* of certain possible paths.

The above analysis of BKS appears to be an appropriate starting point for building a Quantum counterpart of this conceptual scheme. From the homotopical point of view there is nothing impossible or unnatural in the idea that a given particle may follow multiple trajectories simultaneously as this is assumed in the Feynman path integral formulation of Quantum Mechanics:

In the present conceptual framework one may rather inquire into the nature

---

[16]Since we are talking about the Classical Mechanics but not about the Relativistic Mechanics, the relevant notion of spacetime is that of the *Neo-Newtonian* spacetime, see [23, p. 202 ff].

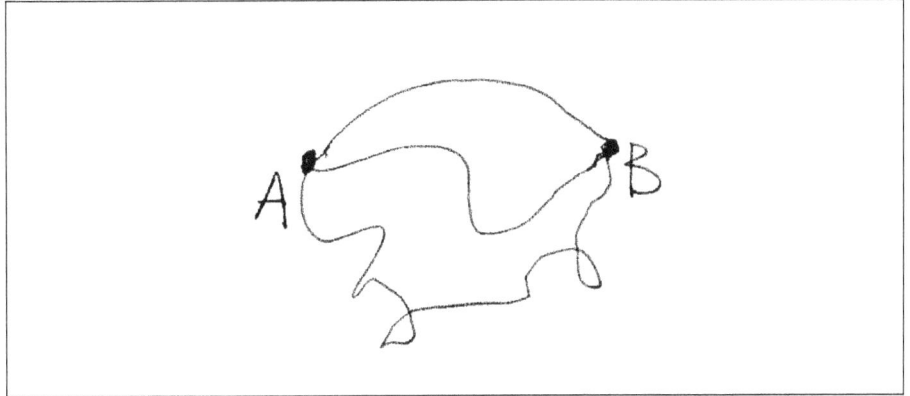

Figure 3: Quantum Paths

of UAP. What is behind the traditional notion according to which the *actual* trajectory of a given particle during its lifetime is necessary unique?

In order to provide a tentative answer let us return to the issue discussed in the last Section. The above analysis of BKS apparently provides an additional evidence in favor of ontic extensionality (OE). The intensional groupoid structure of K represents *possible* trajectories of particles. But since in the real world each particle has its unique worldline the groupoid K is reduced (truncated) to the extensional set $A \simeq Pt$. Conversely, OE in the given context implies UAP. However OE is compatible with BKS only if one understands the modal property of being *possible* (for paths) in purely epistemic terms - say, as a lack of knowledge about the actual trajectories. Alternatively, one may think about possible paths in K as physically real. This latter view violates OE but it is not wholly unreasonable. Quantum Mechanics where UAP does not apply, provides additional reasons for taking it seriously. I stop here and leave an attempt to develop a HoTT-based theory of identity for Quantum Mechanics for a different occasion.

References

[1] D. J. Chalmers. On Sense and Intension. *(J. Tomberlin, ed.) Philosophical Perspectives 16: Language and Mind: Blackwell*, pages 135–182, 2002.

[2] H. Deutsch. Relative Identity (entry in Stanford Encyclopedia of Philosophy). Retrieved September 29, 2016, from https://plato.stanford.edu/entries/identity-relative/, 2002.

[3] G. Frege. *Begriffsschrift: eine der arithmetischen nachgebildete Formelsprache des reinen Denkens.* Halle, 1879.

[4] G. Frege. *Die Grundlagen der Arithmetik.* Verlag von W. Koebner, Breslau, 1884.

[5] G. Frege. Über Sinn und Bedeutung. *Zeitschrift für Philosophie und philosophische Kritik*, 100:25–50, 1892.

[6] G. Frege. *On Sense and Reference in: Translations from the Philosophical Writings of Gottlob Frege, ed. by Geach and M. Black*, pages 56–78. Oxford: Basil Blackwell, 1952.

[7] G. Frege. *Grundgesetze der Arithmetik.* Hildesheim Olms, 1962.

[8] G. Frege. *The Basic Laws of Arithmetic. Exposition of the System / Translated and edited, with an introduction, by M. Furth.* California Press, 1964.

[9] S. French. Quantum Physics and the Identity of Indiscernibles. *British Journal of the Philosophy of Science*, 39:233–246, 1988.

[10] S. French and D. Krause. *Identity in Physics: A Historical, Philosophical, and Formal Analysis.* Oxford University Press, 2006.

[11] P. T. Geach. *Logic Matters.* Basil Blackwell, 1972.

[12] M. Hofmann and T. Streicher. A Groupoid Model Refutes Uniqueness of Identity Proofs. *Proceedings of the 9th Symposium on Logic in Computer Science (LICS), Paris*, 1994.

[13] M. Hofmann and T. Streicher. The Groupoid Interpretation of Type Theory. *G. Sambin and J. Smith (eds.), Twenty-Five Years of Constructive Type Theory, Oxford University Press*, pages 83–111, 1998.

[14] K. C. Klement. *Frege and the Logic of Sense and Reference.* Routledge, 2002.

[15] P. Martin-Löf. *Intuitionistic Type Theory (Notes by Giovanni Sambin of a series of lectures given in Padua, June 1980).* Napoli: BIBLIOPOLIS, 1984.

[16] F. J. Pelletier. Did Frege Believe Frege's Principle? *Journal of Logic, Language, and Information*, 10:87–114, 2001.

[17] Univalent Foundations Project. *Homotopy Type Theory: Univalent Foundations of Mathematics.* Institute for Advanced Study (Princeton); available at https://homotopytypetheory.org/book/, 2013.

[18] W. V. Quine. *Ontological Relativity and Other Essays.* Columbia University Press, 1969.

[19] A. Rodin. *Axiomatic Method and Category Theory (Synthese Library vol. 364).* Springer, 2014.

[20] U. Schreiber. *Quantization via Linear homotopy types.* arXiv: 1402.7041, 2014.

[21] U. Schreiber and M. Shulman. *Quantum Gauge Field Theory in Cohesive Homotopy Type Theory.* arXiv:1408.0054, 2014.

[22] M. Shulman. *Brouwer's Fix-Point Theorem in Real-Cohesive Homotopy Type Theory*. arXiv:1509.07584, 2016.
[23] L. Sklar. *Space, Time, and Spacetime*. University of California Press, 1974.
[24] B. Smith. Truthmaker realism. *Australasian Journal of Philosophy*, 77(3):274–291, 1999.
[25] C. Wright. *Frege's Conception of Numbers as Objects*. Aberdeen University Press, 1983.

Received: 29 September 2016

On a Combination of Truth and Probability: Probabilistic IF Logic

Gabriel Sandu

Department of Philosophy, History, Culture and Arts, University of Helsinki
Unioninkatu 40 A, 00014 Helsinki, Finland
`sandu@mappi.helsinki.fi`

Abstract: I will give a short exposition of Independence-Friendly logic (IF logic), a system of logic which extends ordinary first-order logic with arbitrary patterns of dependent and independent quantifiers. Truth and falsity of IF sentences is defined in terms of the existence of winning strategies in a 2-player win-lose games of imperfect information. One consequence of imperfect information is the existence of indeterminate IF sentences (on finite models). I sketch how indeterminacy may be overcome using von Neumann's Minimax Theorem. My exposition draws on ideas in [6].

Keywords: IF logic, IF games.

Introduction

In a seminal paper [2], Goldfarb points out that "The connection between quantifiers and choice functions or, more precisely, between quantifier-dependence and choice functions, is the heart of how classical logicians in the twenties viewed the nature of quantification." [2, p. 357]. For a less historical but more systematic point of view [8], Terence Tao, notices that we know how to render in first-order logic statements like:

1. For every x, there exists a y *depending on* x such that $B(x, y)$ is true

and

2. For every x, there exists a y *independent of* x such that $B(x, y)$ is true.

The first one can be rendered by

$$\forall x \exists y B(x, y)$$

and the second one by

$$\exists y \forall x B(x,y).$$

(Here $B(x,y)$ is a binary relation holding of two objects x, y). Things become more complicated when four quantifiers and a 4-place relation $Q(x, x', y, y')$ are involved. We can express in first-order logic statements like:

3. For every x and x', there exists a y depending only on x and a y' depending on x and x' such that $Q(x, x', y, y')$ is true

and

4. For every x and x', there exists a y depending on x and x' and a y' depending only on x' such that $Q(x, x', y, y')$ is true

by

$$\forall x \exists y \forall x' \exists y' Q(x, x', y, y')$$

and

$$\forall x' \exists y' \forall x \exists y Q(x, x', y, y')$$

respectively. However, Tao continues, one cannot always express the statement

5. For every x and x', there exists a y depending only on x and a y' depending only on x' such that $Q(x, x', y, y')$ is true.

His conclusion is that

> It seems to me that first order logic is limited by the linear (and thus totally ordered) nature of its sentences; every new variable that is introduced must be allowed to depend on all the previous variables introduced to the left of that variable. This does not fully capture all of the dependency trees of variables which one deals with in mathematics. (Idem)

1. Independence-friendly logic

Independence-friendly logic (IF logic) introduced by Hintikka and Sandu in [4], is intended to represent patterns of dependence and independence of quantifiers like those exemplified by 5 which go beyond those expressible in ordinary first-order logic. More exactly, IF logic contains quantifiers of the form

$$(\exists x/W)\ (\forall x/W)$$

where W is a finite set of variables. The intended interpretation of $(\exists x/W)$ is: the existential quantifier $\exists x$ is independent of the quantifiers which bind the variables in W. The notion of independence involved here is a game-theoretical one and corresponds to the mathematical notion of uniformity. The example 5 above will be rendered in the new formalism by:

$$\forall x \forall x' (\exists y/\{x'\})(\exists y'/\{x,y\}) Q(x, x', y, y').$$

The original interpretation of IF formulas is given by semantical games of imperfect information. An alternative, equivalent interpretation is by skolemization. We shall adopt the latter.

2. Truth in IF logic

Let φ be a formula of IF logic in a given vocabulary L and U a finite set of variables which contains the free variables of φ. We expand the vocabulary L of φ to $L^* = L \cup \{f_\psi : \psi \text{ is a subformula of } \varphi\}$. The skolemized form or skolemization of φ with variables in U is defined by the following clauses, as detailed in [6]:

$Sk_U(\psi) = \psi$, for ψ an atomic subformula of φ or its negation

$Sk_U(\psi \circ \theta) = Sk_U(\psi) \circ Sk_U(\theta)$, for $\circ \in \{\vee, \wedge\}$

$Sk_U((\forall x/W)\psi) = \forall x Sk_{U \cup \{x\}}(\psi)$

$Sk_U((\exists x/W)\psi) = Subst(Sk_{U \cup \{x\}}(\psi), x, f_{\exists x}(y_1, ..., y_n))$

where $y_1, ..., y_n$ enumerate all the variables in $U - W$. We notice that if $W = \varnothing$ the last clause becomes

$$Sk_U((\exists x)\psi) = Subst(Sk_{U \cup \{x\}}(\psi), x, f_{\exists x}(y_1, ..., y_n))$$

where $y_1, ..., y_n$ enumerate all the variables in U. That is, we recover the notion of skolemization for the standard quantifiers. We abbreviate $Sk_\varnothing(\varphi)$ by $Sk(\varphi)$. An interpretation of $f_{\exists x}(y_1, ..., y_n)$ is called a Skolem function.

Example. We skolemize the sentence φ

$$\forall x \forall x' (\exists y/\{x'\})(\exists y'/\{x,y\}) Q(x,x',y,y').$$

We denote $(\exists y'/\{x,y\})Q(x,x',y,y')$ by ψ. $Sk(\varphi)$ is obtained through the following steps:

$$
\begin{aligned}
Sk_{\{x,x',y,y'\}}(Q(x,x',y,y')) &= Q(x,x',y,y') \\
Sk_{\{x,x',y\}}(\psi) &= Q(x,x',y,f_{y'}(x')) \\
Sk_{\{x,x'\}}((\exists y/\{x'\})\psi) &= Q(x,x',f_y(x),f_{y'}(x')) \\
Sk_{\{x\}}(\forall x'(\exists y/\{x'\})\psi) &= \forall x' Q(x,x',f_y(x),f_{y'}(x')) \\
Sk_\varnothing(\forall x \forall x'(\exists y/\{x'\})\psi) &= \forall x \forall x' Q(x,x',f_y(x),f_{y'}(x')).
\end{aligned}
$$

The original vocabulary L receives an interpretation through an L-structure $\mathbb{M}$ in the usual way. We are now ready for the truth-definition.

Definition 1. Let φ be an L-sentence of IF logic and $\mathbb{M}$ an L-structure. We say that φ is true in $\mathbb{M}$, $\mathbb{M} \models^+ \varphi$, if and only if there exist functions $g_1, ..., g_n$ of appropriate arity in M to be the interpretations of the new function symbols $f_{x_1}, ..., f_{x_n}$ in $Sk(\varphi)$ such that

$$\mathbb{M}, g_1, ..., g_n \models Sk(\varphi).$$

3. Falsity in IF logic

In order to deal with falsity, we shall define another translation procedure, $Kr_U(\varphi)$ (we continue to follow [6]):

$Kr_U(\psi) = \neg \psi$, for ψ an atomic subformula or its negation

$Kr_U(\psi \vee \theta) = Kr_U(\psi) \wedge Kr_U(\theta)$,

$Kr_U(\psi \wedge \theta) = Kr_U(\psi) \vee Kr_U(\theta)$

$Kr_U((\exists x/W)\psi) = \forall x Kr_{U \cup \{x\}}(\psi)$

$Kr_U((\forall x/W)\psi) = Subst(Kr_{U \cup \{x\}}(\psi), x, f_{\forall x}(y_1, ..., y_m))$

where $y_1, ..., y_m$ are all the variables in $U - W$. We call the value of interpretation of $f_{\forall x}(y_1, ..., y_m))$ a Kreisel counter-example.

By analogy with the truth definition, we stipulate that an IF sentence φ is false in a structure $\mathbb{M}$, $\mathbb{M} \models^- \varphi$, if and only if there exist functions $h_1, ..., h_m$ of appropriate arity in M to be the interpretations of the new function symbols $f_{x_1}, ..., f_{x_m}$ in $Kr(\varphi)$ such that

$$\mathbb{M}, h_1, ..., h_m \models Kr(\varphi).$$

4. Indeterminacy and signaling

Here is an example of an IF sentence which is neither true nor false in any structure $\mathbb{M}$ which contains at least two elements:

$$\varphi = \forall x (\exists y / \{x\}) x = y.$$

It may be checked that $Sk(\varphi) = \forall x\, x = c$, where c is a new 0-place function (individual constant); and $Kr(\varphi) = \forall y \neg d = y$. Then by the definitions above, we have:

$\mathbb{M} \models^+ \varphi$ iff there is $a \in \mathbb{M}$ such that $\mathbb{M}, a \models \forall x\, x = c$
$\mathbb{M} \models^- \varphi$ iff there is $b \in \mathbb{M}$ such that $\mathbb{M}, b \models \forall y \neg d = y$.

As the structure $\mathbb{M}$ contains at least two elements, none of the assertions on the right side is true. Thus we have both $\mathbb{M} \not\models^+ \varphi$ and $\mathbb{M} \not\models^- \varphi$.

It is interesting to compare the previous example with ψ

$$\forall x \exists z (\exists y / \{x\}) x = y$$

whose skolemization is

$$\forall x\, x = g(f(x)).$$

It may be checked that this sentence is a logical truth. Unlike in ordinary first-order logic, the example shows that inserting a dummy existential quantifier in an IF sentence changes its semantical value. Hodges has discussed this example in [5], and the phenomenon of signaling in IF logic.

5. Expressive power

Example of sentences of the form

$$\forall x \forall x' (\exists y / \{x'\})(\exists y' / \{x, y\}) Q(x, x', y, y')$$

which are mentioned by Tao and which are not first-order definable are not difficult to find. We prefer to use a different example, which will turn up to be useful for other purposes too. There is an IF sentence which expresses the (Dedekind) infinity of the universe M. M is said to be (Dedekind) infinite iff there is a function $h : M \to M$ which is an injection and in addition there is an element in M which is not the image under h of any element of M. The sentence we look for is φ_{inf}

$$\exists w \forall x (\exists y / \{w\})(\exists z / \{w, x\})(x = z \land w \neq y).$$

The Skolem form of φ_{inf} is

$$\forall x (x = g(f(x)) \land c \neq f(x)).$$

It can be checked that φ_{inf} is true in a model iff the function f is an injection which range is not the entire universe. On the other side if $\mathbb{M}$ is finite, it may be shown that we have both $\mathbb{M} \not\models^{+} \varphi_{inf}$ and $\mathbb{M} \not\models^{-} \varphi_{inf}$. Thus we have produced another example of an indeterminate IF sentence.

6. Strategic games

Consider our earlier IF sentence $\varphi = \forall x (\exists y / \{x\}) x = y$ and a finite model $\mathbb{M}$. The set $S_\exists$ of Skolem functions of Eloise in this game reduces to the set of all individuals in M which can be the values of the new function symbols in $Sk(\varphi) = \forall x x = c$. In this case $S_\exists = M$. And the set $S_\forall$ of Kreisel counter-examples of Abelard in this game reduces to the set of all individuals in M which can be the values of the new function symbols in $Kr(\varphi) = \forall y \neg d = y$. Thus $S_\forall = M$. We can now formulate a two-player strategic game in which we let $S_\exists$ be the set of (pure) strategies of Eloise, and $S_\forall$ the set of (pure) strategies of Abelard. The two players choose simultaneously $s \in S_\exists$ and $t \in S_\forall$, respectively. The payoff of the outcome is determined in a very simple way: if s and t satisfy the equation $x = y$, Eloise wins (1 euro). Otherwise Abelard wins. Here is the complete matrix of the game for the case in which $S_\exists = \{1, 2, 3\} = S_\forall = M$:

	1	2	3
1	(1,0)	(0,1)	(0,1)
2	(0,1)	(1,0)	(0,1)
3	(0,1)	(0,1)	(1,0)

The rows represent the strategies of Eloise and the columns the strategies of Abelard. In (m,n), $m \in \{0,1\}$ is the payoff of Eloise, i.e. $u_\exists(m,n) = m$, and n is the payoff for Abelard for the corresponding pair of strategies.

It is interesting to compare this game to the one associated with the IF sentence $\psi = \forall x(\exists y/\{x\})x \neq y$ and $M = \{1,2,3\}$:

	1	2	3
1	(0,1)	(1,0)	(1,0)
2	(1,0)	(0,1)	(1,0)
3	(1,0)	(1,0)	(0,1)

We shall call these games strategic IF games, and denote them by $\Gamma(M,\varphi) = (S_\exists, S_\forall, u_\exists, u_\forall)$. Obviously these games are

- win-lose: Every game has exactly two payoffs, 0 and 1.
- 1 sum: For every $s \in S_\exists$ and $t \in S_\forall$ we have: $u_\exists(s,t) + u_\forall(s,t) = 1$.

Let $\Gamma = (S_\exists, S_\forall, u_\exists, u_\forall)$ be a finite strategic IF game. For $s^* \in S_\exists$ and $t^* \in S_\forall$, the pair (s^*, t^*) is an equilibrium in Γ iff the following two conditions are jointly satisfied:

(i) $u_\exists(s^*, t^*) \geq u_\exists(s, \tau t^*)$ for every strategy s in $S_\exists$. In other words

$$u_\exists(s^*, t^*) = max_s u_\exists(s, t^*)$$

(ii) $u_\forall(s^*, t^*) \geq u_\forall(s^*, t)$ for every strategy t in $S_\forall$. In other words

$$u_\forall(s^*, t^*) = max_t u_\forall(s^*, t).$$

We can check that in our earlier strategic IF games $\Gamma(\mathbb{M}, \forall x(\exists y/\{x\})x = y)$ and $\Gamma(\mathbb{M}, \forall x(\exists y/\{x\})x \neq y)$ where $\mathbb{M} = \{1,2,3\}$, there are no equilibria. Obviously this is a reflection of the fact that these games are undetermined.

6.1. Mixed strategies equilibria in IF games

There is an equilibrium in every IF game if, instead of pure strategies, we switch to mixed strategies. Let $\Gamma = (S_\exists, S_\forall, u_\exists, u_\forall)$ be a finite IF strategic game. A mixed strategy ν for player i in this strategic game is a probability distribution over S_i, that is, a function $\nu : S_i \to [0,1]$ such that $\sum_{\tau \in S_i} \nu(\tau) = 1$. ν is uniform over $S'_i \subseteq S_i$ if it assigns equal probability to all strategies in S'_i and zero probability to all the strategies in $S_i - S'_i$. Obviously we can simulate a pure strategy s with a mixed strategy ν such that ν assigns s probability 1. Given a mixed strategy μ for player $\exists$ and a mixed strategy ν for player $\forall$, the expected utility for player i is given by:

$$U_i(\mu, \nu) = \sum_{s \in S_\exists} \sum_{t \in S_\forall} \mu(s)\nu(t)u_i(s,t).$$

When $s \in S_\exists$ and ν is a mixed strategy for player $\forall$, we let

$$U_i(s, \nu) = \sum_{t \in S_\forall} \nu(t)u_i(s,t).$$

Similarly if $t \in S_\forall$ and μ is a mixed strategy for player $\exists$, we let

$$U_i(\mu, t) = \sum_{s \in S_\exists} \mu(s)u_i(s,t).$$

Von Neumann's well known Minimax Theorem shows that every finite, constant sum, two player game has an equilibrium in mixed strategies. It is also well known that every two equilibria in such a game returns the same expected utility to the two players. Thus we can talk about the expected utility returned to player $\exists$ by an IF strategic game. This justifies the next definition:

Definition 2. Let φ be an IF sentence and $\mathbb{M}$ a finite model. When $0 \leq \varepsilon \leq 1$ we define: $\mathbb{M} \models_\varepsilon^{eq} \varphi$ iff the expected utility returned to player $\exists$ by the strategic game $\Gamma(\mathbb{M}, \varphi)$ is ε.

The above definition gives us the (probabilistic) value of an IF sentence φ on a given finite model $\mathbb{M}$. It can be shown that this interpretation is a conservative interpretation of the earlier interpretation, in the following sense.

Proposition 1. *For every IF sentence φ and finite model $\mathbb{M}$ we have: $\mathbb{M} \models^+ \varphi$ iff $\mathbb{M} \models_1^{eq} \varphi$; and $\mathbb{M} \models^- \varphi$ iff $\mathbb{M} \models_0^{eq} \varphi$.*

The next proposition is often useful for checking that a pair of mixed strategies is an equilibrium.

Proposition 2. *Let μ^* be a is a mixed strategy for player $\exists$ and ν^* is a mixed strategy for player $\forall$ in the strategic IF game Γ. The pair (μ^*, ν^*) is an equilibrium in Γ if and only if the following conditions hold:*

1. $U_\exists(\mu^*, \nu^*) = U_\exists(\sigma, \nu^*)$ for every $\sigma \in S_\exists$ in the support of μ^*;
2. $U_\forall(\mu^*, \nu^*) = U_\forall(\mu^*, \tau)$ for every $\tau \in S_\forall$ in the support of ν^*;
3. $U_\exists(\mu^*, \nu^*) \geq U_\exists(\sigma, \nu^*)$ for every $\sigma \in S_\exists$ outside the support of μ^*;
4. $U_\forall(\mu^*, \nu^*) \geq U_\forall(\mu^*, \tau)$ for every $\tau \in S_\forall$ outside the support of ν^*.

Recall our earlier examples $\Gamma(\mathbb{M}, \forall x(\exists y/\{x\})x = y)$ and $\Gamma(\mathbb{M}, \forall x(\exists y/\{x\})x \neq y)$ where $\mathbb{M} = \{1, 2, 3\}$. In both cases the uniform strategies $\mu^*(1) = \mu^*(2) = \mu^*(3) = \frac{1}{3}$ and $\nu^*(1) = \nu^*(2) = \nu^*(3) = \frac{1}{3}$ form an equilibrium. The value of the first game is $\frac{1}{3}$ and that of the second game is $\frac{2}{3}$. Thus $\mathbb{M} \models^{eq}_{\frac{1}{3}} \forall x(\exists y/\{x\})x = y$ and $\mathbb{M} \models^{eq}_{\frac{2}{3}} \forall x(\exists y/\{x\})x \neq y$.

A more complex argument shows that for M a finite model with n elements we have $\mathbb{M} \models^{eq}_{\frac{n-1}{n}} \varphi_{inf}$. Thus when n grows to infinity the value of φ_{inf} approaches 1, as expected.

Notes

IF logic has been introduced by Hintikka and Sandu in [4]. In [3], Hintikka discusses the foundational role of IF logic in the philosophy of mathematics. The basic model theoretical properties of IF logic from a game-theoretical perspective are described by Mann, Sandu, and Sevenster in [6]. In that work the probababilistic interpretation of IF logic, which is the source of our exposition in section 6, is thoroughly studied. The idea to use von Neumann's Minimax Theorem in the context of partially ordered quantifiers is due to Ajtai as mentioned in [1]. The first systematic investigation of strategic IF games is provided by [7]. Recently, an alternative approach to IF logic has been developed which replaces the independence of quantifiers by the dependence between terms. In this new setting, (5) is rendered by

$$\forall x \forall x' \exists y \exists y' (= (x, y) \wedge = (x', y') \wedge Q(x, x', y, y')).$$

The intended meaning of $' = (x, y)'$ is: y functionally depends on x. The semantical interpretation of this language is based on Hodges' compositional interpretation, introduced in [5]. A self-contained introduction to this logic is [9].

References

[1] A. Blass and Y. Gurevich. Henkin quantifiers and complete problems. *Annals of Pure and Applied Logic*, 32(C):1–16, 1986.

[2] W. Goldfarb. Logic in the twenties: The nature of the quantier. *Journal of Symbolic Logic*, 44:351–368, 1979.

[3] J. Hintikka. *Principles of Mathematics Revisited*. Cambridge University Press, Cambridge, 1996.

[4] J. Hintikka and G. Sandu. Informational independence as a semantic phenomenon. In J. E. Fenstad et al., editors, *Logic, Methodology and Philosophy of Science*, volume 8, pages 571–589. Elsevier, Amsterdam, 1989.

[5] W. Hodges. Compositional semantics for a language of imperfect information. *Logic Journal of the IGPL*, 5:539–563, 1997.

[6] A. Mann, G. Sandu, and M. Sevenster. *Independence-Friendly Logic: A Game-theoretic Approach*. Cambridge University Press, Cambridge, 2011.

[7] M. Sevenster. *Branches of Imperfect Information: Logic, Games and Computation*. Phd thesis, University of Amsterdam, Amsterdam, 2006.

[8] T. Tao. Printer-friendly CSS, and nonfirstorderisability. Retrieved on January 15, 2016, from the website: URL = <https://terrytao.wordpress.com/2007/08/27/printer-friendly-css-and-nonfirstorderizability/>, 2007. What's new: Updates on my research and expository papers, discussion of open problems, and other maths related topics.

[9] J. Väänänen. *Dependence Logic*. Cambridge University Press, Cambridge, 2007.

Received: 13 July 2016

Towards Analysis of Information Structure of Computations

Anatol Slissenko

Laboratory for Algorithmics, Complexity and Logic (LACL)
University Paris East Créteil (UPEC)
61 av. du Général de Gaulle 94010, Créteil France
and
ITMO, St. Petersburg, Russia
`slissenko@u-pec.fr`

Abstract: The paper discusses how one can try to analyze computations, and maybe computational problems from the point of view of information evolution. The considerations presented here are very preliminary. The long-standing goal is twofold: on the one hand, to find other vision of computations that may help to design and analyze algorithms, and on the other hand, to understand what is realistic computation and what is real practical problem. The concepts of modern computer science, that came from classical mathematics of pre-computer era, are overgeneralized, and for this reason are often misleading and counter-productive from the point of view of applications. The present text discusses mainly what classical notions of entropy might give for analysis of computations. In order to better understand the problem, a philosophical discussion of the essence and relation of knowledge/information/uncertainty in algorithmic processes might be useful.

Keywords: computation, problem, partition, entropy, metric.

1. Introduction

The goal of this paper is to discuss along what lines one can look for ways to describe the quantity of information transformed by computations. This may permit to better understand the computations themselves and, possibly, what is practical computation and what is practical algorithmic problem. The considerations presented here are very preliminary, more of philosophical than of mathematical flavor. We consider rather straightforward geometrical and information ideas that come to mind. Usually they are not sufficient taken directly. Making explicit the obstacles may help to devise more productive approaches.[1]

In Introduction we give some arguments that illustrate that the mathematical formulations of computational problems we usually consider, are overgeneralized, and sometimes this hinders the development of practical algorithms or the understanding why certain algorithms for theoretically hard problems work well in practice. In Section 2 we outline some approaches to measuring information in computations, and discuss their weak and strong points. Section 3 is about the structure of problems for which we can presumably develop measures of information along the lines described in the previous section. It contains also a short discussion of the role of linguistic considerations in describing practical problems.

Why traditional mathematical settings look too general for practical computer science? And when it is inevitable and when maybe not?

Most notions used in theoretical computer science either come from mathematics of pre-computer era or are developed along mathematical lines of that epoch. From mathematics of pre-computer era the computational theory borrows logics, logical style algorithms (lambda-calculus, recursive function, Turing machine), general deductive systems (grammars), Boolean functions, graphs. More specific notions like finite automata, Boolean circuits, random access machines etc., though motivated by modeling of computations, are of traditional mathematical flavor. All these concepts played and continue to play fundamental role in theoretical computer science, however other, more adequate concepts are clearly needed.

I can illustrate this thesis by Boolean functions and their realization by circuits. Almost all Boolean functions of n variables have exponential circuit complexity $(2^n/n)$ [9], and there is an algorithmic method to find such an optimal realization for a given 'random' function [6]. But it is clear that even for $n = 64$, that is not so big from practical viewpoint, one cannot construct a circuit with $2^n/n$ gates. So one can state that almost all Boolean functions will never appear in applications. The notion of Boolean function is of evident practical value, but not in its generality. All this does not say that the general notion and the mentioned result on the complexity of realization are useless in theory (moreover, they are known to be useful). But an optimal circuit construction for almost all Boolean functions is not of great value for practical

---

[1] Some of them were developed later to become quite mathematical.

Boolean functions.

Consider another example. We know that the worst-case complexity of the decidability of the theory of real addition is exponential [2]. This theory is a set of valid closed formulas that are constructed from linear inequalities with integer coefficients with the help of logical connectives, including quantifiers over real numbers (in fact, only rational numbers are representable by such formulas, as the only admissible constants are integers). In particular, one can express in this theory the existence of a solution of a system of linear inequalities, and various parametric versions of this problem, e.g., whether such a solution exists for any value of some variable in some interval. The complexity of recognition of validity of the formulas grows up with the number of quantifier alternations.

The mentioned exponential lower bound on the computational complexity of the theory of real addition is proven along the following lines. Denote $\mathbb{B} =_{df} \{0, 1\}$ and denote by $\mathbb{B}^*$ the set of all strings over $\mathbb{B}$. Under some technical constraints for any algorithm f from $\mathbb{B}^*$ to $\mathbb{B}$, whose complexity is bounded by some exponential function φ, and for any its input $x \in \mathbb{B}^*$ one can construct a formula $\Phi(f, x)$ of sufficiently small size (polynomial in the size of f and x) that is valid if and only if $f(x) = 0$.

Within a reasonable algorithmic framework (e.g., for some random access machines, like LRAM from [10]) one can construct a predicate $f : \mathbb{B}^* \to \mathbb{B}$ whose upper bound on computational complexity is φ, and any algorithm that computes this predicate has lower bound $\theta \cdot \varphi$, for some $0 < \theta < 1$. This f is a diagonal algorithm, I do not know other kind of algorithms for this context. Such a diagonal algorithms works like follows. Assume that the complexity of computing $\varphi(|x|)$, where $|x|$ is the length of $x \in \mathbb{B}^*$, is bounded by its value $\varphi(|x|)$. The algorithm f computes $\varphi(|x|)$ and makes roughly $\varphi(|x|)$ steps of simulation of algorithm with the code x applied to input x. If the process ends within less that $\varphi(|x|)$ steps then f outputs the value different from the value computed by the algorithm with the code x, otherwise it outputs say, 0 (in the latter case the value is not important).

Thus, the recognition of the validity of formulas $\Phi(f, x)$ has a high complexity. But they are not formulas that appear in practice. Moreover, practical formulas, that may have a good amount of quantifier alternations, are semantically much simpler, they never speak about diagonal algorithms, though may speak about practical algorithms, e.g., about execution and properties of hard real-time controllers.

The just presented argument is valid for all negative complexity results (undecidability, high lower bounds, relative hardness) with the existing proofs. And here one arrives at another 'incoherence' between theory and practice that can be illustrated by the TAUT problem, i.e., by the problem of recognition of the validity of propositional formulas. This problem is considered as relatively hard (more precisely, coNP-complete) in theory, but existing algorithms solve very efficiently practical instances of this problem, and the problem is considered as an easy one by people working in applications. This is not the only example.

There are similar examples of another flavor, like the practical efficiency of linear programming algorithms. Here one finds mathematically interesting results of their average behavior. However, traditional evaluation of the average or Teng-Spielman smooth analysis [13] deal with sets of inputs almost none of which appears in practice. If one accepts Kolmorogov algorithmic vision of randomness, i.e., a string (or other combinatorial construct) is random if its Kolmogorov complexity is close to the maximal value, then one gets another argument that random constructs cannot appear from physical or human activity.

Many people believe that physical processes may produce truly random data. Many years ago, it was somewhere in the 70th, G. M. Adelson-Velsky[2] told me that M. M. Bongard[3] showed, using not very complicated learnability algorithm, that Geiger counter data, that were considered as truly random, can be predicted with a probability definitely higher that $1/2$. Who else analyzed physical 'random data' in this way? Notice that standard statistical tests that are used to prove randomness can be easily fooled by simple deterministic sequences, e.g., Champernowne's sequence. Happily, in practice 'sufficiently random' sequences suffice.

The practical inputs are always described in a natural language whose constructs are numerous but incomparably less numerous than arbitrary constructs, so they are not so random.

One may refer to the ideology of modern mathematics. Modern mathematics does not study arbitrary functions, nor arbitrary continuous functions, nor even arbitrary smooth functions. It studies particular, often rather smooth, manifolds on which often, though not always, acts a group with some properties modeling

---

[2]Georgy Maximovich Adelson-Velsky (1922–2014) was a well-known Soviet and Israeli mathematician and computer scientist.

[3]Mikhail Moiseevich Bongard (1924–1971) was a well-known Soviet computer scientist.

properties inspired by applications in mind.

It is not so evident how to find a structure to study in algorithmic problems, but it is much simpler to see a structure in computations, namely, in sets of runs (executions). One can try to find geometry in these sets. An intuitive sentiment is that any algorithm transforms information, so we can try to find geometry in computations using this or that concept of information.

It is improbable that one approach will work for all types of algorithms that appear in practice. The frameworks we use to study different types of algorithms are different. For example, reactive real-time systems are studied not as data base queries, computer algebra algorithms are studied not in the same way as combinatorial algorithms etc. In this paper I try to look at off-line 'combinatorial' algorithms without defining this class rigorously. Roughly speaking such an algorithm processes a finite 'combinatorial' input accessible from the very beginning, where each bit is 'visible' except maybe some integers that are treated as abstract atoms or 'short' integers with addition and comparison. Examples are string matching, binary convolution, TAUT, shortest path in graphs with integer weights etc.

But algorithms of this vaguely defined class may be very different from the point of view of their analysis. For example, take diagonal algorithms and compare such an algorithm with an algorithm like just mentioned above. One can see that runs of diagonal algorithms are highly diverse, within the same length of inputs we may see a run that corresponds to an execution of a string-matching algorithm, another run that correspond to solving a linear system etc. In the algorithms mentioned above the runs are more or less 'similar'. My first idea was to say that this distinguishes practical algorithms from non-practical ones. However, E. Asarin immediately drew my attention to interpreters that are quite practical and whose sets of runs are of the same nature that the set of runs of diagonal algorithms. It is interesting that compilers (to which N. Dershowitz drew my attention in the context of a discussion on practical and impractical algorithms some time ago) are in the same class that the mentioned combinatorial algorithms because they do not execute the programs that they transform. But interpreters are not in the same class as the combinatorial algorithms that are under study here. We do not demand that an interpreter diminish the computational complexity of the interpreted algorithm. And the interpretation itself slows down the interpreted algorithm by a small multiplicative constant that we can try to diminish. In some way, the output of the interpreter is a trace of the interpreted algorithm, so their diversity is intrinsic, and the length of their outputs is compared with their time complexity. We consider algorithms whose outputs are 'much shorter' than their time complexity.

2. How to evaluate similarity of computations?

Some syntactic precisions on the representation of runs of algorithms are needed. Suppose that F is an algorithm of bounded computational complexity that has as its inputs some structures (strings, graphs etc.) and whose outputs are also some structures.

By the size of an input we mean not necessarily the length of its bit code but some value that is more intuitive and 'not far' from its bit size. E.g., the number of vertices for a weighted graph, the length of vectors in binary convolution etc. In any case the bit size is polynomially bounded by our size. Thus, for a weighted graph we assume that weights are integers whose size is of the order of logarithm of the number of vertices if the weights are treated as binary numbers or whose size is $O(1)$ if they are treated abstractly.

We mention two very simple examples, namely palindrome recognition and sum of elements of a string over $\mathbb{B}$.

Assume that for the structures under consideration a reasonable notion of size is defined, and the set of all inputs of size n, that are in the domain of F, is denoted by $\boldsymbol{dm}_n(F)$ or $\boldsymbol{dm}$ if F and n are clear from the context. The set of corresponding values of F is denoted $\boldsymbol{rn}_n(F)$ or $\boldsymbol{rn}$. We assume that n is a part of inputs. Below n is fixed and often omitted in the notations.

We look at algorithms from the viewpoint of logic. Though in programming, as well as in logic, any program may be seen as an abstract state machine, there is no terminology that is commonly accepted in logic and programming. For example, what is called variable in programming is not variable in logic; from the point of view of logic it is a function without arguments but that may have different values during the execution of the program. In order to avoid such discrepancy we use logical terminology that was developed by Yu. Gurevich for his Abstract State Machines [3], and may be applicable to any kind of programs. Our framework is not that of Yu. Gurevich machines, we deal with executions of low-level programs seen as some kind of abstract state machines.

An algorithm computes the values of outputs using *pre-interpreted* constants like integers, rational numbers, Boolean values, characters of a fixed alphabet, and pre-interpreted functions like addition, order relations over numbers and other values, Boolean operations. These functions are *static*, i.e., they do not change during the executions of F. The other functions are *abstract* and *dynamic*. The inputs are given by the values of functions (that constitute the respective structure) that F can only read; they are *external* (as well as pre-interpreted functions). The functions that can be changed by F are its *internal functions*, they are subdivided into *output functions* and *proper internal* functions. We assume for simplicity that the output functions are updated only once. Dynamic functions may have arguments, like, e.g., arrays, and we limit ourselves to such functions that have one natural argument. When the argument i in such a function f is fixed, this $f(i)$ can be considered as nullary function, i.e. as a function without arguments. All these functions constitute a vocabulary of the algorithm.

We consider computations only for inputs from a finite set $dm_n(F)$. These computations are represented as sets of traces that we describe below. We can treat such sets abstractly without precise notion of algorithm. However, for better intuitive vision, we describe a simple algorithmic language that gives a general notion of algorithm and that suffices for our examples.

Term is defined as usual, and without loss of generality, we consider non nested terms, i.e., terms whose arguments are only variables if any. *Guard* is a literal. *Update* (assignment) is an expression $g := \theta$, where g is an internal function, and θ is a term. Constructors of a program (algorithm) are: update, sequential composition (denoted ;), branching **if** $guard$ **then** P **else** P', where P and P' are programs, **goto**, **halt**. As delimiters we use brackets.

A *state* is an interpretation of the vocabulary of the algorithm. A state is changed by updates in the evident way. The initial state is common for all inputs, we assume that the initial value of any internal function f is symbol ♮ that represents *undefined*, is never used in updates, and that $f^{-1}(♮) = \emptyset$. A *run* is usually defined as a sequence of states, but we use an equivalent representation of executions as traces.

Given an input $X \in dm_n(F)$ a trace $tr(X)$ is constructed as follows according to the executed operators: update is written as it is in the program; in the case of conditional branching **if** $guard$ **then-else** we put in the trace either $guard$ or its negation depending on what is true in this trace. For simplicity the initial state and **halt** are not explicitly mentioned in traces, neither **goto**. Thus, a trace is a sequence of updates and guards that are called *events*. The tth event in a trace $tr(X)$ is denoted $tr(X, t)$. These events are *symbolic*. An execution gives values to the internal functions, and thus, an interpretation of any event.

Denote by $\mathbf{t}_F^*(X)$ the time complexity of F for input X, and by $\mathbf{t}_F(n)$ the maximum of these values, i.e., the worst-case time complexity of F over $\boldsymbol{dm}_n(F)$.

For an input $X \in \boldsymbol{dm}_n(F)$ and a time instant t, $1 \leq t \leq \mathbf{t}^*(X)$, we denote by $f[X, t]$ the value of a internal function f in $\boldsymbol{tr}(X)$ at t, the value is defined recursively together with the recursive definition of trace given just above. If f is not undated at t then $f[X,t] = f[X, t-1]$. If $\boldsymbol{tr}(X, t)$ is of the form $f := g(\eta)$ then $f[X, t] = g(\eta[X, t-1])[X, t-1]$.

Consider two examples.

Palindrome recognition. Inputs are non empty strings of length n over an alphabet $\mathcal{A}$ with $\alpha \geq 2$ characters. For simplicity assume that n is even and set $\nu =_{df} \frac{n}{2}$. We denote the input by w, and the character in the ith position by $w(i)$. We take a straightforward algorithm φ that compares characters starting from the ends and going to the middle of the input. We use % to mark comments, and we omit **halt** that is evident.

Algorithm φ:

% i is a loop counter, r is the output (0 means non palindrome, 1 palindrome)

1:    $i := 0$;

2:    **if** $i < \nu$ **then** $\Big(i := i + 1$;

                if $w(i) = w(n - i + 1)$ **then goto** 2 **else** r:=0$\Big)$

        else r:=1

Algorithm φ has two types of traces (one with output 0 and the other with output 1):

$i := 0, i < \nu, i := i+1, w(i) = w(n-i+1), \ldots, i < \nu, i := i+1,$
$w(i) = w(n-i+1), i < \nu, i := i+1, w(i) \neq w(n-i+1),$ r:=0

$i := 0, i < \nu, i := i+1, w(i) = w(n-i+1), \ldots, i < \nu, i := i+1,$
$w(i) = w(n-i+1), i \geq \nu,$ r:=1

A trace of the first type may have different lengths starting from 5, but the length of the trace of the second type is always the same.

For a string $aabaaa$ with $a \neq b$, if in the respective trace we replace the internal functions, as well as n, by their values we can write:

$i := 0, 0 < 3, i := 0 + 1, w(1) = w(6), 1 < 3, i := 1 + 1, w(2) = w(5),$
$2 < 3, i := 2 + 1, w(3) \neq w(4),$ r:=0

Sum modulo 2 of bits of a string. Inputs are strings of the set $\mathbb{B}^n$.
Algorithm σ:
% x is input, r is output, i is a loop counter, s is an intermediate value
1:      $i := 0;\ s := 0;$ %Initialization
2:      **if** $i < n$ **then** $i := i + 1;\ s := s + x(i);$ **goto** 2
3:              **else** $r := s$ % case $i \geq n$

All traces of σ are 'symbolically' the same (the algorithm is oblivious), for clarity we put in an event the value of i acquired before this event:
$i := 0,\ s := 0,\ 0 < n,\ i := 0 + 1,\ s := s + x(1),\ 1 < n,\ i := 2,$
$s := s + x(2), \ldots, n - 1 < n,\ i := n,\ s := s + x(n),\ n \geq n,\ r := s$

Remark. For Boolean circuits we can also produce traces that are even simpler, as a Boolean circuit is a non branching oblivious algorithm. Such a trace consists of updates, each one being an application of the Boolean function attributed to a vertex of the circuit, to the values attributed to its predecessors.

Denote by $\boldsymbol{Tr_n}$ the set of all traces for inputs from $\boldsymbol{dm_n}$. The length $|\boldsymbol{tr}(X)|$ of a trace $\boldsymbol{tr}(X)$, $X \in \boldsymbol{dm_n}$, is the number of occurrences of events in it, i.e., the time complexity $\mathbf{t}_F^*(X)$.

2.1. A syntactic similarity of traces

A straightforward way to compare two traces is the following one. We look in $\boldsymbol{tr}(X)$ and $\boldsymbol{tr}(Y)$ for a longest common subsequence (we tacitly assume that some equivalence between events is defined), and take as a measure of similarity the size of the rest. More precisely, if S is the longest common subsequence then we take as measure the value $|\boldsymbol{tr}(X)| + |\boldsymbol{tr}(Y)| - 2|S|$, where $|S|$ is the size (the number of elements) of a sequence S. This measure is something like the size of symmetric difference of two sequences.

We can go further, and to take into account only causal order in what concerns the order of events, and to permit a renaming of proper internal functions and their values. The causal order is defined as follows. If the function updated or used (in the case of guard verification) in an event e depends on a function updated earlier in an event e' then e' *causally precedes* e. Taking a transitive closure of this relation we get *causal order* between events in a given trace. This generalization is too technical (details can be found in [12]), and as I cannot give examples of realistic applications, it is just mentioned as a theoretical possibility.

The measure introduced above gives a pseudo-metric (it is like metric except that two different traces may have zero distance; in our case the zero distance relation is an equivalence) over traces. As the trace space Tr_n is clearly compact, this metric permits to define epsilon-entropy [5] on it. This entropy is defined as follows. For a given ε (in our case it is a natural number) take an ε-net of minimal size such that the ε-balls centered at the points of the net cover all the space. Then $\log s$, where s is the size of this net, is the ε-entropy. It gives the size complexity of the ε-approximation of the space, or to say it differently, how much information one needs to have, in order to describe an element of the space with accuracy ε.

Consider our examples.

Trace space of φ. We define similarity as follows (it is a rather general way to define it). First, in the right-hand side of each update $f := \theta$ replace all proper internal functions of θ by their values. In guards replace all internal functions by their values. We get as transformed events the expressions: $i := m$, where $m \in \mathbb{N}$ and $m = (\ldots((0+1)+1) + \cdots + 1)$, $m < \nu$, $m \geq \nu$, $w(m) = w(n-m+1)$, $w(m) \neq w(n-m+1)$, $r = 0, r = 1$. As similarity (we refer to it as 'weak similarity') we take the syntactic equality of these transformed events.

With this similarity we have $(\nu + 1)$ different (*classes of similar*) traces (ν classes with $r = 0$ at the end, and 1 class with $r = 1$): denote by P_k traces with $(k-1)$ equalities and one inequality in the k comparison, $1 \leq k \leq \nu$, and by P the only trace with $r = 1$. The distance between P_k and P_l is $3|k - l|$, and between P_k and P is $3(\nu - k) + 2$. If we take $\varepsilon = 2$ then ε-net should include all the traces but, however, it is of size $(\log n + \mathcal{O}(1))$. If we take $\varepsilon = 3p$, $p \in \mathbb{N}$, then as an ε-net we can take each pth trace ordered according to their lengths; hence, $3p$-entropy is of size $\lceil \frac{n}{2p} \rceil = \lceil \frac{\nu}{p} \rceil$ (maybe plus 1).

The situation changes if we take stronger similarity. We say that $w(m) = w(n-m+1)$ and $w(m') = w(n-m'+1)$ are similar if $m = m'$ (as before) and the respective values of inputs are the same $w(m) = w(m')$ (for $\neq$ we demand also $w(n-m+1) = w(n-m'+1)$). In this case the trace space becomes of exponential size. We illustrate this kind of similarity for algorithm σ.

Trace space of σ. We define similarity of event of the form $i := i + 1$ and of the form $i < n$ as in the previous case: values of $(i+1)$ in similar events of the form $i := i+1$ and the value of i in similar events of the form $i < n$ should be equal. Two events of the form $s := s + x(i)$ are similar if the values of i, as well as of the acquired values of s, are equal. Any string from $\mathbb{B}^n$ may be a string of consecutive values s starting from $s := 0 + x(1)$ that equals to $x(1)$. Thus the set of traces of σ with this similarity and our metric divided by 2 is isometric to the Boolean cube $\mathbb{B}^n$ with Hamming metric. This space is studied in the coding theory, and I cannot say more than can be found there.

Unfortunately, the metric spaces in the examples above do not say much about the advancement of the algorithm towards the result. If we take spaces of traces up to some time instant and their dynamics with growing time, it does not help much neither. Moreover, the size of the space $\boldsymbol{Tr}_n$ is bounded by $|\boldsymbol{dm}_n|$, and does not depend on the complexity of F, and this is also a shortcoming of this approach.

2.2. Remark on Kolmogorov complexity approach

Why not to measure distance between traces on the basis of Kolmogorov complexity? This question was put by some of my colleagues.

A direct application of Kolmorogov algorithmic entropy [4] to measure similarity of traces does not give results corresponding to our intuition. Indeed, in [4] Kolmogorov defines entropy as conditional complexity $\boldsymbol{K}(\alpha|\beta)$. Similarity of structures α and β may be measured as $\mathfrak{K}(\alpha, \beta) = \boldsymbol{K}(\alpha|\beta) + \boldsymbol{K}(\beta|\alpha)$. This is not a metric, strictly speaking, however, we call this function $\mathfrak{K}$-*distance* as it has a flavor of intuitive distance-like measure.

Denoting by $|F|$ and $|X|$ binary lengths of respectively F and X we get
$$\boldsymbol{K}(\boldsymbol{tr}(X)/\boldsymbol{tr}(Y)) \leq |F| + \boldsymbol{K}(X/Y) + O(1) \leq |F| + |X| + O(1).$$
This formula follows from an observation that X and F are sufficient to calculate the trace $\boldsymbol{tr}(X)$. Thus, whatever be an algorithm F and whatever be its computational complexity, the $\mathfrak{K}$-distance between traces from $\boldsymbol{Tr}_n$ is not greater than $O(|X|)$ that we assume, for simplicity, to be $O(n)$. On the other hand, given a minimal length program G that computes $\boldsymbol{tr}(X)$ from $\boldsymbol{tr}(Y)$ (thus, $|G| = \boldsymbol{K}(\boldsymbol{tr}(X)/\boldsymbol{tr}(Y))$) one can get X from Y as follows: from Y one computes $\boldsymbol{tr}(Y)$ using F (whose size is a constant), then using G one computes $\boldsymbol{tr}(X)$ and finally extracts X from $\boldsymbol{tr}(X)$ with the help of a simple fixed program, say E, whose length is a constant (without loss of generality, we can assume that the input is reproduced at the beginning of each trace). All this gives (we put 'absolute' constants $|F|$, $|E|$ in the last $O(1)$)
$$\boldsymbol{K}(X|Y) \leq |F| + |G| + |E| + O(1) \leq \boldsymbol{K}(\boldsymbol{tr}(X)/\boldsymbol{tr}(Y)) + O(1).$$
We assume that the cardinality of binary codes of $\boldsymbol{dm}_n(F)$ is at least 2^n (hence,

almost all inputs have Kolmogorov complexity $n - o(n)$), then the chain rule for Kolmogorov complexity (e.g., see [4]) for almost all X, Y gives
$$K(X|Y) = K(X,Y) - K(Y) - O(\log K(X,Y)) \geq n - c\log n$$
for some constant $c > 0$.

Together with the previous formula this gives a lower bound for $K(tr(X)/tr(Y))$ that shows that $\mathfrak{K}$-distance is almost always of order of n that can hardly be seem as satisfactory for evaluation of similarity of traces from Tr_n.

What is said above, does not exclude that other types of Kolmogorov style complexity could work better (e.g., a more general notion of entropy [11] is based on inference complexity.). In particular, resource bounded complexity approaches may prove to be productive if we find a 'good' description of information extracted by algorithm as datum (structure); however, this remains an open question.

2.3. Similarity via entropy of partitions

In this subsection we outline another approach to measure similarity of traces. It refers to the classical entropy of partitions. We use partitions of the inputs. For this reason a probabilistic measure over the inputs is needed. Such a measure is a technical means, so there is no evident way to introduce it. We do it taking into account an intuition related to the evolution of the 'knowledge' of the algorithm. When an algorithm F starts its work it 'knows' nothing about its output. So all values from rn_n are equiprobable.

Let $M = |rn_n(F)|$. As any of these M values is equiprobable (imagine that an input is given by an adversary who plays against F), we set $P_n(F^{-1}(Y)) = \frac{1}{M}$ for all $Y \in rn_n(F)$, and inside $F^{-1}(Y)$ the measure is uniform as the algorithm a priori has no preferences. In particular, if F is a 2-valued function, say $rn(F) = \mathbb{B}$, then its domain is partitioned into two sets $F^{-1}(0)$ and $F^{-1}(1)$ with the same measure $1/2$ of each set. E.g., for palindromes we the measure of a palindrome is $\frac{1}{2a^\nu}$ and that of a non palindrome is $\frac{1}{2(a^n-a^\nu)}$. There is nothing random in the situation we consider, we wish only to model the evolution of the knowledge of an algorithm during its work. So this way to introduce a measure may be not the best one.

Suppose that f is updated at t and $f[X,t] = v$. How to describe the knowledge acquired by F via this event at t that gives $v = f[X,t]$? This value v may be acquired by f in different traces, even several times in the same trace, and at different time instants. The traces are not 'synchronized' in time, however, we can compare events, as in subsection 2.1, due to this or that similarity relation, that is determined by our goal and our vision of the situation. Notice that formally speaking similarity is a relation between pairs (X, t), where $X \in dm(F)$ and $1 \leq t \leq \mathbf{t}_F^*(X)$. Similarity can be defined not only along the lines described in subsection 2.1. One may think about quite different ways. Just to give an idea, one can, for example, consider as similar events corresponding to the kth execution of the same command of the program with or without demanding equality of these or that values. Or one can permit renaming of internal function as it was mentioned at the beginning of subsection 2.1.

Suppose that some similarity relation $\sim$ is fixed.

To compare traces we attribute to each event of a trace a partition of inputs. Thus, to each trace there will be attributed a sequence of partitions. Taking into account that the set of inputs is a space with probabilistic measure we can define a distance between partitions and furthermore a distance between sequences or sets of partitions.

For any input X and an instant t, $1 \leq t \leq \mathbf{t}^*(X)$, denote by $\boldsymbol{sm}(X, t)$ all the inputs X' such that $(X, t) \sim (X', t')$ for some t'. Clearly, $X \in \boldsymbol{sm}(X, t)$. Denote by $\boldsymbol{pt}(X, t)$ the partition of $\boldsymbol{dm}_n$ into $\boldsymbol{sm}(X, t)$ and its complement that we denote $\boldsymbol{sm}(X, t)^c =_{df} \boldsymbol{dm}_n \setminus \boldsymbol{sm}(X, t)$.

Thus, each input X determines a *sequence* $(\boldsymbol{pt}(X, t))_t$ or a set $\{\boldsymbol{pt}(X, t)\}_t$ of *partitions* of $\boldsymbol{dm}_n(F)$. These constructions, namely sequence or set, provide different opportunities for further analysis, e.g., we can define distance between metric spaces, e.g., see [1, ch. 7].

For measurable partitions of a probabilistic space $\mathcal{P} = (\Omega, \Sigma, P)$ one can define entropy (no particular technical constraints are needed in our case of finite sets), see [8] or books like [7].

Let $\mathcal{A}$ and $\mathcal{B}$ be measurable partitions of $\mathcal{P}$ (in our situation all the sets are measurable).

Entropy $H(\mathcal{A})$ and conditional entropy $H(\mathcal{A}/\mathcal{B})$ are defined as

$$H(\mathcal{A}) = -\sum_{A \in \mathcal{A}} P(A) \log P(A), \quad H(\mathcal{A}/\mathcal{B}) = -\sum_{B \in \mathcal{B}, A \in \mathcal{A}} P(A \cap B) \log \frac{P(A \cap B)}{P(B)}. \tag{1}$$

The conditional entropy permits to introduce Rokhlin metric [8] between partitions:
$$\rho(\mathcal{A},\mathcal{B}) = H(\mathcal{A}/\mathcal{B}) + H(\mathcal{B}/\mathcal{A}) = 2H(\mathcal{A} \vee \mathcal{B}) - H(\mathcal{A}) - H(\mathcal{B}),$$
(here $\mathcal{A} \vee \mathcal{B}$ is common refinement of partitions $\mathcal{A}$ and $\mathcal{B}$, that is the partition formed by all pairwise intersection of sets of $\mathcal{A}$ and $\mathcal{B}$).

There are other ways to introduce distance between partitions, e.g., see [7, 4.4], so one can take or invent maybe more productive metrics or entropy-like measures.

Unfortunately, the combinatorial difficulties of estimating such distances are discouraging, they do not justify what we get form them. We illustrate this for the palindrome recognition algorithm φ.

Denote $w^=(1..k) =_{df} \{w : \bigwedge_{1 \le i \le k} w(i) = w(n-i+1)\}$ (the set of words whose prefix of length k permits to extend it to a palindrome), denote by $w^{\ne}(1..k)$ the complement of $w^=(1..k)$; in particular, $w^=(k..k) = \{w : w(k) = w(n-k+1)\}$ and $w^{\ne}(k..k) = \{w : w(k) \ne w(n-k+1)\}$. Probabilities are easy to calculate (we use them in the next subsection), here $1 \le k < m \le \nu$:

$$\boldsymbol{P}(w^=(1..k) \cap w^{\ne}(k+1..m)) = \frac{\alpha^{\nu-m}(\alpha^{m-k}-1)}{2(\alpha^\nu-1)}, \quad (2)$$

$$\boldsymbol{P}(w^=(1..k)) = \frac{1}{2} + \frac{\alpha^{\nu-k}-1}{2(\alpha^\nu-1)}, \quad \boldsymbol{P}(w^{\ne}(1..k)) = \frac{\alpha^{\nu-k}(\alpha^k-1)}{2(\alpha^\nu-1)}. \quad (3)$$

(We omit technical details, the role of the formulas is illustrative.)

However, when we try to calculate the distance between partitions, take for example $\rho(\pi^=(k), \pi^=(m))$, where $\pi^=(s) = (w^=(1..s), w^{\ne}(1..s))$, we arrive at a formula that is a sum of several expressions like $\left(\frac{1}{2} + \frac{\alpha^{\nu-s}-1}{2(\alpha^\nu-1)}\right) \log \left(\frac{1}{2} + \frac{\alpha^{\nu-s}-1}{2(\alpha^\nu-1)}\right)$, that is hard to evaluate. And what is worse the result is not very instructive, e.g.,

$$\rho(\pi^=(1), \pi^=(\nu)) \approx \begin{cases} 0.9 & \text{if } \alpha = 2 \\ 0.67 & \text{if } \alpha = 3 \\ 0.6 & \text{if } \alpha = 4 \end{cases}$$

Technical combinatorial difficulties do not discard the idea of geometry of spaces of events or traces, the point is to find a geometry and its interpretation that really deepens our understanding of algorithms and problems.

2.4. The question of information convergence

Now we discuss how similarity of events may serve to evaluate the rate of convergence of a given algorithm towards the result.

Among the first ideas that come to mind is the following one. The result $F(X)$ for an input X is represented in terms of a partition of $dm_n(F)$ into $F^{-1}(F(X))$ and its complement $F^{-1}(F(X))^c$. The current knowledge of F at an instant t is in its current event that also defines a partition denoted above $pt(X,t)$.

How this local knowledge represented by $pt(X,t)$, is related to the partition $(F^{-1}(F(X)), F^{-1}(F(X))^c)$ mentioned just above? A possible answer is: compare $pt(X,t)$ (the local knowledge at an instant t in terms of partitions) with the partition $(F^{-1}(F(X)), F^{-1}(F(X))^c)$. This idea can be a priori implemented differently, for example, in terms of conditional probabilities or in terms of conditional entropies.

If we try to apply this idea to any of our examples, we find that the commands that control the loops give trivial partition $(dm, \emptyset)$ because they are in all traces, and these events give nothing useful. So we take only events that process inputs.

Consider φ (the example of palindromes). Using (2), (3) we get, omitting technicalities and taking sufficient approximations:

$$P(r=1|w^=(1..k)) \approx \frac{1}{1+A(k)}, \quad P(r=0|w^=(1..k)) \approx \frac{A(k)}{1+A(k)}, \quad (4)$$

where $A(k) = \alpha^{-k} - \alpha^{-\nu}$. The probabilities (4) do not reflect our information intuition that φ converges to the result when $k \to \nu$ as one goes to 1, and the other to 0. But if we take the respective entropy

$$-\frac{1}{1+A(k)} \log \frac{1}{1+A(k)} - \frac{A(k)}{1+A(k)} \log \frac{A(k)}{1+A(k)}, \quad (5)$$

we see that it goes to 0, thus, to total certainty.

Consider σ (sum modulo 2). The similarity that we used for the trace space of σ in subsection 2.1 we call here *weak similarity*. Denote $\sigma^{-1}(a)$ by $\sigma = a$. Clearly, $|\sigma = 0| = |\sigma = 1| = 2^{n-1}$, $P(\sigma = a) = \frac{1}{2}$, and P is a uniform distribution over $\mathbb{B}^n$.

Denote by $S_k(a)$, where $a \in \mathbb{B}$, the set $\{x : s+x(k) = a\}$; it is a set of type

$sm(X,t)$. For $k < n$ and all $a, b \in \mathbb{B}$ we have

$$P(S_k(a)) = \frac{|S_k(a)|}{2^n} = \frac{2^{n-1}}{2^n} = \frac{1}{2}, \quad P(S_n(a) \cap S_k(b)) = \frac{1}{4} \quad (6)$$

$$P(\sigma = a | S_k(b)) = \frac{P(S_n(a) \cap S_k(b))}{P(S_k(b))} = \frac{1}{2} \quad (7)$$

We see that nothing changes with advancing of time, i.e., with $k \to n$. If we apply formula (1) for conditional entropy, it gives a constant. Hence, with this similarity, we do not see any convergence of σ to the result.

Let us try a stronger similarity: we say that a event $s := s + x(k)$ is (strongly) similar to $s := s + x'(k)$ if $x(i) = x'(i)$ for all $1 \leq i \leq k$. Denote by $Z(\chi)$, where $\chi \in \mathbb{B}^k$, $1 \leq k < n$, the set of inputs x such that for event $s := s + x(k)$ there holds $x(i) = \chi(i)$ for $1 \leq i \leq k$; this set describes the set of inputs of strongly similar events. We have $|Z(\chi)| = 2^{n-k}$, and $|(\sigma = a) \cap Z(\chi)| = 2^{n-k-1}$, thus, $P(Z(\chi)) = 2^{-k}$ and $P((\sigma = a) \cap Z(\chi)) = 2^{-k-1}$. So the measure of the space of continuations of the known part of the input diminishes. The respective term in conditional entropy (1) gives $-2^{-k-1} \log \frac{1}{2} = 2^{-k-1}$ that is encouraging but the term related to $Z(\chi)^c$ (notice that $P(Z(\chi)^c) = 1 - 2^{-k}$ and $P((\sigma = a) \cap Z(\chi)^c) = \frac{1}{2} - 2^{-k-1}$) bring us back to values that practically do not diminish. All this means only that the classical entropy does not work, and we are to seek for entropy-like measures that truly reflect our intuition.

The partition based measures of convergence look promising. However, one can say that the number of partitions is limited by an exponential function of $|dm|$. So if the complexity is very high, e.g., hyper-exponential, then there is 'not enough' of partitions to represent the variety computations. In fact, we think about certain class of problems that are outlined in the next section for which there seems to be 'enough' of partitions. As for high complexity problems, another interpretation of input data is needed. Some hints are given in the next section 3.

3. On the structure of problems

Here are presented examples of problems together with a reference to their inner structure that may be useful for further study of information structure of computations and that of problems themselves along the lines discussed in the paper. The examples below concern only simple 'combinatorial problems'. The instances of these problems are finite graphs (in particular, strings, lists, trees etc.) whose edges and vertices may be supplied with additional objects that are either abstract atoms with some properties or strings. As examples of problems that are not in this class one can take problems with exponential complexity like theory of real addition or Presburger arithmetics. The problems in the examples below are divided into 'direct' and the respective 'inverse' ones.

Direct Problems

(A1) *Substring verification.* Given two strings U, W over an alphabet with at least two characters and a position k in W, to recognize whether $U = W(k, k+1, \ldots, k+|U|-1)$, i.e., whether U is a substring of W from position k.

(A2) *Path weight calculation.* Given a weighted (undirected) graph and a path, calculate the weight of the path.

(A3) *Evaluation of a Boolean formula for a given value of variables.* Given a Boolean formula Φ and a list X of values of its variables, calculate the value $\Phi(X)$ for these values of variables.

(A4) *Permutation.* Given a list of elements and a permutation, apply the permutation to the list.

(A5) *Binary convolution (or binary multiplication).* For simplicity we consider binary convolution that represents also the essential difficulties of multiplication. Given 2 binary vectors or strings $x = x(0) \ldots x(n-1)$ and $y = y(0) \ldots y(n-1)$ calculate
$$z(k) = \sum_{i=0}^{i=k} x(i)y(k-i), \quad 0 \leq k \leq (2n-2),$$
assuming that $x(i) = y(i) = 0$ for $n - 1 < i \leq (2n-2)$.

Inverse Problems

(B1) *String matching.* Given two strings W and U over an alphabet with at least two characters, to recognize whether U is a substring of W.

(B2) *Shortest path.* Given a weighted (undirected) graph G and its vertices u and v, find a shortest path (a path of minimal weight) from u to v.

(B3) *Propositional tautology TAUT* Given a propositional formula Φ, to recognize whether it is valid, i.e., is true for all assignment of values to its variables. A variant that is more interesting in out context is MAX-SAT: given a CNF (conjunctive normal form), to find the longest satisfying assignment of variables, i.e. an assignment that satisfies the maximal number of clauses.

(B4) *Sorting.* Given a list of elements of a linearly ordered set, to find a permutation that transforms it into an ordered list.

(B5) *Factorization.* Given z, to find x and y whose convolution or product (in the case of multiplication) is z.

Examples (A1)–(A4) give algorithmic problems whose solution, based directly on their definitions, is practically and theoretically the most efficient. Each solution consists in a one-directional walk through a simple data structure making, again rather simple, calculations – something that is similar to scalar product calculation.

In (A1) the structure is a list $(k, k+1, \ldots, k+|U|-1)$, and while walking along it, we calculate conjunction of $U(i) = W(i)$ for $k \leq i < (k+|U|)$ until i reaches the last value or $false$ appears.

Example (A2) is similar, where the list of vertices constituting the linear structure is explicitly given, and the role of conjunction of (A1) is played by addition.

The structures used in (A3) depend on the representation of Φ and of the distribution of values of its variables. In any case one simple linear structure does not suffice here. Suppose Φ is represented in DNF (Disjunctive Normal Form), i.e., as a disjunction of conjunctions. This can be seen as a list of lists of literals, and a given distribution of values is represented as an array corresponding to a fixed order of variables. So given a variable, its value is immediately available. Thus, the representation of values is a linear structure, and DNF is a linear structure of linear structures. It is more interesting to suppose that Φ is a tree. Then we deal with the representation of values and with a walk, again without return, through a tree with calculating the respective Boolean functions at the vertices of the tree. So we see another simple basic structure, namely a tree.

In example (A4), while walking through two given lists, namely a list of elements and a permutation, a third list (a list of permuted elements) is constructed.

Example (A5) is more complicated, and the definition of problem does not give an algorithm that may be considered as the best; it is known that the direct algorithm for convolution is not the fastest one. Here there is no search, and for this reason this problem is put in the class of direct ones, but there is a non-trivial intermixing of data. One may see the description of the problem as a code of data structures to extract, and then to calculate the resulting values by simple walks through these data structures. The number of the data structures to extract is quadratic. In order to find a faster algorithm, one should ensure the same intermixing but using different data structures and operations.

Examples (B1)–(B5) give algorithmic problems of search among substructures coded in inputs. The number of these substructures, taken directly from the definition, is quadratic for (B1), and exponential for (B2)–(B5). The substructures under search should satisfy conditions that characterize the corresponding direct problem. More complicated problems code substructures not so explicitly as in examples (B1)–(B5). To illustrate this, take e.g., quantifier elimination algorithm for the formulas of the theory of real addition, not necessarily closed formulas. Here it is not evident how to define the substructures to consider. The quantifier elimination by any known algorithm produces a good amount of linear inequalities that are not in the formula. So the formula codes its expansion that is more than exponentially bigger as compared with the initial formula itself.

Whatever be the mentioned difficulties, intuitively the substructures and constraints generated by a problem may be viewed as an extension of the set of inputs. And in this extended set one can introduce not only measure but also metrics that give new opportunities to analyze the information contents and the information evolution. One can see that the cardinality constraints on the number of partitions that was mentioned in subsections 2.3 and 2.4 is relaxed. This track has not been yet studied, though one observation can give some hint to how to proceed. When comparing substructures it seems productive to take into account its context, i.e., how it occurs in the entire structure. For example, we can try to understand the context of an assignment A of values to variables of a propositional formula Φ in the following manner. Pick up a variable x_1 and its value v_1 from A and calculate the result $\Phi(x_1, v_1)$ of the standard simplification of Φ where x_1 is replaced by Boolean value v_1. This resulting residue formula gives some context of (x_1, v_1). We can take several variables and look at the respective residue as at a description of context. This or that set of residues may be considered as a context of A. It is just an illustration of what I mean here by 'context'.

A metric over substructures may distinguish 'smooth' inputs from 'non-smooth' ones, and along this line we may try to distinguish practical inputs from non practical ones. Though it is not so evident.

For some 'simple' problems such a distinction is often impossible. It looks hard to do for numerical values. The set of such values often constitutes a variety with specific properties that may represent realistic features but almost all elements of such varieties will never appear in practical computations. An evident example is binary multiplication. Among 2^{128} possible inputs of multiplication of 64-bit numbers most of them will never be met in practice.

A remark on the usage of linguistical frameworks

One more way to narrow the sets of inputs to take into account, is a language based one. Inputs describing human constructions, physical phenomena, and their properties, when they are not intended to be hidden, have descriptions in a natural language. Encrypted data are not of this nature. So for input data with non hidden information, we have a grammar that generates these inputs. Such a grammar dramatically reduces the number of possible inputs and, what is more important, defines a specific structure of inputs. The diminishing of the number of generated inputs is evident. For example, the number of 'lexical atoms' of the English is not more than 250 thousands, i.e., not more than 2^{18}. On the other hand, the number of strings with at most, say, 6 letters is at least $26^2 = 2^{6 \cdot \log 26} > 2^{6 \cdot 4.7} > 2^{28}$ (here 26 is the number of letters in English alphabet). The set of cardinality 2^{18} is tiny with respect to the set of cardinality 2^{28}. If one tries to evaluate the number of phrases, the difference becomes much higher.

But this low density of 'realistic' inputs does not help much without deeper analysis. The particular structure of inputs may help to devise algorithms more efficient over these inputs than the known algorithms over all inputs; there are examples, however not numerous and mainly of more theoretical value. So if one wishes to describe practical inputs in a way that may help to devise efficient algorithms, one should find grammars well aimed at the representation of particular structures of inputs. This point of view does not go along traditional mathematical lines when we look for simple and general descriptions, that are usually too general to be adequate to the computational reality.

The grammar based view of practical inputs may influence theoretical vision of a problem. For example, consider the question of quality of encryption. The main property of any encryption is to be resistant to cryptanalysis. Notice that linguistic arguments play an essential role in practical cryptanalysis. In reality the encryption is not applied to all strings, it mostly deals only with strings produced by this or that natural language, often rather primitive. Thus, there are relations defined over plain texts. E.g., some substrings are related as subject-predicate-direct compliment, etc. A good encryption should not leave traces of these relations in the encrypted text. What does it mean? Different precisions come to mind. A simple example: let P be a predicate of arity 2 defined over plain texts, and its arguments be of small bounded size. Take concrete values A and B of arguments of P. Assume that we introduced a probabilistic measure on all inputs (plain texts), and hence we have a measure of the set S^+ of inputs where $P(A, B)$ holds and of its complement S^-. Now suppose that we have chosen a predicate Q over 'substructures' of encrypted texts (I speak about 'substructures' to underline that the arguments of Q are not necessarily substrings, as for P), again simple to understand. Denote by E^+ the set of encrypted texts for which Q is true for at least one argument and by E^- its complement. The encryption well hides $P(A, B)$ if the measures of all 4 sets $(S^\alpha \cap E^\beta)$, where $\alpha, \beta \in \{+, -\}$, are very 'close'. This example gives only an idea but not a productive definition.

However, in order to find grammars that help to solve efficiently practical problems 'semantical' nature of sets of practical inputs should be studied.

Conclusion

The considerations presented above are very preliminary. The crucial question is to define information convergence of algorithms, not necessarily of general algorithms, but at least of practical ones.

One can imagine also other ways of measuring similarity of traces. We can hardly avoid syntactical considerations when keeping in mind the computational complexity. However semantical issues are crucial, and may be described not only in the terms chosen in this paper.

The analysis of philosophical question of relation of determinism versus uncertainty in algorithmic processes could clarify the methodology to choose. Here algorithmic process is understood at large, not necessarily as executed by a computer. Though the process is often deterministic, and if we adhere to determinism then it is always deterministic, at a given time instant, when it is not yet accomplished, we do not know with certainty the result, though some knowledge has been acquired. The question is: what is or how to formalize the knowledge (information) that the algorithm acquires after each step of its execution?

Acknowledgments

I am thankful to Edward Hirsch and Eugene Asarin for their remarks and questions that stimulated this study. I am also grateful to anonymous referees for their remarks and questions that considerably influenced the final text.

The research was partially supported by French *Agence Nationale de la Recherche* (ANR) under the project EQINOCS (ANR-11-BS02-004) and by Government of the Russian Federation, Grant 074-U01.

References

[1] D. Burago, Yu. Burago, and S. Ivanov. *A Course in Metric Geometry*, volume 33, Graduate Studies in Mathematics. Americal Mathematical Society, Providence, Rhode Island, 2001.

[2] M. Fischer and M. Rabin. Super-exponential complexity of presburger arithmetic. In *Complexity of Computation, SIAM-ASM Proceedings, vol. 7*, pages 27–41, 1974.

[3] Yu. Gurevich. Evolving algebra 1993: Lipari guide. In E. Börger, editor, *Specification and Validation Methods*, pages 9–93. Oxford University Press, 1995.

[4] A. N. Kolmogorov. On the Logical Foundations of Information Theory and Probability Theory. *Probl. Peredachi Inf.*, 5(3):3–7, 1969. In Russian. English translation in: Problems of Information Transmission, 1969, 5:3, 1–4, or in: Selected Works of A.N. Kolmogorov: Volume III: Information Theory and the Theory of Algorithms (Mathematics and its Applications), Kluwer Academic Publishers, 1992.

[5] A. N. Kolmogorov and V. M. Tikhomirov. ε-entropy and ε-capacity of sets in function spaces. *Uspekhi Mat. Nauk*, 14(2(86)):3–86, 1959. In Russian. English translation in: Selected Works of A.N. Kolmogorov: Volume III: Information Theory and the Theory of Algorithms (Mathematics and its Applications), Kluwer Academic Publishers, 1992.

[6] O. B. Lupanov. The synthesis of contact circuits. *Dokl. Akad.Nauk SSSR*, 119:23–26, 1958.

[7] N. F. G. Martin and J. W. England. *Mathematical Theory of Entropy*. Addison-Wesley, Reading, Massachusetts, 1981.

[8] V. A. Rokhlin. Lectures on the entropy theory of measure-preserving transformations. *Russian Math. Surveys*, 22(5):1–52, 1967.

[9] C. Shannon. The synthesis of two-terminal switching circuits. *Bell System Technical Journal*, 28(1):59–98, 1949.

[10] A. Slissenko. Complexity problems of theory of computation. *Russian Mathematical Surveys*, 36(6):23–125, 1981. Russian original in: *Uspekhi Matem. Nauk*, 36(2):21–103, 1981.

[11] A. Slissenko. On Measures of Information Quality of Knowledge Processing Systems. *Information Sciences: An International Journal*, 57–58:389–402, 1991.

[12] A. Slissenko. On Entropy in Computations. In *Proc. of the 8th Intern. Conf. on Computer Science and Information Technology (CSIT'2011), September 26–30, 2011, Yerevan, Armenia. Organized by National Academy of Science of Armenia*, pages 25–30. National Academy of Science of Armenia, 2011. ISBN 978-5-8080-0797-0.

[13] D. A. Spielman and S.-H. Teng. Smoothed analysis of algorithms: Why the simplex algorithm usually takes polynomial time. *J. ACM*, 51(3):385–463, 2004.

Received: 17 May 2016

Horizons of Scientific Pluralism: Logics, Ontology, Mathematics

Vladimir L. Vasyukov

Institute of Philosophy, Russian Academy of Science
12/1, ul. Goncharnaya, Moscow 109240, Russia

vasyukov4@gmail.com

Abstract: Discussions on the scientific pluralism typically involve the unity of science thesis, which has been first advanced by Neo-Positivists in the 1930-ies and later widely criticized in the late 1970-ies. In the present paper the problem of scientific pluralism is examined in the context of modern logic, where it became particularly pertinent after the emergence of non-Classical logics. Usual arguments in favor of a unique choice of "the" logical system are of an extralogical nature. The conception of Universal Logic as a theory of mutual translatability and combination of alternative logical systems allows for a more constructive approach to the issue. Logical pluralism gives rise not only to the ontological pluralism but also to non-Classical mathematics based on various non-Classical logics. Our analysis of ontological pluralism rises the following question: is our mathematics globally Classical and locally non-Classical (i.e. having non-Classical parts) or rather, the other way round, is globally non-Classical and only locally Classical? We conclude that in the context of post-non-Classical science the logical pluralism justifies one's freedom to chose logical tools in conformity with one's aims, norms and values.

Keywords: unity of science, scientific pluralism, logical pluralism, universal logic, ontological pluralism, non-classical mathematics.

1. An issue of the unity of science

Presently many philosophers and scientists are inclined to take a pluralistic position regarding scientific theories or methods. It is a common wisdom that the totality of natural phenomena cannot be possibly explained with a single theory or a single approach. (cf. [14]). Current debates on the scientific pluralism usually involve the 'Unity of Science' thesis first advanced by Neo-Positivists in the 1930-ies. According to this thesis

> Laws and concepts of particular sciences have to belong to the one system and be reciprocally related. They have to form certain unified science

with a common system of concepts (common language), separate sciences are just the members of it and their languages are parts of the common language. [15, p. 147–148]

In 1978 Patrick Suppes [27] in his presidential address to the Philosophy of Science Association claimed that the time for defending science against metaphysics (which he took to be the original rationale for the unity of science movement) had passed. Suppes argued that neither the languages of scientific disciplines nor their subject matters were reducible to one language and one subject matter. Nor was there any unity of method beyond the trivially obvious such as use of elementary mathematics.

The majority of philosophers of science were not particularly enthusiastic about Suppes's ideas. A noticeable exception was Nancy Cartwright and her collaborators who stressed the irreducible variety of scientific disciplines involved in solving concrete scientific problems. Later Cartwright [7] elaborated a pluralistic account of a 'dappled world' composed of a number of separate areas. Each particular area of this world is ruled by its own laws, so that this system laws form a loose patchwork, which does not reduce to a single compact system of fundamental laws. A similar view has been put forward by John Dupré [11] who also supports a pluralist metaphysical position called the "promiscuous realism".

One has to distinguish between the pluralism *in* science and the pluralism *about* science. At any stage of their development sciences typically use a variety of different approaches corresponding to different aspects of studied phenomena. They use various representational or classificatory schemes, various explanatory strategies, various models and theories, etc. This is a pluralism *in* science. The pluralism *about* science is a view according to which such a plurality of approaches in science is ineliminable as a matter of principle, and that it does not constitute any deficiency in knowledge. According to this view, an analysis of meta-scientific concepts (such as theory, explanation, evidence) should take into consideration the possibility that in the long run the explanatory and investigative aims of science can be best achieved with a pluralistic science.

Modern scientific monism can be described as follows [14, p. x]:

- the ultimate aim of a science is to establish a single, complete, and comprehensive account of the natural world (or the part of the world investigated by the science) based on a single set of fundamental principles;

- the nature of the world is such that it can, at least in principle, be completely described or explained by such an account;

- there exist, at least in principle, methods of inquiry that if correctly pursued will yield such an account;

- methods of inquiry are to be accepted on the basis of whether they can yield such an account;

- individual theories and models in science are to be evaluated in large part on the basis of whether they provide (or come close to providing) a comprehensive and complete account based on fundamental principles.

Notice that the above description does not imply that the wanted complete theory of everything is necessarily unique. Nevertheless such the uniqueness assumption is often taken for granted.

The Vienna's Circle's thesis of the Unity of Science describes this unity in ontological terms. As Alan Richardson notes, when Rudolf Carnap claims to establish the unity of 'the object domain of science' he

> does this by presenting a language in which all significant scientific discourse can be formulated. Putative metaphysical things such as essences, however, cannot be constructed – that is, they cannot be defined in the language – and this is the fact that Carnap uses to expunge metaphysical talk. Metaphysics does not speak of things in the object domain of science; there is only one such domain, and it contains all the objects that can be referred to, so metaphysics strictly does not speak of anything at all. [23, p. 6]

Carnap adds that

> we can, of course, still differentiate various types of objects if they belong to different levels of the constructional system, or, in case they are on the same level, if their form of construction is different. [6, p. 9]

He gives an example of synthetic geometry where complex constructions are built from basic elements such as points, straight lines, and planes. Such constructions may involve several different layers but all statements about these constructions are ultimately the statements about their basic elements. So we have here different types of objects and yet a unified domain of objects from which they all arise.

The question arises: how big and how independent can be such complexes? It turns out that the "global" monism in the sense of the above definition allows, after all, for a pluralistic picture if one splits it into a number of "local" monisms based on independent complexes. A good example is a situation in today's non-Classical logics to which we now turn.

2. Logical pluralism and logical monism

The Tower of Babel is a cultural pattern, which recurs again and again. The first attempt of its erection, as it is well known, ended up with a catastrophe and produced multiple languages and the lack of understanding between the builders of this monster. However this was not the end of the story. A new Babel Tower dating back to Aristotle and the Stoics was the project of developing a unique and uniform logic supposed to provide rules of correct reasoning for all. This attempt seemed successful throughout the last two thousand years but eventually it failed as a result of the development and proliferation of the so-called non-Classical logics. Some thinkers including Aristotle himself considered certain deviations from the Classical logic earlier but only in the beginning of the 20-th century researches began to explore this new territory systematically. As a result many today's logicians hold a view according to which there exist many alternative systems of logic rather than a single "right" logic. This view is known under the name of *logical formalism*. Although the philosophical analysis of logical pluralism is still in its infancy the soundness of this view is hardly any longer questionable. It is possible that the logical pluralism will point to ways out of some deadend of modern logic and determine a strategy for developing logic in the 21st century. Implications of logical pluralism for the modern also still wait to be studied. In what follows we shall consider some problems of logical and metalogical pluralism and explore their implications for ontology and foundations of mathematics.

It may appear that the logical monism does not need an argumentative defense because it is supported by more then two thousand years of the history of logic. However the situation is not so is simple. Does the Classical logic in some sense imply the logical monism? Or perhaps some non-Classical logic can play the same role of the only "right" logic common for all? The Intuitionistic logic at certain point of history was considered as a candidate for this role. Later were considered some other candidates such as the Relevant logic, which allows one to avoid certain paradoxes appearing in the Classical logic. According to Stephen Read [21], the only purpose of logic is to distinguish between valid and invalid inferences. Hence, the argument goes, there is only one "true" logic, which can be nothing but the Relevant logic.

However if one takes into account how the concept of relevance has been modified in the course of the 20-th century, one can hardly accept this and similar arguments of logical monists. All such arguments are ultimately ethical or aesthetic arguments rather than properly logical. They call for the "lost paradise", from where logics and logicians have been earlier expelled. The existing experience of metalogical researches indicates that there is no logical system satisfying all wanted metalogical properties and free from all paradoxes. As a matter of fact, it is difficult to single out even a short list of universal meta-properties which the ideal logical system of logical monists should necessarily possess.

Earlier R. Carnap [5] put forward the Principle of Tolerance in logic according to which logic should justify conclusions rather than establish some bans. There is no moral in logic and everyone has a liberty of building his or her own system of logic. As a matter of fact Carnap talks about the choice of formal language rather than the choice of logic. As it has been shown by G. Restall [22] one and the same language may admit for different logical consequence relations. So the distinction between language and logic is essential in this context.

Beall and Restall point to the following problem of logical pluralism:

> Which of these many logics governs your reasoning about how many logics there really are? In other words, which logic ought to govern your reasoning about the nature of logic itself? And indeed, which logic ought to govern your reasoning about the nature of logic itself? [1, p. 6]

Indeed, a goal of logical pluralist is to study mutual relationships between the known logical systems. These logical systems can be seen either as a list of candidates for the same role of "the" unique "true" logic or as a friendly "logic community" providing different answers to the same questions. The builders of the Babel Tower eventually lost a common language and a mutual understanding. Does the existing logical community await the same fate?

A basic problem of logical pluralism is the problem of relationships between different systems of logic. How such systems can be compared and evaluated? If we recall that a logical theory is always a theory of some individual domain then the logical pluralism can be understood as the thesis according to which the one and the same domain, generally, admits for several alternative logics.

Logical rules do not depend on empirical reliability, they cannot be cancelled because of empirical observations: logic is aprioristic by its very nature. Hartry Field argues [12] that a system of logic accepted a priori can be eventually replaced by an alternative logical system, equally designed a priori, under the pressure of facts. This view qualifies as a sort of *fallibilistic apriorism* (borrowing the term from the philosophy of science). However such a revision of logic can be possibly viewed as a mere recognition of the fact that the old logic simply did not correspond to the studied individual domain. As notices Ottavio Bueno [4] this possibility cannot be ruled out a priori.

3. Logical eclecticism and logical relativism

The logical monism is a dogmatic position. The logical eclecticism, in its turn, is a variety of logical pluralism, which makes a choice of the best logical system from a list of such systems and aims at harmonizing competing approaches. On the other hand it operates like logical monism when it rejects certain moments of known logical systems as "erroneous".

A problem of logic eclecticism, as well as of any other sort of eclecticism, is the arbitrariness of choices: one chooses and uses certain principles without having any general theory justifying the choice. However the choice between logical systems becomes interesting when one translates problems formulated in some given logical framework into a different logical framework. This allows one to look at the given problem from a different viewpoint and sometimes helps to find an unexpected solution.

The same feature belongs to the position called *logical relativism*. Roy Cook describes it as follows: one qualifies as a relativist about a particular phenomenon if and only if one thinks that the correct account of it is a function of some distinct set of facts [9, p. 493]. How many similar correct accounts of the same set of facts can exist in principle? If the answer is that such accounts are multiple then this position reduces to a version of pluralism; of one assumes that there is only one such account then it reduces to monism. In this context Cook distinguishes between the *dependent* and *simple* varieties of pluralism. While former variety of pluralism is based on the relativism the latter is not. It may appear that an obvious example of the dependent pluralism is given by the *Tarskian Relativism* [29] according to which every term in a formal language can be equally treated either as logical or non-logical. But, as

Varzi rightly notices, the Tarskian Relativism implies a stronger form of logical relativism according to which different ways of specifying the semantics of terms are equally admissible. It is possible, for example, that you and I agree that identity is a logical constant but you may think that it stands for a transitive relation whereas I may not accept this assumption.

4. Metalogical relativism as the consequence of logical pluralism

Varzi's paper referred to above makes it clear that Tarskian Relativism adds to the logical pluralism a new dimension related to the choice of logical semantics. Each variant of logical semantics comes with its own conception of logical consequence. Indeed, the usual definition of logical consequence – the conclusion follows from the given premises when in every case where the premises are true the consequence is also true – only looks neutral. In fact it involves the concept of truthfulness which depends on the chosen semantics of logical terms. Alternatively one may use in this definition a metaimplication opening thus yet a further dimension of pluralism.

Should be one's metalogic necessarily Classical? Graham Priest, considering Tarski's theory of truth and his T-construction, writes that

> sometimes it is said that Tarskian theory must be based on Classical logics: this logic is required for the construction to be performed. Such a claim is just plain false. It can be carried out in intuitionistic logics, paraconsistent logics, and, in fact, most logics. [20, p. 45]

Thus the Tarskian Relativism turns into the metalogical relativism and the metalogical pluralism. It allows for considering various alternative definitions of logical consequence such as: "the conclusion follows from premises if and only if any case in which each premise is true is also a case in which conclusion is relevantly true" (a case of Relevant metalogic), "the conclusion follows from premises if and only if any case in which each premise is true is also a case in which conclusion is intuitionistically true" (a case of Intuitionistic metalogics), "the conclusion follows from premises if and only if any case in which each premise is true is also a case in which conclusion is paraconsistently true" (a case of Paraconsistent metalogic), "the conclusion follows from premises if and only if any case in which each premise is true is also a case in which conclusion is quantum logically true" (a case of quantum metalogic), etc.

Moreover, apparently nothing prevents one from correlating one's concept of logical consequence with a non-Classical logic. Then the above definition can be modified as follows: "the conclusion intuitionistically follows from premises if and only if any case in which each premise is intuitionistically true is also a case in which conclusion is intuitionistically true" (the case of Intuitionistic logic and metalogic), "the conclusion relevantly follows from premises if and only if any case in which each premise is relevantly true is also a case in which conclusion is relevantly true" (the case of Relevant logic and metalogic), "the conclusion intuitionistically follows from premises if and only if any case in which each premise is relevantly true is also a case in which conclusion is relevantly true" (the case of Relevant metalogic for Intuitionistic logic), "the conclusion relevantly follows from premises if and only if any case in which each premise is intuitionistically true is also a case in which conclusion is intuitionistically true" (the case of Relevant metalogic for Intuitionistic logic), etc. Here the choice may be limited by certain specific properties of these 'cases' [24, p. 396].

Thus we can formulate a "metalogical" definition of logical consequence as follows:

A conclusion is valid in the given logic if in the corresponding metalogic the validity of premises implies the validity of the conclusion.

On this basis it is possible to construe two further different versions of the above metalogical definition:

(i) a conclusion is valid in some logic if in some metalogic the validity of premises implies the validity of the conclusion.

(ii) a conclusion is valid in some logic if in all metalogics the validity of premises implies the validity of the conclusion.

The second version is hardly realistic since all possible metalogics can be hardly taken into account. One may also suspect that the choice of metalogic may depend on the existence of 'translation' from certain logic to the given logic. Indeed, all "mixed" principles arise via a meddling or substituting semantics of one logic to another. These semantic operations may provide grounds for further arguments pro or contra the monistic (when logic always coincides with the metalogic) and (when logic and metalogic may differ conceptually) points of view.

A non-Classical metaimplication gives rise to a meta-metalogical definition of logical consequence as follows:

- A conclusion follows from premises iff the truth of the conclusion follows from the truth of premises iff in all cases the truth of premises implies the truth of the conclusion.

On the one hand, this is a bad infinity. But on the other hand, this situation can be described in terms of S. Kripke's theory of truth [16]:

- A conclusion logically follows from premises if and only if the truth of the conclusion follows from the truth of premises if and only if the truth of the truth of the conclusion follows from the truth of the truth of premises.

- *Mutatis mutandis* in case of the 'mixed' principle. In this case in addition to Kripke's considerations of cases of the truth or falsity at corresponding meta-levels we need also to construe the truth on pluralistic variants of meta-levels.

5. Logical pluralism and universal logics

How statements of the form 'A follows from B iff B is true implies A is true in metalogic M' can be compared in the case of different metalogics? Some authors suggest that this can be done with a theory of *Universal Logic* that would provided criteria for such a comparison (see [32, 33]). The Universal Logic (UL) is a theory of translatability and combination of logical systems. The above statements can be compared with UL as follows. First one constructs a translation F from (meta)logic Y_1 to (meta)logic Y_2. Then
'A follows from B iff B *is true* implies in Y_1 A *is true*'
translates under F into
'A follows from B iff $F(B$ *is true*) implies in Y_2 F (A *is true*)'.

If such translations between different metalogics exist then we can speak about a local metalogical monism: the translatability gives us an invariant kernel preserved through translations.

Instead of linking by means of translation we would consider, using methods of universal logic, the combinations of two formulations, e.g. join of two formulations. In this case join of two logics gives us the uniform logic possessing properties of both initial logics. In particular, in union $Y_1 \oplus Y_2$ of two metalogics Y_1 and Y_2 "joint" consequence relation is defined by means of a condition:

if from *A is true* in one metalogic (Y_1 or Y_2) follows *B is true* in the same metalogic then *B jointly* follows from *A* (i.e. within the framework of the metalogic $Y_1 \oplus Y_2$).

To put it more precisely "jointly follows" gives us that

- *A* follows from *B* iff *A is true* implies in $Y_1 \oplus Y_2$ *B is true*.

Instead of unions of metalogics one can also use their product (taking pairs of metaformulas as new metaformulas), so the definition becomes

if *B multiplicatively* follows from *A* (i.e. within the framework of the metalogic $Y_1 \otimes Y_2$) then from *A is true* in both metalogics (Y_1 and Y_2) follows *B is true*.

"Multiplicativeness" gives us that

- if *A is true* follows in metalogic $Y_1 \otimes Y_2$ from *B is true* then *A* follows from *B*.

Similarly one can consider the *exponential* and *co-exponential* local metalogical monism combining metasystems Y_1, Y_2 into $Y_1 \Rightarrow Y_2$ and $Y_1 \Leftarrow Y_2$ respectively and then use the "implications" of these combined metasystems in the definition of logical consequence of the same form (provided such combinations as allowed in UL).

An obstacle for this project is the omniscience problem: we cannot explicitly describe all possible logics in advance and hence cannot accomplish all possible combinations of logics. The above types of combinations of (meta)logical systems do not exhaust all possible combinations being only the most common ones.

6. From logical to ontological pluralism

According to J. Bocheński, the modern logic is "a most abstract theory of objects whatsoever" or a "physics of the object in general". Thus "logic, as it is now constituted, has a subject matter similar to that of ontology" [3, p. 288].

In effect, ontology is a prolegomenon to logic. While ontology is an informal, intuitive inquiry into the basic properties and basic aspects of entities in general, logic is the systematic, formal, axiomatic elaboration of these ontological intuition. While ontology as it is usually practiced is the most abstract theory of real entities, logic in its present state is the general ontology of both real and ideal entities [3, p. 290].

Thereby logical pluralism is 'dangerous' because it implies the ontological pluralism. Since any logical theory is always a theory of some domain of individuals, the acceptance of this or that logic compels to certain assumptions, hypotheses about the cognizable objects inhabiting this area and described by our theory. It is a good thing if we are in a position to control these assumptions; too often such assumptions remain tacit.

Ontological assumptions are specific to languages – artificial or natural. The term "ontological commitment" that denotes this phenomenon can be understood either as an ontological assumption, or an ontological obligation or as an ontological hypothesis. Scientific artificial languages, which are always designed for a definite purpose, may enforce certain ontological commitments not intended by their designers.

Such troubles are rooted in the fact that formal languages designed for the scientific purposes should cope with two different ontologies, one of which represents the domain of scientific inquiry while the other belongs to the language itself and depends on its formal properties. The history of science of the 20-th century makes it clear that interactions between these two ontological layers cannot be ignored.

How ontological assumptions of a given formal language can be identified? An answer is given by A. Church's criterion: a language carries an ontological commitment associated with every sentence, which is analytic in this language, i.e., of every true sentence whose truth is granted by the semantics of this language. The distinction between analytical and synthetic sentences is made here as follows:

> One can single out two types of propositions: propositions, whose truth or falsity should be established on the basis of semantic rules of the system, and propositions, whose truth or falsity cannot not be seen from them. Such division of statements of language in respect to fixed semantic system, division on analytical and synthetic in this sense, in our opinion, is indisputable. The question consists in their exact definition and interpretation. [25, p. 88]

The usual semantics of the first-order Classical logic is given in terms of its Tarskian models. The *universe of all sets* and the related set theory provide in this case the proper ontology for this language. Thus in the case of this particular language the 'theory of objects in general' coincides with some version of set theory (possibly with urelements and empirical predicates, see [8]).

However the set theory is itself an elementary theory, i.e., a set of formal statements deduced from a conservative axiomatic extension of predicate logic with certain non-logical axioms, which describe formal properties of predicate $\in$. By modifying the logical part of this theory one can obtain a new theory based on some non-Classical logic: Paraconsistent, Relevant, Quantum, Fuzzy etc. Thus one obtains a class of non-Classical set-theoretic universes associated with their non-Classical underlying logics.

There is another simple argument supporting the claim that logical pluralism implies the pluralism of universes. Consider usual definitions of operations of join $\cup$, meet $\cap$ and complement $/$ on sets

$$x \cup y =_{def} \{a : a \in x \vee a \in y\},$$

$$x \cap y =_{def} \{a : a \in x \wedge a \in y\},$$

$$x/y =_{def} \{a : a \in x \wedge \neg(a \in y)\}.$$

A pluralist may ask: what type of connectives $\vee$ (or), $\wedge$ (and), $\neg$ (it is incorrect, that) are used in these definitions? If these are Classical connectives then the algebra of subsets of a given set is Boolean.

But what happens, if one modifies the operations on sets using non-Classical logic connectives $\vee$, $\wedge$, $\neg$ and then construes an algebra for the obtained new operations? Since in Tarskian models set-theoretic operations are responsible for truth values of formulas this provides us with an interpretation of a non-Classical logic in the Classical universe. In this way one can interpret in the given Classical universe as many non-Classical logics as one wants. One can also use a non-Classical universe and introduce in it Classical set-theoretic operations. So one gets an interpretation of Classical logic (along with non-Classical ones) in a non-Classical universe.

Is there a way to check whether "our" universe is Classical or non-Classical? Logical pluralism gives an answer in negative. One can assume the existence of a global underlying logic for a given universe but this global logic does not determine any set of local logics, which this universe may admit. Of course, we talk about global and local logics in this context only metaphorically as markers fixing a state of affairs.

7. Non-Classical mathematics: as many logics as mathematics

The 20-th century has witnessed how the original intuitionist and constructivist renderings of set theory, arithmetic, analysis, etc. were later accompanied by those based on relevant, paraconsistent, non-contractive, modal, and other non-Classical logical frameworks. This development led to the ongoing scientific program of "Non-Classical Mathematics". At the conference "Non-Classical Mathematics 2009" (June 2009, Hejnice, Czech Republic) the non-Classical Mathematics 2009 has been defined as a study of mathematics which is formalized by means of non-Classical logics. The Program of this conference included the following sections:

- Intuitionistic mathematics: Heyting arithmetic, Intuitionistic set theory, topos-theoretic foundations of mathematics;

- Constructive mathematics: constructive set or type theories, pointless topology;

- Substructural mathematics: Relevant arithmetic, non-contractive naive set theories, axiomatic fuzzy set theories;

- Inconsistent mathematics: calculi of infinitesimals, inconsistent set theories;

- Modal mathematics: arithmetic or set theory with epistemic, alethic, or other modalities, modal comprehension principles, modal treatment of vague objects, modal structuralism.

It is obvious, that there is not one but many true mathematics. But it remains unclear how these different mathematics interact. Are they complementary or mutually exclusive? This situation resembles that with non-Euclidean geometries. This analogy suggests questions like this: is our mathematics globally Classical, and only locally non-Classical or, on the contrary, it is globally non-Classical and locally Classical?

Gaisi Takeuti develops a *quantum set theory*, which involves a quantum-valued universe. It remains however unclear whether the

> mathematics based on quantum logic has a very rich mathematical content. This is clearly shown by the fact that there are many complete Boolean

algebras inside quantum logic. For each complete Boolean algebra $\mathcal{B}$, mathematics based on $\mathcal{B}$ has been shown by our work on Boolean valued analysis to have rich mathematical meaning. Since mathematics based on $\mathcal{B}$ can be considered as a sub-theory of mathematics based on quantum logic, there is no doubt about the fact that mathematics based on quantum logic is very rich. The situation seems to be the following. Mathematics based on quantum logic is too gigantic to see through clearly. [28, p. 303]

Robert Meyer proposes a construction of Relevant arithmetic built along the same 'pluralistic' line on a basis of Relevant logic [18]. Recall that Peano Arithmetic (PA) is based on the first-order Classical logic (FOL) and involves a number of non-logical axioms. Relevant Peano arithmetic R# according to Meyer is obtained from PA via a replacement of FOL by a system of Relevant logic R, leaving the non-logical axioms unchanged.

One more instance of a non-Classical mathematical theory is given by K. Mortensen in his book *Inconsistent Mathematics* [19]. Claiming that "philosophers have hitherto attempted to understand the nature of contradiction, the point however is to change it", Mortensen describes the mathematics based on the Paraconsistent logic.

In a more sophisticated way a non-Classical logical basis is used in theories of formal topology. A topological structure is usually specified via a specification of set of opens closed under the set-theoretical intersection. By modifying the concept of intersection one obtains a family of new topologies. In particular the set-theoretic intersection can be replaced by the operation of monoidal multiplication. Such constructions can be made with a non-Classical set theory interpreted in a Classical universe.

When one accepts logical pluralism and allows for various logical foundations formal topological properties can be equally taken into account. An example of such an account can be found in the Quantum theory (QT). Garrett Birkhoff and John von Neumann demonstrated an equivalence between experimental statements of QT and subspaces of Hilbert spaces. The set-theoretic intersection of two given experimental statements (represented as the closed vector subspaces of Hilbert space) is also an experimental statement (i.e., a closed vector subspace of Hilbert space). Whence one easily defines a topological structure using the standard definition of boundary.

However when one takes into account the fact that the negation of an experimental statement is its orthogonal complementation, one obtains a formal topology, which differs from its Classical counterpart.

Today's mathematics is going through a paradigm shift in its foundations from the set-theoretic paradigm to the category-theoretic one. From a logical point of view Category theory like Set theory is an elementary theory based on the Classical first-order calculus with equality.

Following N. C. A. da Costa, O. Bueno and A. Volkov [10] one can build *Paraconsistent* elementary theory of categories using the paraconsistent logic $C_1^=$. The axioms of the Paraconsistent category theory include all usual axioms with the Classical negation and some new axioms with the paraconsistent negation. One can also construct a Paraconsistent category theory [33] using axioms for category theory proposed by G. Blanc and M.-R. Donnadieu [2].

Recall that *topos* is a category of a special kind in which there exists a special object bearing a structure of Heyting algebra. The above algorithm for developing non-Classical mathematical theories allows one to build various 'quasi-toposes' by replacing Heyting algebra with some other algebras of logics. For example, the replacement of Heyting algebra by the paraconsistent da Costa algebra brings a 'potos' (aka da Costa topos). A potos is a paraconsistent universe in which one can develop paraconsistent mathematical theories just as in the case of the Intuitionistic mathematics. While in the usual topos the paraconsistency features only in special constructions and in this sense remain local artefacts, in a potos the paraconsistency is organic and underlies all further constructions. In the paraconsistent universe the Classical mathematics features as an artefact, i.e. as a local deviation from the paraconsistent regularities.

Similarly one can replace Heyting algebra with the Relevant one and thus obtain a category called 'reltos' which interprets the Relevant logic and allows for developing the Relevant mathematics [34]. This short list does not exhaust all possibilities for developing the non-Classical mathematics.

Toposes, generally, are non-Classical constructions, namely, constructive intuitionistic universes.

> By imposing natural conditions on a topos (extensionality, sections for epics, natural numbers object), we can make it correspond precisely to a model of Classical set theory. Thus, to the extent that set theory provides a foundation for mathematics, so too does topos theory. [13, p. 344]

What a "natural condition" means precisely in this context?

In a topos-theoretic context a Classical universe is a local construction (being a special case of general topos) while the nature of general topos is purely intuitionistic, i.e., essentially non-Classical. Thus the general topos serves as a global non-Classical foundation of mathematics, which can be Classical locally.

Other kinds of non-Classical mathematics can be similarly obtained locally in the same global intuitionistic context. This can be achieved with Lawvere's 'variable sets' aka intensional sets aka "set-theoretical concepts" (R. Goldblatt's terminology). According to Goldblatt, the intension or meaning of a given expression, is an "individual concept expressed by it". For example, if $\varphi(x)$ is the statement 'x is a finite ordinal' then the intension of φ is the *concept* of a finite ordinal. In the categorical language this concept is represented by a functor that assigns to each $p \in P$ a set of things known "at stage p" to be finite ordinals [13, p. 212].

By varying p, one can impose different "natural restrictions" on given sets of individuals and thus obtain set-theoretic concepts, which describe non-Classical sets. In particular, such a variation can be used for interpreting quantum logics in toposes; in this case the obtained set-theoretic concepts characterize quantum sets.

Likewise it is possible to use functor category Set^A from the so-called CN-category (which is a category-theoretic equivalent of da Costa algebra) to category Set. This category is a topos. Notice that the completeness of da Costa C^1 paraconsistent system has been proved with respect to a similar topos [30]. A similar approach can be used in the case of Relevant logic R [31].

Presently only a small minority of mathematicians expresses an interest in the non-Classical mathematics (beyond its intuitionistic and constructive varieties, which are related to the theory of computability). There are two reasons for this. First, the non-Classical mathematics so far did not bring anything interesting for the viewpoint of mathematical novelty. Researches in this field still focus on mathematical characteristics of non-Classical logics and their models. This common tendency is evident in spite of some noticeable exceptions (e.g. Kris Mortensen's book 'Inconsistent Mathematics', an attempt by K. Piron to reformulate quantum mechanics on quantum logic foundations). Perhaps the development of interactive non-Classical provers and decision-making systems will be able to make this research filled more vivid. The effectiveness and the convenience of the human-machine interaction may serve as a strong argument in favour of this or that non-Classical mathematics.

Second, there is a danger for non-Classical mathematician to become a 'hero of deserted landscapes'. Polish science-fiction writer Stanisław Lem distinguished between three kinds of genius [17, p. 89]. A genius of the third kind is an ordinary genius who is beyond the intellectual scope of his age. A genius of the second kind is a hard nut, which his contemporaries cannot crack. Such a genius usually gets a postmortem recognition. Geniuses of the first and the highest kind remain wholly unknown – both during their lifetimes and after their deaths. Their intellectual impact is so revolutionary that no one can evaluate it. Lem provides a fictitious historical example of a manuscript by an anonymous Florentian mathematician of 18-th century, which *prima facie* appeared to be a work in Alchemy but at a closer examination turned out to be a project of alternative mathematics, which differed drastically from our mathematics as we know it. Checking whether this alternative mathematics is better or worse than the usual one would require a lifetime work of hundreds of scientists working on the manuscript by the Florence Anonymous in a way similar to which Bolyai, Lobachevsky and Riemann worked on Euclid. In reality most mathematicians simply avoid developing any 'parallel' mathematics.

8. Conclusion

Recent developments in logic support a pluralistic logical picture of the world. Besides, it should not be expected that such situation is true only for logics. The emergence of non-Classical mathematics should not be seen as a supporting evidence for logicism. It should be rather understood as a natural consequence of the internal pluralism of logic which has been made explicit in recent developments. Having in mind D. Hilbert's view according to which logic is a metamathematics one can see that logical pluralism implies the plurality of mathematics, i.e., the plurality of mathematical pictures of the world.

Describing the Classical science Kant famously remarked that "each science is as much a science as much there is mathematics in it". Can one really expect a 'pluralization' of such scientific disciplines as physics and biology along with the pluralization of mathematics? From the Classical point of view the answer should be affirmative. However, we are living in the epoch of post-non-Classical rather than Classical science. For this reason scientific pluralism is limited with a variety of systems of social values and goals, which dictate choices of our research strategies. According to V. S. Stepin

> The post-non-classical type of scientific rationality broadens the field of reflection over activity. It takes into account correlation of obtained knowledge of the object not only with specificity of means and operations of

activity, but also with value-goal structures. Here we explicate the connection between intrascience goals and extra-scientific, social values and goals. [26, p. 634]

So the pluralism of the modern logic is rather a precondition of freedom in our choices of logical toolkits, which determines directions of our researches.

The development of logic in the 20-th century made clear that certain metalogical characteristics which were earlier believed to be universal were actually not universal. This concerns, in particular, the completeness and the consistency of logical systems, which make no sense in the case of paraconsistent logical systems (albeit they have such properties as paraconsistency and paracompleteness). Notice that Relevant logics can be paraconsistent and at the same time consistent and complete. Such facts provide an additional evidence in favour of the post-non-Classical view according to which a logician or a mathematician should select his or her formal toolkit on the basis of certain goals, values and norms.

References

[1] J. C. Beall and G. Restall. Defending Logical Pluralism. In John Woods and Bryson Brown, editors, *Logical Consequence: Rival Approaches. Proceedings of the 1999 Conference of the Society of Exact Philosophy*, pages 1–22. Hermes, Stanmore, 2001.

[2] G. Blanc and M.-R. Donnadieu. Axiomatisation de la catégorie des catégories. *Cah. Topol. Géom. Différent.*, XVII, 2:1–38, 1976.

[3] J. M. Bocheński. Logic and Ontology. *Philosophy East and West* 24, VII(3):275–292, 1974.

[4] O. Bueno. Is Logic A Priori? *The Harvard Review of Philosophy*, XVII:105–117, 2010.

[5] R. Carnap. *The Logical Syntax of Language*. Adams and Co., Littlefield, 1959. Translated by Amethe Smeaton.

[6] R. Carnap. *The Logical Structure of the World and Pseudoproblems in Philosophy*. Open Court Publishing, Chicago and La Salle, Illinois, 2005.

[7] N. Cartwright. *The Dappled World*. Cambridge University Press, Cambridge, 1999.

[8] N. B. Cochiarella. Predication Versus Membership in the Distinction between Logic as Language and Logic as Calculus. *Synthese* 77:37–72, 1988.

[9] R. T. Cook. Let a thousand flowers bloom: A tour of logical pluralism. *Philosophy Compass*, 5(6):492–504, 2010.

[10] N. C. A. da Costa, O. Bueno, and A. Volkov. Outline of a Paraconsistent Category Theory. In P. Weingartner, editor, *Alternative Logics. Do Science Need Them?*, pages 95–114. Springer, Berlin, Heidelberg, New York, 2004.

[11] J. Dupré. *The Disorder of Things: Metaphysical Foundations of the Disunity of Science*. Harvard University Press, Cambridge, Mass., 1993.

[12] H. Field. Epistemological Nonfactualism and the A Prioricity of Logic. *Philosophical Studies* 92:1–24, 1998.

[13] R. Goldblatt. *Topoi. The Categorial Analysis of Logic*. North-Holland, Amsterdam, New York, Oxford, 1984.

[14] S. H. Keller, H. E. Longino, and C. K. Waters. Introduction: The Pluralist Stance. In S. H. Keller, H. E. Longino, and C. K. Waters, editors, *Scientific Pluralism (Minnesota studies in the philosophy of science; 19)*, pages vii–xxix, University of Minnesota Press, Minneapolis, 2006.

[15] V. Kraft and Der Wiener Kreis. *Der Ursprung des Neopositivismus*. Springer-Verlag, Wien, New York, 1968.

[16] S. Kripke. Outline of Truth Theory. *The Journal of Philosophy*, 72(9):695–717, 1975.

[17] S. Lem. *Doskonała próżnia. Wiełkość urojona*. Wydawnictwo Literackie, Krakow, 1974.

[18] R. K. Meyer. Relevant arithmetic. *Bulletin of the Section of Logic*, 5:133–137, 1976.

[19] K. Mortensen. *Inconsistent Mathematics*. Kluwer, Dordrecht, 1995.

[20] G. Priest. *Doubt Truth to be a Liar*. Clarendon Press, Oxford, 2008.

[21] S. Read. Monism: the one true logic. In D. De Vidi and T. Kenyon, editors, *A Logical Approach to Philosophy: Essays in Honour of Graham Solomon*, pages 193–209, Springer, 2006.

[22] G. Restall. Carnap's Tolerance, Meaning and Logical Pluralism. *Journal of Philosophy* 99:426–443, 2002.

[23] A. W. Richardson. The Many Unities of Science: Politics, Semantics, and Ontology. In *Scientific Pluralism (Minnesota studies in the philosophy of science; 19)*, pages 1–25. University of Minnesota Press, Minneapolis, 2006.

[24] R. Routley and R. Meyer. Semantics of Entailment. In H. Leblanc, editor, *Truth, Syntax and Modality*, pages 199–243. Amsterdam London, 1973.

[25] E. D. Smirnova. Analiticheskaya istinnost' (An Analytical Truth). *Metodologicheskiye aspekty kognitivnykh protsessov (Vychislitel'nye sistemy, 172)*, pages 74–134. Novosibirsk, 2002. In Russian.

[26] V. S. Stepin. *Theoretical Knowledge*. Synthese Library, volume 326. Springer, 2005.

[27] P. Suppes. The Plurality of Science. In Peter Asquith and Ian Hacking, editors, *PSA 1978: Proceedings of the 1978 Biennial Meeting of the Philosophy of Science Association*, vol. 2, pages 3–16. Philosophy of Science Association, East Lansing, Mich., 1978.

[28] G. Takeuti. Quantum Set Theory. In S. Beltrametti and B. van Fraassen, editors, *Current Issues on quantum logic*, pages 303–322. Plenum, New York London, 1981.

[29] A. C. Varzi. On Logical Relativity. *Philosophical Issues*, 10:197–219, 2002.

[30] V. L. Vasyukov. Paraconsistency in Categories. In D. Batens, C. Mortensen, G. Priest, and J.-P. van Bendegem, editors, *Frontiers of Paraconsistent Logic*, pages 263–278, Research Studies Press Ltd., Baldock, Hartfordshire, England, 2000.

[31] V. L. Vasyukov. Paraconsistency in Categories: Case of Relevant Logic. *Studia Logica*, vol. 98, 3:429–443, 2011.

[32] V. L. Vasyukov. Structuring the Universe of Universal Logic. *Logica Universalis*, vol. 1, 2:277–294, 2007.

[33] V. L. Vasyukov. Logicheskiy pluralism i neklassicheskaya teoriya kategoriy (Logical Pluralism and Non-Classical Category Theory). *Logicheskiye Issledovaniya*, issue 18, pages 60–76. Tsentr gumanitarnych initsiativ, Moscow St. Petersburg, 2012. In Russian.

[34] V. L. Vasyukov. Reltoses Semantics for Relevant Logic. In *Devyatye Smirnovskiye chteniya po logike. Materialy mezhdunarodnoy nauchnoy konferentsiyi, 17–19 iyunia 2015*, pages 13–15, Sovremennye tetradi, Moscow, 2015.

Received: 29 May 2016

www.ingramcontent.com/pod-product-compliance
Lightning Source LLC
Chambersburg PA
CBHW050126170426
43197CB00011B/1729